Advances in
MARINE BIOLOGY

Advances in MARINE BIOLOGY

ADVANCES IN
MARINE BIOLOGY

Edited by

GERAINT A. TARLING
British Antarctic Survey
Natural Environment Research Council
High Cross, Cambridge
United Kingdom

AMSTERDAM • BOSTON • HEIDELBERG • LONDON
NEW YORK • OXFORD • PARIS • SAN DIEGO
SAN FRANCISCO • SINGAPORE • SYDNEY • TOKYO
Academic Press is an imprint of Elsevier

Academic Press is an imprint of Elsevier
32 Jamestown Road, London NW1 7BY, UK
Radarweg 29, PO Box 211, 1000 AE Amsterdam, The Netherlands
Linacre House, Jordan Hill, Oxford OX2 8DP, UK
30 Corporate Drive, Suite 400, Burlington, MA 01803, USA
525 B Street, Suite 1900, San Diego, CA 92101-4495, USA

First edition 2010

Notice
No responsibility is assumed by the publisher for any injury and/or damage to persons or property as
a matter of products liability, negligence or otherwise, or from any use or operation of any methods,
products, instructions or ideas contained in the material herein. Because of rapid advances in
the medical sciences, in particular, independent verification of diagnoses and drug dosages should
be made.

ISBN: 978-0-12-381308-4
ISSN: 0065-2881

For information on all Academic Press publications
visit our website at elsevierdirect.com

Printed and bound in UK

10 11 12   10 9 8 7 6 5 4 3 2 1

Working together to grow
libraries in developing countries
www.elsevier.com | www.bookaid.org | www.sabre.org

ELSEVIER   BOOK AID
            International   Sabre Foundation

Contributors to Volume 57

Cornelia Buchholz
Alfred Wegener Institute for Polar and Marine Research, Bremerhaven, Germany

Friedrich Buchholz
Alfred Wegener Institute for Polar and Marine Research, Bremerhaven, Germany

Janine Cuzin-Roudy
Observatoire Océanologique, Université Pierre et Marie Curie (Paris)-CNRS, Villefranche-sur-Mer, France

Natalie S. Ensor
British Antarctic Survey, Natural Environment Research Council, High Cross, Cambridge, United Kingdom

Torsten Fregin
Universität Hamburg, Zoologisches Institut, Hamburg, Germany

Peter Fretwell
British Antarctic Survey, Natural Environment Research Council, High Cross, Cambridge, United Kingdom

Edward Gaten
Department of Biology, University of Leicester, Leicester, United Kingdom

William P. Goodall-Copestake
British Antarctic Survey, Natural Environment Research Council, High Cross, Cambridge, United Kingdom

Michel Harvey
Maurice Lamontagne Institute, Fisheries and Oceans Canada, Mont-Joli, Québec, Canada

Magnus L. Johnson
Centre for Environmental and Marine Sciences, University of Hull, Scarborough, United Kingdom

Stein Kaartvedt
King Abdullah University of Science and Technology, Thuwal, Saudi Arabia

Chiara Papetti
Department of Biology, University of Padova, Italy

Tomaso Patarnello
Department of Public Health, Comparative Pathology, and Veterinary Hygiene, University of Padova, Italy

Reinhard Saborowski
Alfred Wegener Institute for Polar and Marine Research, Bremerhaven, Germany

Katrin Schmidt
British Antarctic Survey, Natural Environment Research Council, High Cross, Cambridge, United Kingdom

Yvan Simard
Maurice Lamontagne Institute, Fisheries and Oceans Canada, Mont-Joli, Québec, Canada and Marine Science Institute, University of Québec at Rimouski, Rimouski, Québec, Canada

John I. Spicer
Marine Biology and Ecology Research Centre, School of Marine Sciences and Engineering, University of Plymouth, Plymouth, United Kingdom

Geraint A. Tarling
British Antarctic Survey, Natural Environment Research Council, High Cross, Cambridge, United Kingdom

Konrad Wiese✠

Lorenzo Zane
Department of Biology, University of Padova, Italy

✠ Deceased

Contents

4. **Physiology and Metabolism of Northern Krill**
   (*Meganyctiphanes norvegica* Sars)   **91**

   John I. Spicer and Reinhard Saborowski

5. **Food and Feeding in Northern Krill**
   (*Meganyctiphanes norvegica* Sars)   **127**

   Katrin Schmidt

6. **Growth and Moulting in Northern Krill**
   (*Meganyctiphanes norvegica* Sars)   **173**

   Friedrich Buchholz and Cornelia Buchholz

SERIES CONTENTS FOR LAST FIFTEEN YEARS*

*The full list of contents for volumes 1–37 can be found in volume 38.

Preface

Northern krill (*Meganyctiphanes norvegica*) is a member of the euphausiids, a crustacean group of 85 species, which all share a relatively uniform morphology and have a pelagic, principally filter-feeding, lifestyle. This group has been estimated to contain a worldwide biomass in excess of 300 million tonnes, and is notably dominant in the pelagic communities of the Southern Ocean, North Pacific and North Atlantic. Consequently, their role in oceanic food webs is pivotal as a major link between primary productivity and consumption by higher trophic levels, particularly that of fish, birds and mammals. Furthermore, they are an important contributor to the carbon cycle through their pronounced diel vertical migration, which can facilitate rapid carbon sequestration to deeper layers.

The last major synthesis of euphausiid research was published 30 years ago by John Mauchline (1980). That review cited just over 800 works. Since then more than 3500 scientific articles on various aspects of euphausiid biology have been published. The breadth of modern day euphausiid research makes it difficult to emulate the achievement of Mauchline (1980), and the preceding review by Mauchline and Fisher (1969), in providing a synthesis that covers all euphausiid species comprehensively. The remit of the present synthesis was therefore focussed onto one species and our choice of Northern krill was made on two grounds. Firstly, Northern krill has an extraordinary physiological and behavioural adaptability, which facilitates one of the widest distributions of any euphausiid species, covering the entire North Atlantic basin from 30°N to 80°N, across which surface temperature ranges from 2 to 25 °C. Antarctic krill, by contrast, has a comparatively stenothermic physiology which probably reflects its exposure to a much a narrower range of conditions, with a latitudinal range of $\sim 15°$ and temperature limits between 0 and 6 °C. Most euphausiid reviews since 1980 have focussed on various aspects of Antarctic krill biology (e.g. Everson 2000—fisheries, Atkinson *et al.*, 2008—distribution and habitats), while species with a contrasting pattern of distribution and physiology have not received similar levels of attention. A review of Northern krill presents a different and complimentary perspective with which to reveal the full capacities of this crustacean group. Secondly, the present is an opportune time to reflect on the achievements made in Northern krill research over the past decades. Bodies of work from a number of euphausiid researchers, perhaps originally instructed by the syntheses of Mauchline, have now reached maturity. The past decade has also seen the completion of a number of relevant major research programmes, notably GLOBEC (*Global Ocean Ecosystem Dynamics programme*) and the Northern krill focussed European programme, PEP (*Physiological Ecology of a Pelagic Crustacean, EU Mast III*). The progress that has been made by individuals, teams and programmes on Northern krill is now substantial and requires distillation.

This volume contains 10 chapters on various aspects of Northern krill biology in an attempt to capture the breadth of recent enquiry. Comparisons with other euphausiid species are made throughout this volume even though Northern krill remains the primary focus. Chapter 1 considers aspects fundamental to the study of Northern krill biology, starting with an overview of its morphology, taxonomy and phylogeny. It discusses different means of sampling euphausiids and considers the recent commercial interest in their exploitation. The chapter also provides an updated synthesis of the pattern of Northern krill distribution, as well as estimates of standing stock biomass and abundance. Chapter 2 (Patarnello *et al.*) adds further detail to the distribution of the species, specifically, the division into different sub-populations, as revealed through population genetics. It also examines the species' evolutionary history and estimates periods of expansion in population size. Chapter 3 (Tarling) considers the population dynamics of *M. norvegica* and how it varies across the species' distributional range, particularly in terms of size distribution and life-cycle pattern.

The majority of Northern krill research over the past 30 years has focussed on physiological rates, processes and cycles, and these are dealt with in the four subsequent chapters, starting with Chapter 4 (Spicer and Saborowski) on metabolism, followed by Chapter 5 (Schmidt) on feeding, Chapter 6 (Buchholz and Buchholz) on moulting and growth and Chapter 7 (Cuzin-Roudy) on reproduction. Each chapter synthesises recent findings in these fields and then provides an up-to-date assessment of how these physiological functions operate. The chapters also consider the responsiveness of the species to environmental clines across its distributional range.

The behaviour of Northern krill has been reviewed from two different perspectives, firstly from observations made through laboratory-based experiments (Chapter 8, Gaten *et al.*) and secondly, from *in situ* observations (mainly nets and acoustics) with a particular focus on diel vertical migration (Chapter 9, Kaartvedt). These contrasting approaches reveal different but complimentary insights into the behavioural traits of Northern krill, and the cues that trigger them.

The volume concludes with a consideration of the role of Northern krill as prey to various predators (Chapter 10, Simard and Harvey). It includes a comprehensive survey of the presence and relative importance of Northern krill in the diets of a wide range of predator species. It also describes some of the characteristic ways in which Northern krill aggregate and how these aggregations are subsequently exploited by predators.

The generation of this synthesis volume has been a multi-author effort and I acknowledge the dedicated and enthusiastic way the authors have applied themselves to the task. I particularly acknowledge the contribution of Konrad Wiese, who sadly passed away during the production of this volume and whose fascination in krill behaviour was an inspiration to us all. The original initiative for this volume derives from Fred Buchholz and Jack Matthews in providing a new thrust to European research on Northern krill via the EU Mast III PEP programme. This volume has benefitted from discussions and correspondence with PEP colleagues as well as other researchers of Northern krill, namely Jean-Phillipe Labat, Patti Virtue, Patrick Mayzaud, Steve Nicol, Stig Falk-Petersen, Padmini Dalpadado, Webjørn Melle, Thor Klevjer, Bo Bergström, Mike Thorndyke, Jalle Strömberg, Ryan Saunders and Inigo Everson. Robin Ross, Langdon Quetin, Peter Wiebe, Angus Atkinson, Eugene Murphy, So Kawaguchi, Evgeny Pakhomov, Volger Siegel, and Jon Watkins have also been generous in sharing their insights into euphausiid biology. John Mauchline remains an enthusiastic supporter of krill research and was an invaluable source of information regarding historical records of Northern krill distribution. Jo Milton, Mandy Tomsett and Joanna Rae of the British Antarctic Survey ably assisted in tracking down references, as did Jane Stephenson and Adrian Burkett of Library and Information Services at the National Oceanography Centre, Southampton and Linda Noble at the National Marine Biological Library, Plymouth. I also thank the production team of Zoe Kruze, Narmada Thangavelu and Sujatha Thirugnanasambandam who have been helpful and understanding throughout. Finally, this volume would not have been achieved without the unconditional support of Denise, Don, Joanne, Angharad, Arwen and Eirig.

GERAINT A. TARLING

CHAPTER ONE

An Introduction to the Biology of Northern Krill (*Meganyctiphanes norvegica* Sars)

Geraint A. Tarling,* Natalie S. Ensor,* Torsten Fregin,[†]
William P. Goodall-Copestake,* *and* Peter Fretwell*

Contents

* British Antarctic Survey, Natural Environment Research Council, High Cross, Cambridge, United Kingdom
[†] Universität Hamburg, Zoologisches Institut, Hamburg, Germany

Advances in Marine Biology, Volume 57
ISSN 0065-2881, DOI: 10.1016/S0065-2881(10)57001-9

1

Abstract

This chapter provides a background to research on Northern krill biology, starting with a description of its morphology and identifying features, and the historical path to its eventual position as a single-species genus. There is a lack of any euphausiid fossil material, so phylogenetic analysis has relied on comparative morphology and ontogeny and, more recently, genetic methods. Although details differ, the consensus of these approaches is that *Meganyctiphanes* is most closely related to the genus *Thysanoessa*. The light organs (or photophores) are well developed in Northern krill and the control of luminescence in these organs is described. A consideration of the distribution of the species shows that it principally occupies shelf and slope waters of both the western and eastern coasts of the North Atlantic, with a southern limit at the boundary with sub-tropical waters (plus parts of the Mediterranean) and a northern limit at the boundary with Arctic water masses. Recent evidence of a northward expansion of these distributional limits is considered further. There have been a variety of techniques used to sample and survey Northern krill populations for a variety of purposes, which this chapter collates and assesses in terms of their effectiveness. Northern krill play an important ecological role, both as a contributor to the carbon pump through the transport of faecal material to the deeper layers, and as a key prey item for groundfish, squid, baleen whales, and seabirds. The commercial exploitation of Northern krill has been slow to emerge since its potential was considered by Mauchline [Mauchline, J (1980). The biology of mysids and euphausiids. *Adv. Mar. Biol.* **18**, 1–681]. However, new uses for products derived from krill are currently being found, which may lead to a new wave of exploitation.

1. INTRODUCTION

In 1960, John Mauchline published a seminal work entitled 'The biology of the Euphausiid Crustacean, *Meganyctiphanes norvegica* (M. Sars)', which provided detailed descriptions of a number of characteristic traits of Northern krill, including diel vertical migration (DVM), feeding, growth mortality rates, maturation and the function of the luminescent organs. The study was carried out in the semi-enclosed Scottish loch, Loch Fyne (Clyde Sea), which offered ready sampling-access and the capacity to follow a population reliably over time (2 years in total). The work was a major step forward in the understanding of ecology and behaviour of this species, to add to what was already known in terms of its distribution (Einarsson, 1945; Ruud, 1936) and feeding habits (Fisher and Goldie, 1959; Macdonald, 1927). Mauchline went on to publish the first major review of euphausiids, with Leonard Fisher, in 1969, which included sections on vertical distribution and migration, food and feeding, and growth, maturity, and mortality that were mainly based on Mauchline (1960). Mauchline updated this review in 1980.

Fifty years on from Mauchline (1960) and 30 years from the updated review of euphausiid biology (Mauchline, 1980), we find that the field has advanced considerably. Since 1980, more than 1000 articles have been published with a euphausiid species name in the title and more than 3500 with a euphausiid species name in the Abstract or as a Keyword (Table 1.1). Of these, around 90% have considered some physiological process, more than 40%, abundance and biomass, and ~10%, the role of krill as a prey item. Furthermore, entire new fields have developed over this time, such as population genetics and genomics. New acoustic- and video-enabled instrumentation has allowed unprecedented insights into euphausiid behavioural ecology. State-of-the-art acoustics can remotely identify aggregations of euphausiids and rapidly assess their distribution and biomass, from local- to basin-scales. Physiological cycles have now been well described, as has the capacity of Northern krill to cope with change in temperature and levels of oxygen saturation.

Everson (2000a) reviewed some of these advances, from a krill (mainly *Euphausia superba*) fisheries perspective. Nevertheless, there is still much recent progress in the understanding of euphausiid biology to be synthesised. A review of the biology of Northern krill, in particular, is timely given the considerable change in its northern distributional limit. Arctic summer-ice is now at a historical low and Atlantic species may potentially incur into higher latitudes. Northern krill is being increasingly reported at latitudes as high as the Spitsbergen fjords (78°56′ N, 12°04′ E, Buchholz *et al.*, 2010). The potential for the establishment of this species further north and the implications this may have on the Arctic ecosystem can be fully considered only through the study of all aspects of its biology which the present volume aims to cover.

As a background, we describe in the present chapter some fundamental details of Northern krill biology. We begin with a description of the general body plan, highlighting the species morphological characteristics. We then review its phylogeny and geographic distribution. Sampling methodology is considered in terms of the diversity of equipment used and its purpose. Finally, we assess the wider ecological role and the commercial relevance of Northern krill in the modern world.

2. Morphology and Taxonomy

2.1. General morphology

The morphology of *M. norvegica* (Northern krill) was first described by M. Sars (1857) as follows:

Table 1.1 ISSN-referenced articles published on major euphausiid species/genera since 1980

	Meganyctiphanes norvegica			Euphausia superba			Euphausia pacifica			All Thysanoessa spp.		
	1980–1989	1990–1999	2000–2010	1980–1989	1990–1999	2000–2010	1980–1989	1990–1999	2000–2010	1980–1989	1990–1999	2000–2010
Number of ISSN articles published												
1. Number with respective species name in title	26	28	54	403	221	257	25	31	41	39	24	22
2. Number with name in title, abstract or keywords	86	164	323	599	696	1071	60	79	136	130	145	208
Percentage of ISSN articles dealing with the following specific disciplines												
3. Genetics	12%	14%	13%	5%	6%	9%	8%	3%	2%	10%	8%	5%
4. Population dynamics	27%	68%	46%	15%	32%	34%	4%	32%	56%	31%	54%	64%
5. Biomass and distribution	38%	54%	50%	20%	45%	50%	16%	52%	59%	49%	54%	41%
6. Physiology and process studies	92%	86%	96%	80%	85%	74%	92%	100%	90%	95%	92%	91%
7. Behaviour	15%	54%	52%	8%	16%	18%	24%	32%	29%	10%	17%	23%
8. Krill predators	4%	7%	22%	2%	10%	13%	4%	10%	22%	0%	8%	0%

Audit carried out through applying search terms to an on-line scientific database (Web of Knowledge, wok.mimas.ac.uk). 1, involved a search for all articles in which the species or common name was contained in the *title*. 2, species or common name in the *title*, *abstract* or *keywords*. 3–8, species or common name in the *title* and discipline related key words in the *title*, *abstract* or *keywords*. Percentages relate to the number of articles relative to category 1. An article may fit more than one of the categories 3–8.

Whitish translucent, spotted above with red. Cephalothorax equalling a third to three-eighths of the whole body in length; with the forehead very short and truncated, not reaching the slender stalks of the very large eyes, with the median anterior edge angular (rostrum rudimentary), with a sharp spine on both sides of the triangle. . ..only seven pairs of gills with none on the first pair of feet, very large on the hindmost. The gills consist of short double curved branches, with a single row of simple tufts in the six anterior pairs, covered with secondary tufts of pinnate embellishments in the last pair (Translation from Latin; Mauchline and Fisher, 1969)

M. norvegica is one of the largest of the 86 described species of euphausiid, and reaches a total body length of between 40 and 50 mm (Baker *et al.*, 1990; Falk-Petersen and Hopkins, 1981), which is surpassed only by a small number of *Thysanopoda* species and *E. superba* (Antarctic krill). The basic body plan of all euphausiids is relatively similar, being divided into two main regions—the cephalothorax and the abdomen (Fig. 1.1). The cephalothorax contains the head and thoracic segments and is completely covered by a carapace. It extends for about one-third of the animal. The abdomen consists of six segments terminated by a telson (Mauchline and Fisher, 1969). The head contains one pair of stalked eyes and two pairs of antennae, which have tactile and olfactory functions. Excretory organs open to the exterior at the bases of the second antennae. The mouthparts consist of a single labrum, paired mandibles, labia, maxillules, and maxillae which filter, macerate, and manipulate food to the mouth. There are only seven pairs of biramous thoracic limbs, the eighth being absent in Northern krill. The exopodites of the thoracic limbs produce currents for the oxygenation of the external gills associated with the thoracic limb bases. The endopodites of these limbs form a food basket that filters particles out of the water. The first five abdominal segments each have one pair of biramous pleopods, with which the animal swims. The sixth abdominal segment bears one pair of biramous uropods. The uropods along with the

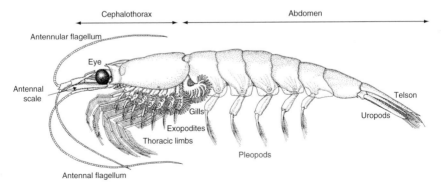

Figure 1.1   Adult *Meganyctiphanes norvegica* (total body length around 40 mm) indicating the main morphological features (Illustration by J. Corley).

telson form a 'tail fin'. The first pair of abdominal limbs of the adult male develops at maturity to form male organs called petasmae (Fig. 1.2). The female organ for collecting sperm, the thelycum (Fig. 1.3), is located ventrally on the sixth thoracic segment.

Euphausiids have two distinct spawning strategies (Gómez-Gutiérrez *et al.*, 2010); about 26 of the species are sac-spawners, where the just-spawned eggs are held into an ovigerous sac supported by the eighth periopod until they hatch. The others are broadcast-spawning species, where eggs are released freely into the water column at the point of spawning. *M. norvegica* belongs to the second of these categories. Euphausiids can also be divided between those species with spherical eyes and those with bilobed eyes and 1 or 2 pairs of elongated thoracic limbs, of which *M. norvegica* belongs to the former. Features that uniquely distinguish *M. norvegica* from other euphausiid species have been provided by Baker *et al.* (1990) as follows: 'Seventh thoracic legs consisting of two elongated joints, with the seventh thoracic exopod being present. Strong post-ocular spines and long recurved antennular (first antennae) lappets are present'.

2.2. Taxonomy

M. norvegica was originally described by Michael Sars in 1857 (Sars 1857) as *Thysanopoda norvegicus*, from four specimens from Florö and Söndfjord in Norwegian waters and a further nine specimens from the guts of two Norway

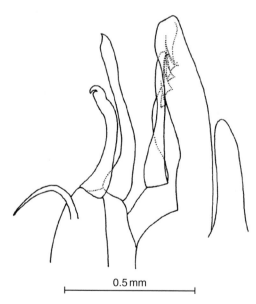

0.5 mm

Figure 1.2   Petasma of an adult male *Meganyctiphanes norvegica*, which is a modification of the internal ramus of the first pleopod (Illustration by J. Corley).

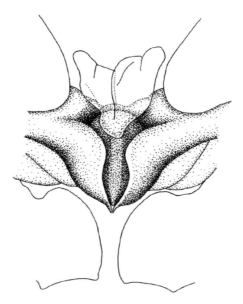

Figure 1.3 Thelycum of an adult female *Meganyctiphanes norvegica* with 2 attached spermatophores. The view is from the ventral side of the cephalothorax (6th thoracic sternite). The thelycum is approximately 1 mm across (Illustration by J. Corley).

haddocks (*Sebastes norvegicus*). The genus *Nyctiphanes* was erected by G. O Sars report in 1883 (Sars, 1883) which reported on the Schizopoda collected during the HMS Challenger expeditions of 1873–1876. In their report of the schizopoda of the north-east Atlantic slope, Holt and Tattersall (1905) decided that the species differed sufficiently from the others of the genus *Nyctiphanes* to warrant the erection of a separate genus *Meganyctiphanes* to contain it. Hansen (1908) accepted Holt and Tattersall's new genus *Meganyctiphanes* and put forward a generic difference detectable in the males, namely the form of the clasping organs (petasma) on the first pair of pleopods. Those of *M. norvegica* are essentially similar to those of *Thysanopoda* and differ from those of *Nyctiphanes*.

Meganyctiphanes is presently a single-species genus. However, Colosi (1918) categorised specimens from the Mediterranean as *Meganyctiphanes calmani* based on differences in the structure of the petasma. This was dismissed as natural variation resulting from the state of maturation by Ruud (1936) in his report on the Euphausiacea collected in the Mediterranean and adjacent seas. No further specimens of *M. calmani* have since been reported.

2.3. Phylogeny

The Euphausiacea have no unequivocal fossil representatives (Jarman, 2001; Maas and Waloszek, 2001). Indeed, the Malacostraca, in which Euphausiacea is placed, is also poorly represented in terms of fossils compared to many other major arthropod groups. Consequently, phylogenetic relationships between the Euphausiacea and other orders of Malacostraca have been investigated using morphological (Medina *et al.*, 1998) and molecular (Jarman *et al.*, 2000) characters found among extant taxa. Medina *et al.* (1998) examined the sperm ultrastructure of *M. norvegica* and found there to be a thick two layered capsule in close contact with the plasma membrane, similar to that found in *Euphausia* (Jamieson, 1991). This resembles the sperm of the decapods *Sergestes arcticus*, *Aristaeomorpha foliacea* and *Stenopus hispidus*. Medina *et al.* (1998) concluded that Euphausiacea are close to primitive decapods, in line with the assertions of previous phylogenetic studies (Gordon, 1955; McLaughlin, 1980; Schram, 1986). Jarman *et al.* (2000) used nuclear ribosomal (28S) DNA sequences to examine Malacostracan phylogeny and concluded that the mysidacean suborder Mysida was the sister taxon of the Euphausiacea with the Anaspidacea being another closely related but more basal group. Decapoda were found to be more distantly related to krill. A recent total evidence (morphology and molecules) Malacostracan phylogeny by Jenner *et al.* (2009), which included *M. norvegica*, revealed conflict between morphological and molecular evidence. The best supported higher-level clade to emerge from their analysis grouped Euphausiacea with the Anaspidacea, Decapoda, and Stomatopoda. Jenner *et al.* (2009) stressed the importance of obtaining new molecular data sources to better resolve the phylogeny of the Malacostraca.

Relationships among genera within the Euphausiacea have been examined using morphological and ontogenetic characters (Casanova, 1984; Maas and Waloszek, 2001) and through molecular investigations (Bargelloni *et al.*, 2000; D'Amato *et al.*, 2008; Jarman, 2001; Patarnello *et al.*, 1996; Zane and Patarnello, 2000). Casanova (1984) examined 13 morphological characters including the shape of the eye, endopodite structure in the second and third thoracic limbs, the degree of differentiation between the male and female antennules, the level of instar variation during development, and the reduction in the sixth and seventh thoracic limbs. From this, Casanova (1984) deduced a pattern of morphological change and constructed a phylogenetic tree (Fig. 1.4a). At the base of the tree was the most archaic taxa, the genus *Bentheuphausia*. Euphausiidae were defined at the next stage up, and comprised five main lineages that ranged from the most primitive (1) *Thysanopoda*, through (2) *Euphausia*, (3) *Meganyctiphanes*, *Nyctiphanes*, *Pseudeuphausia*, and *Thysanoessa*, (4) *Tessarabrachion* to (5) *Nematoscelis*, *Nematobrachion*, and *Stylocheiron*, the latter of which was considered the most advanced.

Maas and Waloszek (2001) used literature data and scanning electron microscope (SEM) examinations of selected ontogenetic stages of *E. superba* to determine phylogenetic relationships within the Euphausiacea. The SEM analysis revealed morphological characters that are either missing in, or significantly changed towards, the adult. Together with adult features,

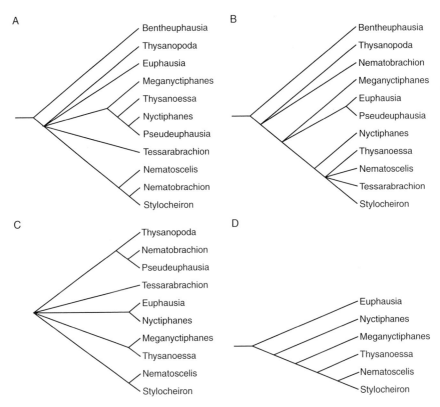

Figure 1.4 Phylogenetic trees of Euphausiacea genera derived from phylogenies and descriptions in (A) Casanova (1984), (B) Maas and Waloszek (2001), (C) Jarman (2001) and (D) Patarnello *et al.* (1996), Zane and Patarnello 2000, Bargelloni *et al.* 2000 and D'Amato *et al.* 2008. With the exception of a few molecular trees, support values were not presented in the original phylogenies. Apart from tree (C), all of the trees are rooted.

these ontogenetic characters were used in cladistic analysis to generate a phylogenetic hypothesis for the Euphausiacea (Fig. 1.4b). The results obtained confirmed many of the traditional taxonomic groups and, additionally, were used to propose three novel groups named 'Euphausiini' (*Euphausia, Pseudeuphausia*), 'Nematoscelini' (*Nyctiphanes,* Nematoscelina) and 'Nematoscelina' (*Nematoscelis, Thysanoessa, Tessarabrachion, Stylocheiron*). *M. norvegica* formed an unresolved trichotomy with the Euphausiini and Nematoscelini. Maas and Waloszek (2001) also questioned the monophyly of *Euphausia, Nematoscelis, Pseudeuphausia,* and *Tessarabrachion* and went as far as to suggest paraphyly for *Thysanopoda* and inferred paraphyly for *Thysanoessa.*

In terms of recent molecular studies, *M. norvegica* has been included in molecular phylogenies in (at least) five publications. These have concerned the evolutionary history of *E. superba* and closely related species (Bargelloni et al., 2000; Patarnello *et al.*, 1996; Zane and Patarnello, 2000), the biogeography of species in the genus *Nyctiphanes* (D'Amato *et al.*, 2008) and the evolutionary history of major Euphausiacea lineages (Jarman, 2001). Probably, the best resolved study with respect to *M. norvegica* was that carried out by Jarman (2001) which considered nuclear ribosomal (28S) DNA analysed using maximum parsimony and maximum likelihood methods of phylogenetic analysis. Jarman (2001) found *M. norvegica* to be a sister to *Thysanoessa macrura* and more distantly related to the 'archaic' *Thysanopoda* (Fig. 1.4c). Other relationships among major Euphausiacea lineages were less clear due to differences between the results obtained from the parsimony and likelihood analyses. In the other molecular phylogenies, which were all based on mitochondrial DNA (16S and ND1 gene sequences) analysed using parsimony, likelihood and Bayesian methods, *M. norvegica* did not form a terminal pairing with *Thysanoessa* but was resolved as sister to a group containing the genera *Thysanoessa*, *Nematoscelis* and *Stylocheiron* (Fig. 1.4d). Taxon sampling was narrower in these mitochondrial DNA studies than in the nuclear DNA study by Jarman (2001), thus comparisons at deeper phylogenetic levels were not possible.

On the whole, there is support from both morphological and molecular studies that *M. norvegica* represents a distinct Euphausiacea lineage. This appears to be more closely related to *Thysanoessa* than to the more 'archaic' genus of *Thysanopoda*. However, the position of *M. norvegica* within Euphausiacea, like that of Euphausiacea within Malacostraca, is far from fully resolved. In order to better resolve the phylogenetic position of *M. norvegica*, it will be necessary to follow the approach advocated by Jenner *et al.* (2009) for higher-level, Malacostraca phylogenetics. This involves looking beyond the most commonly used sources of data, namely nuclear ribosomal and mitochondrial DNA sequences, to new sources of data such as nuclear coding genes.

2.4. Photophores and their function

M. norvegica translates as 'Norwegian bright shiner' signifying its notable capacity to bioluminesce spontaneously. Its bioluminescence is produced by 10 separate light organs (photophores): one on each eyestalk, two pairs on the ventral thorax and four separate organs under the abdomen (Herring and Locket, 1978; Fig. 1.5). At a microscopic level, the photophores appear as highly organised, bell-shaped structures (Grinnell *et al.* 1988; Herring & Locket, 1978; Petersson, 1968). Light is produced *via* a luciferin–luciferase-type of biochemical reaction that is under hormonal and neural control in the light-generating cells (Herring & Locket, 1978).

Serotonin (5-hydroxytryptamine, 5-HT) (Kay, 1962) and several analogues (Herring and Locket, 1978) are stimulants for bioluminescence in krill

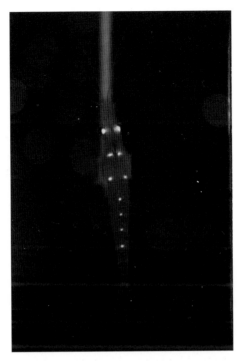

Figure 1.5 The pattern of photophore illumination in Northern krill (*Meganyctiphanes norvegica*) as viewed ventrally. The image was photographed in darkness, exposure time 40 s/400 ASA. The natural colour, a narrow band blue of 480 nm wavelength, is visible to the human eye with background irradiation dimmer than 10^{-2} μW/cm^2. Note that parts of the appendages are illuminated. (See Colour Insert.)

when added to water in aquaria. Light production can continue for several hours after single injection of Serotonin into the haemolymph and can continue for many days on repeated injection of the substance (Fregin and Wiese, 2002).

Krönström *et al.* (2007) showed that the interaction between nitric oxide and Serotonin plays a modulatory role at several levels in the control of light production in *M. norvegica*. Inside the photophores, numerous capillaries drain haemolymph into the light-producing structure (lantern). Filamentous materials around these capillaries act as sphincters. The sphincters are controlled by nerves containing 5-HT. When exposed to muscle-relaxing substances (papaverine and verapamil), krill respond with luminescence, suggesting that the sphincter structures are functionally involved in the control of light production (Krönström *et al.*, 2009), probably through restricting the flow of oxygenated haemolymph to the light-generating cells.

Light emission is directional, with a dark pigment masking emission in most directions, and a highly reflective lining focusing it. By means of small muscles, the organs can be pointed in different directions and the intensity of the bioluminescence can be regulated to fit the current ambient situation (Fregin and Wiese, 2002; Hardy and Kay, 1964; Herring and Locket, 1978; Kay, 1965; Mauchline, 1960).

In terms of patterns of bioluminescence, there is a lack of meaningful *in situ* observations and most studies have been based on the reaction of the organisms to capture or post-capture manipulation. Macdonald (1927) stated that freshly caught *M. norvegica* spontaneously flash during the first 12–24 h but cease to do so quickly after catch. Hardy and Kay (1964) wrote that the number of flashes was highest in the first 10 h after capture, after which it decreased quickly but could be observed up to 50 h after capture. Mauchline (1960) noticed that exposing krill to high intensities of white light for 20 min (room lights) and then transferring the animals to total darkness ('light–dark-experiments') will trigger the bioluminescence in 60–80% of animals. Hardy and Kay (1964) used this effect to collect the most responsive animals for their experiments. The first onset of the light production becomes visible after 20 s and lasts for up to 120 s (Fregin and Wiese, 2002).

Little has been concluded about the true function of bioluminescence in krill. During vertical migration experiments within plankton wheels, Körte (1964) reported that the krill glowed for around 40–80% of the time, suggesting that bioluminescence may be related to specific behaviours. Clarke (1963) considered that, under well-lit surface waters, the visual contrast of krill to an upward looking predator may be countershaded by the photophores. Light production may also be a form of communication between individuals. Mauchline (1960) found that individuals can stimulate each other to flash in a kind of 'chain reaction' if he placed individual krill in adjacent jars. Fregin and Wiese (2002) report that artificial light flashes presented to a group of krill evoked a signalling behaviour, in which the animals pointed the beams of light from their photophores for a fraction of a second at conspecifics at the same depth level. There was a fixed delay between the signal and the response. Such light signals can probably be observed at a distance of 10 m or more in clear water (Fregin and Wiese, 2002). Through observations made *in situ* by a submersible, Widder *et al.* (1992) found that the scattering layers, in which *M. norvegica* was one of the main emitters of bioluminescence, rarely emitted light spontaneously but did so once disturbed. They proposed that the light displays were a defensive function that may distract or blind a predator, warn conspecifics, or expose primary predators to secondary predators. Increased numbers of *in situ* observations are likely to provide better information with which to ascertain the full functional attributes of bioluminescence in Northern krill.

3. GEOGRAPHIC DISTRIBUTION

3.1. General remarks

Zoogeographically, *M. norvegica* is classified as a boreal species (Dalpadado and Skjoldal, 1991; Einarsson, 1945; Mauchline and Fisher, 1969). It characteristically inhabits shelf–slope regions and waters between coastal banks and deep basins (Melle *et al.*, 2004). Within these habitats, it usually occurs in areas with bathymetries greater than 100 m (Hjort and Ruud, 1929; Melle *et al.*, 1993) probably because its preferred depth during the day is between 100 and 500 m from which it performs a DVM to the surface layers at night (Mauchline and Fisher, 1967).

   M. norvegica has been reported to alter its distribution seasonally. Glover (1952) found that it moved towards coastal areas during the period January to May, whereas it spreads towards more oceanic areas between June and December. Mauchline (1960) reports a similar winter dispersion in the Loch Fyne (Clyde Sea) population. The population movements may be associated with breeding, with the spring aggregations being a precursor to mating and spawning. Once these have been effected, dispersion takes place (Mauchline and Fisher, 1967).

   Einarsson (1945) proposed that temperature is the factor that mainly limits the distributional extent of breeding, with the 5 °C isotherm at 100 m in May limiting the northern breeding extent and the 15 °C surface isotherm in February, the southern breeding extent. Most subsequent studies have supported these proposed limits, although a functional link between breeding and temperature remains to be elucidated. The spawning area is generally more restricted than the overall distribution of the species. Fowler *et al.* (1971) found that 18 °C approximated a lethal upper temperature limit for large adults and that the species altered its vertical distribution in waters off Monaco as a consequence. In the north, the species has been reported in waters as cold as 2 °C (Einarsson, 1945; Hollingshead and Corey, 1974; Mauchline and Fisher, 1969). Einarsson (1945) considered the younger stages (the adolescents) may be relatively stenothermic, since they are generally not found in waters less than 6 °C. However, he also considered the larvae to be more eurythermic, extending well beyond the spawning limits. As well as temperature, Einarsson (1945) noted that the breeding area is also a function of the winter distribution, which is generally more limited than that in the summer. Adult Northern krill are considered to be euryhaline, with a lower salinity tolerance limit of around 24 PSU, while the larvae are even more tolerant of low salinities (Buchholz and Boysen, 1988).

   The following section discusses some of the more regional limits within the wider distributional range of Northern krill (Fig. 1.6).

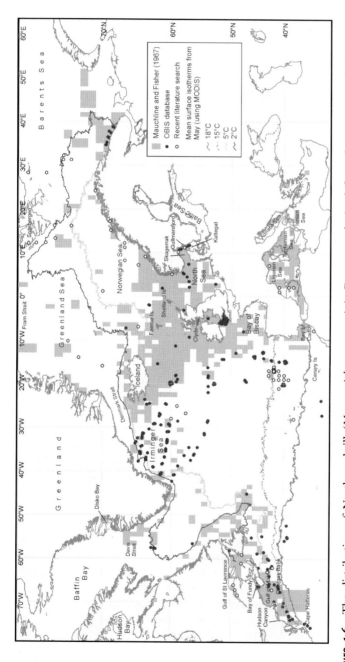

Figure 1.6 The distribution of Northern krill (*Meganyctiphanes norvegica*). Data was obtained from three sources: shaded areas were transposed from Mauchline and Fisher (1967), including any records of larvae or point samples. Blue dots represent records of *M. norvegica* present on the Ocean Biogeographic Information System (OBIS—www.iobis.org) as of January 2010. Red dots represent further novel records obtained in a recent literature search by the present authors. Surface isotherms were extracted from data provided by the NASA Moderate Resolution Imaging Spectroradiometer (MODIS) representing a mean value for May (2003–2010): the 2 and 18 °C isotherms approximate the distributional limits, with a few exceptions; the 5 and 15 °C isotherms represent the notional breeding limits, following Einarsson (1945). (See Colour Insert.)

3.2. North Atlantic west coast

Along the North American coast, the southern limit of distribution is about Cape Hatteras (35° N; Hansen, 1915; Leavitt, 1938; Bigelow and Sears, 1939). It occurs in great abundance in the Gulf of Maine and Bay of Fundy (Bigelow, 1926; Fish and Johnson, 1937; Hansen, 1915; Soulier, 1965; Whiteley, 1948). In the Pasamaquoddy area (Bay of Fundy), *M. norvegica* is known to seek deeper warmer water during the winter to avoid the chilling down to < 1 °C that occurs in the upper water layers (Bigelow, 1926; Hollingshead and Corey, 1974). It does not occur in great numbers in the inner Bay of Fundy, nor onto the southwest Nova Scotia shelf (Kulka *et al.*, 1982). In the Gulf of Maine, the species is usually found in waters of between 3 and 4 °C. It may congregate around the 100 m contour further offshore, around Georges Bank, but with seasonal shifts in its distribution (Bigelow, 1926; Whiteley, 1948). The species is common in the estuary of the St Lawrence river (Préfontaine and Brunel, 1962), the Gulf of St Lawrence and Chaleur Bay (Lacroix, 1961), where there has been detailed descriptions of the interactions between its vertical migration behaviour and the prevalent 3D advective regimes (Chapter 10). The species is absent in Hudson Bay and the Hudson Strait marks the northern limit of distribution along this coast.

3.3. Arctic limits

Disko Bay is the northernmost point that *M. norvegica* has been recorded on the west Greenland coast (Dunbar, 1942), those being larvae taken at the end of May. Nevertheless, there are no records of breeding adults and it is unlikely that breeding takes place in this region (Einarsson, 1945; Mauchline and Fisher, 1967). Saunders *et al.* (2007) found juvenile and adult size classes in the central and northern Irminger Sea and also in the east Greenland coastal region, although the relative number of larger individuals was comparatively less in the Irminger Sea. Larval stages were relatively scarce although they may have been distributed below the maximum net-depth of 400 m. Einarsson (1945) considered that no considerable spawning takes place in the cold less saline water of the East Greenland Current and that spawning is restricted almost entirely to Atlantic water, which can reach to the slope of the Greenland continental shelf. This current may be the origin of small numbers of larvae and adults caught further to the north and east, in the Denmark Strait (Mauchline and Fisher, 1967). Records are rare in the Greenland Sea, mainly restricted to larval stages (Ussing, 1938). Damas and Koefoed (1909) record *M. norvegica* at 80°03' N, 2°47' E. Hirche *et al.* (1994) found *M. norvegica* to be the most numerous euphausiid species in the Fram Strait, although absolute numbers were very low.

Around Iceland, adults can occur in numbers comparable to other well-populated regions (Astthorsson and Gislason, 1997). However, breeding is almost entirely restricted to the outer parts of the southern continental shelf, especially on the western side (Einarsson, 1945). For instance, Astthorsson (1990) found no records of larvae in Ísafjord, northwest Iceland, and concluded that recruitment came from outside. Astthorsson (1990) also found that the abundance of adults in the fjord increased during the autumn and winter, probably as a result of autumn water exchange.

M. norvegica has been found mainly in the southwestern part of the Barents Sea where its occurrence depends on transport with inflowing water from the west (Dalpadado and Skjoldal, 1991; Drobysheva, 1979). Dalpadado and Skjoldal (1991) did not find the species in the central or northern Barents Sea during their investigations in May–June nor at any stations during an August cruise to the northern Barents Sea. Drobysheva (1957) found *M. norvegica* specimens with spermatophores on both sexes in the Barents Sea, but larvae were registered only in the western part in the warm summer of 1954; she therefore assumed that *M. norvegica* does not reproduce in the Barents Sea and that recruitment occurs partly from the warmer Norwegian Atlantic Current. Bjorck (1916) did not record any *M. norvegica* in the fjords of Spitsbergen but recently, Buchholz *et al.* (2010) found younger specimens of *M. norvegica* in Kongsfjord, west Spitsbergen. They found no sign of reproducing adults.

Further to the east, *M. norvegica* has been recorded in the White Sea and also off the south-western part of Nova Zemlya and along the Murmansk coast, although breeding is not believed to occur in this region (Boldovsky, 1937; Drobysheva, 1957, 1961; Zelickman, 1958, 1961; Mauchline and Fisher, 1967). No one has found *M. norvegica* east of 52° E, which appears to be its north-easterly limit apart from one isolated record by Sars (1900) from 81°24′ N 125° E.

3.4. North-east Atlantic and fringes

M. norvegica becomes increasingly common heading from the northern to the southern reaches of the Norwegian coast and the adjacent Norwegian Sea. At the northern extremities, Falk-Petersen and Hopkins (1981) found a *M. norvegica* population in Balsfjorden (69°32′ N 19°06′ E) containing adults with spermatophores but no evidence of any larvae. They concluded that the population in this region depended entirely on recruitment from external coastal waters, possibly associated with the northward flow of the Norwegian Coastal Current. On the mid Norwegian shelf (Møre plateau, Halten, Sklinna, Træna banks), the abundance of *M. norvegica* reaches >20 individuals per square meter Dalpadado (2006), with the main distribution being within Atlantic waters of the shelf region. The species is common off the southwestern Norwegian coast, and in the Skagerrak and Kattegat and adjacent fjords such as Gullmarsfjord (Heegard, 1948). Heegard (1948) considered that the species bred in all of these environments apart from the Kattegat, but breeding and recruitment in deep basins within the Kattegat has since been reported by Boysen and Buchholz (1984). It is absent from the North Sea, probably because of its shallow bathymetry (Kramp, 1913).

There have been studies on various life-stages on the Iceland-Farøe ridge, around the Faerøes and in the Faerøe-Shetland Channel (Glover, 1952; Lindley, 1982a; Lucas et al., 1942; Ruud, 1926). Further detailed studies on the behaviour and life-history of the species have been carried out on populations resident along the Scottish west coast (Mauchline, 1960; Tarling, 2003); Williamson (1956) has studied the local distribution in the Irish Sea and it is also common off the west and south coasts of Ireland (Frost, 1932). Although a contributor to the euphausiid community in the south-west reaches of the British Isles, the species becomes increasingly less common moving into the open North Atlantic, south of 55° S (Lindley, 1982a).

3.5. Mediterranean and south-eastern limits

M. norvegica is located throughout the Bay of Biscay and in the area westward of it (Ruud, 1936). Ruud (1936) also records the species off the Portuguese coast and southwards to Gibraltar as well as in the Bay of Cadiz. Bouvier (1907) presents records of the species between the Cape Verde Islands and Tenerife (the Canaries), while Mauchline (1980) reported a record from A de C Baker that the species was found westwards of the Azores at 37°34.3′ N 25°22.0′ W and also off the West African coast at 28°04.8′ N 14°04′ W. Thiriot (1977) reported that M. norvegica, along with Euphausia krohni, dominated the euphausiids in the oceanic zone to the north of Casablanca (Morocco). The species was also found in a region of active upwelling from Casablanca down to Cape Juby (27°57′ N 12°56′ W).

M. norvegica occurs throughout the western Mediterranean (Alvarino, 1957; Dion and Nouvel, 1960; Furnestin, 1960; Jespersen, 1923; Ruud, 1936), particularly between 5° and 10° E (Jespersen, 1923; Labat and Cuzin-Roudy, 1996). In the eastern Mediterranean, it has neither been found south of 35° N nor east of 29° E (Mauchline and Fisher, 1967). It has been recorded in the Sea of Marmara and the Aegean Sea, in small numbers (Jespersen, 1923; Ruud, 1936). It is more common in the Adriatic and Ionian seas, particularly in deep waters (Guglielmo, 1979), and also in the Tyrrhenian Sea (Brancato et al., 2001).

3.6. Distributional trends and changes

There is little information regarding any recent changes in the southerly limits of this species, which may partially reflect the paucity of appropriate sampling in these regions. However, there has been particular interest in the recent extent of the northerly distributional limits, along with increased incursion of water with Atlantic origin into Arctic waters (Cottier *et al.*, 2005; Cottier *et al.*, 2007). In particular, inflowing coastal and Atlantic waters from the southwest are cooled and gradually transformed into Arctic waters in the Barents Sea (Loeng, 1989; Midttun, 1989). This makes the Barents Sea a zoogeographical transition zone between the Atlantic-boreal and the Arctic and its pelagic communities are strongly influenced by immigration from inflowing currents (Dalpadado and Skjoldal, 1991; Skjoldal and Rey, 1989).

The first remarks regarding the changing distribution of *M. norvegica* in this region were made by Mauchline and Fisher (1967) who stated that "in the early part of this century, [*M. norvegica*] was rare but became common in the southern Barents Sea and extended eastwards up to Nova Zemlya by the late 1930s. It now appears to be always present, although no breeding has yet been recorded." Siegel (2000a) did not consider that this trend was directly related to increasing temperature but rather to an increasing number of years in which a 'stronger drift of the North Atlantic Current carry krill to northern waters'.

Zhukova *et al.* (2009) examined *M. norvegica* abundance in the Barents Sea relative to sea-water temperature using more than 50 years of net-sample data. They found that the portion of *M. norvegica* varied considerably over the time period, from 10 to 20% of the total euphausiids population in the warm 1950s and 1960s to almost completely disappearing in the 1970–1990s. The peak of species occurrence (18–26%) took place in the beginning of the warm period (1999–2000) after a succession of cold years. Nevertheless, there was a subsequent decline in the proportion of *M. norvegica* to 7% in 2004–2005 despite the fact that sea-water temperatures remained abnormally warm. This was believed to be the result of fish predation, particularly by cod fingerlings. The study of Zhukova *et al.* demonstrates that the establishment of a species such as *M. norvegica* outside of its previous boundaries depends not only on suitable physical conditions but also on the food-web dynamics of the new communities into which they are introduced.

4. THE SAMPLING OF NORTHERN KRILL

Historically, much of what we have learnt about Northern krill has been obtained through the deployment and analysis of trawled-nets. Net sampling remains the standard method with which to determine the size-structure of krill populations and to obtain specimens for incubation and experimentation. Nevertheless, net sampling alone has limitations in ascertaining distribution patterns and true levels of biomass since Northern krill can avoid on-coming nets quite effectively given their fast swimming speed and high visual acuity. Active acoustics are now frequently used alongside net-sampling deployments in carrying out such surveys, giving the added benefits of greater coverage and minimal avoidance. Direct visual observations *via* manned-submersibles or remotely operated vehicles (ROVs) are also providing useful insights into the behavioural ecology of krill species. The use of each of these techniques and their intercomparison are discussed in the following sections.

4.1. Net sampling

Nets that have achieved the greatest success in obtaining samples of Northern krill have incorporated a number of attributes, of which the main ones are:

- Being as large as possible
- Being propelled relatively fast at a constant speed
- Remaining at least 85% efficient throughout the haul
- Being free of obstructions forward of the mouth, such as tow lines
- Generally being dark coloured with no shiny metal components

Some of the earliest nets were simple ring nets, with diameters generally between 0.5 and 3 m (e.g. Einarsson, 1945; Mauchline, 1960; Table 1.2). They either contained meshes of between 200 μm and 7 mm, or stramin, a coarse canvas of hemp, with 500 threads to 100 cm. Oblique tows at speeds of $0.5-1$ m s^{-1} have been the most common deployment method. Kaartvedt *et al.* (2002) customised this simple device with a strobe light to temporarily startle the krill and decrease avoidance. The plummet net, which is a ring net packed with lead shot around its rim, decreases avoidance through a rapid vertical approach to krill from above (Hovekamp, 1989). The rapidity of the deployments and the fact that little pressure is placed on the cod-end makes plummet nets effective at maintaining captured specimens in a healthy state for further process work. A commonly used alternative for this purpose is the Isaacs-Kidd Midwater Trawl (Isaacs and Kidd, 1953) which has a mouth opening of 1 m^2 and a mesh size of 1.5 mm. The net is easy to deploy and capable of rapid retrieval, which minimises the amount of time captured animals remain in the cod-end.

To examine the depth distribution and behaviour of krill populations, multinets have been mainly deployed, which are systems based on the principle of opening and closing a series of individual plankton nets in succession. Two widely used types are the BIONESS (Sameoto *et al.*, 1980), and the MOCNESS (Wiebe *et al.*, 1985). The systems are towed horizontally or obliquely with the nets being incremented at desired intervals. Best results have been achieved through sampling mainly during the upward trajectory. The BIONESS is towed at higher speeds

Table 1.2 A synthesis of the different types of nets deployed to capture Northern krill (*Meganyctiphanes norvegica*) categorised according to the purpose of the net deployment

Net sampling device	Purpose of net deployment					Net and deployment parameters					
	Abundance/biomass geographic surveys	Population dynamics and biochemistry	Horizontal and vertical distribution	Behaviour and feeding	Live specimens	Accompanying acoustic data	Mouth	Mesh	Number of nets	Towing speed (knots)	Haul Speed
Simple nets WP2/Conical net/Nansen Net/ring trawl	4	1, 2, 4, 5, 8, 9	4	6 (with strobe light), 10	3, 7, 11		*Area*—0.5 m^2[1], 2 m^2[5], 1.2 m^2[8] *Diameter*—0.6 m[2], 1 m[2,11,6,7,10], 2 m[4], 3 m[4]	200 μm[2,3], 300 μm[1], 500 μm[2,7], 7 mm[4], 2 mm[5,9,10]	1	4[2]	1 m s^{-1}[1] 0.75 m s^{-1}[10]
Bongo nets	12, 15	13, 14	12, 16	12		120 kHz echosounder. [16]	*Area*—0.2 m^2[13], 0.05 m^2[13] *Diameter*—0.6 m[12,14,15], 0.75 m[16]	200 μm[13], 243 μm[16], 335 μm[14,15], 505 μm[12]	2	2.5[14,15]	10 m min^{-1} [14,15]
Stramin Net (conical)	4	4, 17	4, 17	17			*Diameter*—1 m[17], 1.5 m[4], 2 m[4]	*Coarse canvas of hemp, 500 threads to 100 cm*	4 (surface, 50 m, 100 m and 150 m)[17]	4[17]	
Isaacs-Kidd Midwater Trawl Isaacs and Kidd (1953)		5, 8, 19, 18, 20, 31, 32	26, 33	6, 23, 25	23, 27, 28, 29, 30, 34, 35	300 kHz *ADCP.* [21] [22] *38 and* 120 kHz echosounder. [24, 25]	*Area*—0.6 m^2[20,21,22,23] [24,25,28,29] 2 m^2[18] 9 m^2[26] *Diameter*—3-feet[6]	500 μm[25,30] 1500 μm[20,21,22] [23,27,28,29] 2 mm[8] 4 mm[18] 1 cm[26]	1	1.75[18] 2.0[22,26] 2.0-2.5[19]	
Ori net Omori (1965)		19							1	2.0-2.5	
Beyer's Low Speed Midwater Trawl		13					*Area*—0.7 m^2	1 mm	1		

Multiple nets											
BIONESS electronic mulitnet Sameoto et al. (1980)	39, 40	41	36, 37, 38, 42	36, 37, 42	3	12, 50, 122 and 200 kHz echosounder:[41] 38 and 120 kHz echosounder:[39, 40, 42] 153 and 300 kHz ADCP:[42]	Area—1 m²	500 μm[36,38] 250, 333 μm[37,39]	12	2–3[37] 4[38]	1.5–2 m s⁻¹[36]
MOCNESS Wiebe et al. (1985)	44, 57	5, 9, 44, 45, 51, 54	2	45, 48, 49, 52, 53, 56, 58	50, 55	153 kHz ADCP:[46, 47, 52, 58] 300 kHz ADCP:[49]	Area—1 m²	333 μm[2,9,44,45,46,47,48,49,52,54,56] 2 mm[9,45,47,48,49,51,52,53,54,55,56]	9	1.5–5[44] 2.5[45,55] 3[46]	0.5 m s⁻¹[46]
Methot net	59	59		60	59		Area—1.5 m²[59] Diameter—2 m[60]	500 μm/ 1.2 mm[60] 1.5 mm[59]	2[59] 1[60]	1.5–2.0[59]	
Combination Rectangular Midwater Trawl Baker et al. (1973), Roe et al (1980)		61, 62					1 m², 8 m²	0.33 μm, 3 mm main mesh, 1.5 mm cod end	2		
Benthic nets											
Macroplankton net attached to bottom trawl	63		63	63			0.2 m²	564 μm	1		
Trawls											
Pelagic trawls: 'Young-fish'[25] 'Åkra'[57,64]	4, 57	4	25, 64, 65, 66		4	38 and 120 kHz echosounder:[25,66]	Area— 100 m²[25,57]	20 cm at opening and 1 cm at			

(continued)

Table 1.2 (continued)

Net sampling device	Purpose of net deployment						Net and deployment parameters				
	Abundance/biomass geographic surveys	Population dynamics and biochemistry	Horizontal and vertical distribution	Behaviour and feeding	Live specimens	Accompanying acoustic data	Mouth	Mesh	Number of nets	Towing speed (knots)	Haul Speed
'Fipec'[65] 'Petersen's young-fish'[4]						18, 38, 70, 120, and 200 kHz echosounders:[64] 38, 120 and 200 kHz echosounders:[65]	*Vertical opening*—30 m[57,64]	rear end[25,57] 3200 mm at front to 20 mm in cod-end[64] 5 mm[65] Stramin (500 threads to 100 cm)[4] 38 mm mesh cod end net with a 20 m long inner-net of 16 mm mesh[57]			
High speed samplers											
Continuous plankton recorder	68	67						280 μm	Continuous spool	Variable ship speed–	3 m³ of water filtered per 16 km of tow
Gulf III sampler	68	68									

Information on net deployment parameters (where detailed in the study) are also provided. Italicised numbers relate to the following articles: 1, Timofeev (2002); 2, Boysen and Buchholz (1984); 3, Mayzaud et al. (2005); 4, Einarsson (1945); 5, Albessard and Mayzaud (2003); 6, Kaarvedt et al. (2002); 7, Torgersen (2001); 8, Albessard et al. (2001); 9, Cuzin–Roudy and Buchholz (1999); 10, Onsrud and Kaartvedt (1998); 11, Conti et al. (2005); 12, Kulka et al. (1982); 13, Falk-Petersen and Hopkins (1981); 14, Astthorsson (1990); 15, Astthorsson and Gislason (1997); 16, Sameoto (1976); 17, Mauchline (1960); 18, Hassan (1999); 19, Labat and Cuzin–Roudy (1996); 20, Thomasson (2003); 21, Liljebladh & Thomasson (2001); 22, Strömberg et al. (2002); 23, Spicer et al. (1999); 24, Everson et al. (2007); 25, Kleyjer & Kaartvedt (2006); 26, Sardou et al. (1996); 27, Schmidt et al. (2004); 28, Cuzin–Roudy et al. (2004); 29, Thomasson et al. (2003); 30, Bämstedt and Karlson (1998); 31, Hollingshead and Corey (1974); 32, Matthews (1973); 33, Guglielmo (1979); 34, Spicer and Strömberg (2002); 35, Salomon et al. (2000); 36, Brancato et al. (2001); 37, Harvey et al. (2009); 38, Sardou and Andersen (1993); 39, Simard and Lavoie (1999); 40, Simard and Sourisseau (2009); 41, Cochrane et al. (2000); 42, Sourisseau et al. (2008); 43, Sourisseau et al. (2008); 44, Dalpadado and Skjoldal (1991); 45, Tarling (2003); 46, Buchholz et al. (1995); 47, Tarling et al. (1999a); 48, Tarling et al. (1999b); 49, Tarling et al. (2002); 50, Saborowski et al. (2002); 51, Tarling and Cuzin–Roudy (2003); 52, Tarling et al. (2001); 53, Lass et al. (2001); 54, Virtue et al. (2000); 55, Saborowski et al. (2000); 56, Tarling et al. (1998); 57, Dalpadado et al. (1998); 58, Zhou et al. (2005); 59, Saunders et al. (2007); 60, Dalpadado et al. (2008); 61, Mauchline (1985); 62, Lindley et al. (1999); 63, Zhukova et al. (2009); 64, Kaartvedt et al. (2005); 65, Cotté and Simard (2005); 66, Kleyjer and Kaartvedt (2003); 67, Lindley (1982a, 1982b); 68, Dalpadado (2006)

(3–4 knots) than the MOCNESS (1.5–2 knots) and may be more efficient at capturing faster moving organisms such as krill. Nevertheless, the MOCNESS has been used effectively in a number of research efforts on euphausiids (Table 1.2). Most such systems collect a suite of environmental parameters during deployment, particularly temperature, salinity, down-welling irradiance and fluorescence.

Some studies have employed trawl nets, designed for larval fish surveys, to capture Northern krill (Table 1.2). In many instances, the surveys were oriented towards the capture of fish and have obtained krill as a by-catch. Nevertheless, such catches have provided a great deal of useful information on distribution and population structure. Similarly, samples of krill obtained from the epibenthic layer have mainly been obtained from surveys with other sampling priorities. For instance, Zhukova et al. (2009) performed a valuable study on Northern krill using a macroplankton net attached to a bottom trawl.

4.2. Active acoustics

Acoustics has perhaps now overtaken net sampling as the preferred means of investigating krill distribution and abundance (Hewitt and Demer, 2000; Macaulay, 2000). Some view nets as secondary tools, whose main purpose is to verify the results of acoustic studies (Pauly et al., 2000).

Acoustic surveying has the advantages of being non-invasive and capable of covering large areas rapidly. Nets nevertheless remain the primary tool for obtaining krill specimens for population and experimental work (Nicol and Brierley, 2010).

Modern day acoustic systems consist of a number of different frequency echosounders, ranging between 18 and 500 kHz. The difference in scattering strength of targets across these frequencies gives information on their body size and tissue density which, through modelling and *in situ* validation, has become a means of inferring target identity remotely. Technologies and analytical techniques continue to develop (e.g. Korneliussen and Ona, 2002) but one well-established approach towards locating aggregations of Antarctic and Northern krill is to compare the scattering strength of aggregations at 38 kHz and 120 kHz (Madureira et al., 1993a,b). All acoustic approaches to estimating numerical density and biomass of krill rely on an estimation of the target strength (TS) of individual specimens. There has been relatively little consideration of Northern krill TS compared to Antarctic krill. Nevertheless, there is much overlap in morphology, size, and tissue density to allow TS estimates to be interchanged. For instance, in an acoustic survey of a Northern krill population in Gullmarsfjord, Sweden, Everson et al. (2007) used a TS of −76.1 dB (range 6.2) at 120 kHz and −85.1 (range 6.5) at 38 kHz which were values reported for *E. superba* by Foote et al. (1990). Foote et al. (1990) estimated *E. superba* TS from a cage experiment where the size range of specimens (30–39 mm) was almost identical to that of the fjordic Northern krill. TS has been estimated during *in situ* observations of Northern krill using a moored upward looking split-beam echosounder by Klevjer and Kaartvedt (2006). Although their mean values differed little from those used by Everson *et al.* (2007), of particular interest was the distinct variability in TS they observed as a result of the size and behaviour of the krill, which could account for a change in TS of the order of ± 10 dB.

It is well recognised that there is often a discrepancy between net-catch and acoustic survey estimates of krill biomass, with the general belief that net sampling may underestimate true biomass by orders of magnitude because of escapement (Hamner and Hamner, 2000; Watkins, 2000). Echosounders also sample much larger volumes than the nets, for example, 50-fold more at 150 m range. Nevertheless, over broad temporal and spatial scales, Atkinson *et al.* (2009) found there to be a general agreement in biomass estimates of Antarctic krill made by nets and acoustics. Sameoto *et al.* (1993) used a combination of echosounders, an optical particle counter, cameras attached to ROVs and a BIONESS net equipped with a strobe light to examine the level of net avoidance by Northern krill. They concluded that net-avoidance was significant but reduced by the strobe lighting. Simard and Sourisseau (2009) carried out a comparison on fine temporal and spatial scales during a 3-day experiment at a fixed location, using strobe-equipped BIONESS and echosounders, and found that diel effects altered both net catch and acoustic estimates of Northern krill biomass. Night-time catches in the krill scattering layer (SL) were 15 times the acoustic estimates, but the situation was reversed during daytime, when the acoustic estimates in the SL were 5 times larger than the catches. Net avoidance explained the daytime anomaly while, at night time, it was believed that a change in krill orientation, affecting TS, led to an underestimation of biomass by the echosounder. The effect of orientation is a major consideration in krill TS model development (Conti and Demer, 2006).

There has been comparatively little use of echosounders to perform large-scale biomass surveys of Northern krill. By contrast, efforts to survey biomass distributions of Antarctic krill have involved simultaneous multi-ship international efforts (Hewitt *et al.*, 2004). This may reflect the greater interest in stock sizes of the latter species. Echosounders have nevertheless revealed interesting details with regards local-scale distribution patterns of Northern krill. In the Gulf of St Lawrence, for instance, combined net and echosounder surveys found a number of vertical and horizontal distributional features of Northern krill scattering layers in relation to upwelling coastal currents, 3D circulation patterns as well as evidence of predator–prey interactions, particularly with regards the feeding behaviour of whales (Simard *et al.*, 1986; Sourisseau *et al.*, 2006, 2008). Moored echosounders have provided revealing insights into vertical migration behavioural patterns (Onsrud *et al.*, 2005) as well as estimates of swimming speed (Klevjer and Kaartvedt, 2003) and TS (Klevjer and Kaartvedt, 2006). Moored acoustic Doppler current profilers (ADCPs) have obtained time-series information at coarser resolution levels. Tarling *et al.* (2002) for instance, was able to detect a normal DVM in Northern krill plus an additional secondary midnight sinking behaviour in its prey, *Calanus finmarchicus*. ADCPs are principally designed to measure current speeds and are particularly suited to documenting the interaction between krill distributions and the horizontal advective regimes (Liljebladh and Thomasson, 2001; Zhou *et al.*, 2005). The devices also provide an estimate of the vertical velocity of particles and Tarling *et al.* (2001) used this parameter to estimate the relative vertical migration speed of individuals within Northern krill swarms.

4.3. Alternative technologies for sampling and observation

Nets and acoustics remain the most tried and tested sampling technologies for the study of Northern krill. In their review, Nicol and Brierley (2010) consider that this will remain so in the study of all krill species into the future. Nevertheless, they also advise that new technologies be embraced and used in parallel, with their effectiveness assessed through cross-comparisons. This is particularly so when krill occupy areas where sampling by traditional methods is limited.

Ship-based acoustics systems have trouble in sampling the very surface of the water column, the 3–15 m of water which resides above the downward looking, ship-borne transducers. Surface reflectance means that this zone is also unresolved by upward looking moored systems. Northern krill is one of several euphausiid species that notably aggregate within surface swarms, either through the action of tides and upwelling or as a result of their own motivation to mate and spawn (Nicol, 1986). Surface swarms offer unique sampling opportunities because the population is concentrated in the surface layer and is often highly visible (Nicol and Brierley, 2010). The swarms can be adequately viewed from ships, as appreciated by the first International krill biomass experiment (FIBEX) in the Southern Ocean, which developed standard protocols to measure surface-swarm parameters such as area, shape, spacing, and even colour (Hampton, 1982). More synoptic data can be obtained from cameras mounted to aircraft, as reported by Nicol (1986) for Northern krill in the Bay of Fundy. Satellite imagery is now approaching similar levels of resolution which may greatly facilitate data coverage and acquisition (Nicol and Brierley, 2010). At smaller scales, it is feasible for divers to enter surface swarms and examine their structure first hand, as has been achieved for *Euphausia pacifica* by Hanamura *et al.* (1984) and Endo *et al.* (1985), and for *E. superba* by Hamner (1984). These studies have reported swarm densities that are orders of magnitude higher than estimated through either nets or acoustics (Hamner and Hamner, 2000).

The epibenthic layer is another region that is poorly sampled by pelagic nets and badly resolved by acoustics. In shelf and upper slope areas, euphausiids can have access to the seafloor and *M. norvegica* has been caught in these regions by hyperbenthic sledges (Beyer, 1992) and bottom trawls containing additional macroplankton nets (Zhukova *et al.*, 2009). This layer may be a daytime residence to escape predation as well as a feeding environment (see Chapter 10 and Chapter 5). Our knowledge of the biology of krill in this layer is further supplemented by observations from manned submersibles and ROVs. For instance, Youngbluth *et al.* (1989) observed Northern krill feeding on epibenthic particles while Jonsson et al. (2001), Hudson and Wigham (2003) and Rosenberg *et al.* (2005) described the swimming behaviour of Northern krill in the epibenthic layer and their interaction with benthic fauna. The potential for these kinds of observations is continuously increasing, as ROVs become more common and increasingly capable of carrying out deeper and longer dives. For instance, from ROV observations, Clarke and Tyler (2008) were recently able to report on *E. superba* feeding benthically at 3500 m. More such studies in Northern krill habitats would be valuable.

5. Ecological Role

M. norvegica can make up a significant part of the biomass of the euphausiid community over large parts of its distributional range (see above). This makes it both an important consumer of productivity at lower trophic levels (Chapter 5) and an important prey item for many higher predators (Chapter 10). Beyond these well-considered roles, however, Northern krill has a number of other ecological functions that have impacts on other communities and contribute to wider biogeochemical processes.

5.1. Benthic interactions

The behaviour and biomass of *M. norvegica*, particularly in near shore waters, may have a profound impact on benthic and epibenthic communities. The species can be potentially a major source of biogenic material (faecal strings, moults, carcasses) for these communities, can act as a bioturbator and transporter of resuspended and flocculent sediments, and can be a major prey item for demersal predators. This contribution to benthic communities was examined in detail by Youngbluth *et al.* (1989), who made observations with a submersible within the Wilkinson and Jordan basins in the Gulf of Maine and in canyons south of Georges bank. The submersible observed high densities of krill faecal strings in a 5–24 m thick layer coincident with the pycnocline (at depths of around 15–30 m) which were collected and later examined for sinking rate and organic-matter content. Youngbluth *et al.* (1989) calculated sinking rates to be $\sim$204 m d^{-1} $\pm$ 24 S.E and organic matter transport to the bottom to be 7–12 mg C m^{-2} d^{-1}. This represented 1–6% of the daily primary production in the surface mixed layer of that region. The packaging of organic material within fast sinking faecal strings not only provides biogenic material to the benthic community but also means that carbon is rapidly exported out of the surface mixed layers, so increasing its likelihood of ultimate carbon sequestration. This effect was also examined in *E. superba* in the Southern Ocean by Tarling and Johnson (2006), where krill were estimated to transport 2.3 × 10^{13} g C to the ocean interior each year.

Nevertheless, in the case of Northern krill, the transport of carbon may not always be downwards. Youngbluth *et al.* (1989) observed particularly dense aggregations of *M. norvegica* (up to 10^4 individuals per cubic meter) close to benthic boundary layer. Consumption of biogenic detritus in the epibenthic region by these krill may account for considerable repackaging and bioturbation. Assuming that these krill feed more or less continuously, individuals that then migrate vertically each night may introduce recycled biogenic and inorganic materials back into the surface mixed layer when they release faecal strings.

The dense epibenthic aggregations may be particularly targeted by groundfish such as cod, hake, pollack as well as squid. The density of these aggregations may also be sufficient to supply the daily food rations of fin whales. Sourisseau *et al.* (2006) considered how the predictable nature of krill to form high density aggregations in certain locations influences the behaviour of whales. The sea bottom boundary may limit the potential for escapement by krill during feeding forays by whales, so increasing whale foraging efficiency.

5.2. Krill production and predator consumption

For understanding the quantitative role of Northern krill in the transfer of organic material from primary production to higher trophic levels, consideration of its productivity must be made. Siegel (2000b) considered the production to biomass (P/B) ratios of *M. norvegica* was relatively low compared to temperate and warm water species. Typical values were around 3:1–4:1. *E. pacifica* in Japan and *E. superba* in the Southern Ocean have similarly low values, ranging between 1:1 and 4:1 (Siegel, 2000b).

There is notable regional and temporal variability in levels of productivity in *M. norvegica*. For instance, Lindley (1982a) estimated that its annual net production at 10 m depth in the North Atlantic varied between 0.80 and 4.31 mg m^{-3} y^{-1} in 1966 and between 1.62 and 18.74 mg m^{-3} y^{-1} in 1967. The range in values, particularly in the second year, was mainly explained by the comparatively high but variable levels of production in Norwegian waters. In other North Atlantic regions, productivity was lower and less variable. This variation does not necessarily translate into accompanying fluctuations in the P/B ratio, which were all between 1.3:1 and 6.3:1, mainly because biomass and production fluctuated in the same direction.

In the Norwegian Sea, the total biomass of euphausiid stocks was estimated at 42 million tonnes (Mt) by Melle *et al.* (2004). According to Lindley (1982b), between 40 and 75% of this stock is made up of Northern krill. Although the Norwegian Sea is a major hotspot for Northern krill, it represents only a fraction of its distributional range (see above) meaning that the total stock biomass may be equivalent to that of Antarctic krill, presently estimated at around 380 Mt (Atkinson *et al.*, 2009). Through the use of three independent methods, Atkinson *et al.* (2009) estimated gross post-larval production in Antarctic krill as 342–536 Mt y^{-1}. Such a level of production may also be achieved by Northern krill, given the similarity in its P/B ratio to that of Antarctic krill (Siegel, 2000b). Drobysheva and Panasenko (1984) calculated that fish consumed about three quarters of the average yearly production of krill in the Barents Sea. Assuming that a similar ratio applies across all Northern krill populations, around 200–400 Mt y^{-1} of krill is potentially consumed by fish predators, with additional amounts being taken by birds and whales. Given the value of these parameters both to biogeochemical models and fishery managers, much greater effort is required to determine their bounds and levels of variability.

6. Commercial Exploitation of Northern Krill

The most established fishery for euphausiids is within the Southern Ocean where presently around 150,000 t of *E. superba* is being harvested per year (CCAMLR, 2010). This represents a fraction of the total allowable limit of 3.47 Mt y^{-1}, set by the Convention for the Conservation of Antarctic Marine Living Resources (CCAMLR), as part of the Antarctic Treaty System, although there is a lower trigger level of 620,000 t at which further consultation is required. This relatively low harvest to stock ratio reflects the logistical and economic difficulties in catching and processing krill at commercially viable levels. Compared to Antarctic krill, the fishery for Northern krill is even more underdeveloped. There has been some commercial interest shown in the krill on the eastern Canadian seaboard, restricted to the Nova Scotia shelf and specifically the St Lawrence estuary. Northern krill is amongst four species there, the others being *Thysanoessa longicaudata*, *Thysanoessa raschii* and *Thysanoessa inermis*. The Maritime Region of Department of Fisheries and Ocean Canada (DFO) in 1991 issued its first scientific permit to harvest zooplankton—for both krill and copepods in the Gulf of St Lawrence. Exploratory fishing commenced in the St Lawrence estuary in 1993. A total allowable catch (TAC) was set in 1994 of 100 t for krill and 50 t for copepods—reported catches for that season were 6.3 t and 0.4 t, respectively. In 1995, TACs were increased to 300 t for krill and 2000 t for *Calanus* but reported catches were 2 t for krill and 1 t for *Calanus* (Harding, 1996). There has been little further interest in this fishery (Everson, 2000b).

Nevertheless, demand for krill may be reawakened as new uses are found. One potential user is the salmonid aquaculture industry, where the demand for alternative protein sources to fishmeal has increased greatly in recent years in parallel with increases in total production (Waagbø *et al.*, 2001). Further growth in the carnivorous fish aquaculture industry depends on new feed resources becoming available. Krill meal has been proposed as one potential new feed.

Ringo *et al.* (2006) assessed the utility of krill meal as a salmon feed in a trial where it was substituted for fishmeal in farms of the Atlantic salmon (*Salmo salar* L.) for 46 days. They found no differences in weight gain, length gain, feed conversion or specific growth rate between the groups that could have been attributed to a change in diet. The dietary switch did alter the hindgut microbiota. Furthermore, when fish were fed the krill diet, hindgut enterocytes were replete with numerous irregular vacuoles, which were not observed in fish fed with fishmeal. The implications of this difference remain unclear. Problems with a diet of krill meal were more clearly determined by Yoshitomi *et al.* (2006), who found that rainbow trout fed krill meal, accumulated fluoride in bones, adversely affecting growth of the vertebrae and decreasing growth. Since over 99% of the fluoride in krill is concentrated in its exoskeleton, the advice from this study was to exclude the exoskeleton from krill meal preparations.

There have also been recent developments in promoting krill products as dietary supplements. It has been proposed that concentrated krill oil can assist health during pregnancy and during treatments for cancer and for immunodeficiency and nutritional disorders, and other benefits are continually being put forward. The main active components are polyunsaturated ω-3 fatty acids, which make up 19% of the total fatty acids in krill. These are relatively stable during processing and can be stored for long periods. Krill are also a source of chitin. Chitin and chitosan, which is derived from chitin, have a range of potential uses in membrane production, hair products and cholesterol-lowering drugs. The pharmaceutical industry is further exploring krill enzymes in the treatment of inflammation and to assist in the healing of wounds.

The means of harvesting krill have also developed dramatically in recent years. A major hurdle is to maintain caught krill in sufficiently good condition to extract the valuable biochemical components. One solution is to pump krill directly onboard through a tube connected to a net cod-end (the part of the net where the krill are retained). The krill arrive on board still alive and fresh, increasing the chances that subsequent extraction procedures for biochemical components are successful. The FV *Saga Sea*, owned by Aker BioMarine, employs such 'suction' harvesting methods and has the capacity to fish up to about 120 000 tonnes of krill annually. The method of capture suits a species, such as Antarctic krill, that aggregates into well defined and detectable swarms. However, Northern krill are often found in more dispersed layers and may not be so amenable to such an approach. Furthermore, Northern krill distribution centres are far more widespread that those of Antarctic krill, which concentrate in the southwest Atlantic sector and Antarctic Peninsula regions of the Southern Ocean (Atkinson *et al.* 2008). Viable commercial exploitation of Northern krill may necessitate a different approach, probably operating at a smaller scale and for local processing and utilisation.

ACKNOWLEDGEMENTS

We thank Andrew Fleming (British Antarctic Survey) for the provision of MODIS data to plot the isotherms on Fig. 1.6. Juliet Corley provided the illustrations for Figs 1.1–1.3. John Mauchline was a source of encouragement and provided good leads to track down old reports and manuscripts. Katrin Schmidt, Janine Cuzin-Roudy, and Fred Buchholz gave valuable comments on drafts of this chapter. The initial inspiration to undertake this work was provided by Fred Buchholz and Jack Matthews and the team-members of the EU PEP programme on Northern krill physiology and ecology. GT and WPGC carried out this review as a contribution to the Ecosystems programme at the British Antarctic Survey.

REFERENCES

Albessard, E., and Mayzaud, P. (2003). Influence of tropho-climatic environment and reproduction on lipid composition of the euphausiid *Meganyctiphanes norvegica* in the Ligurian Sea, the Clyde Sea and the Kattegat. *Mar. Ecol. Prog. Ser.* **253,** 217–232.

Albessard, E., Mayzaud, P., and Cuzin-Roudy, J. (2001). Variation of lipid classes among organs of the Northern krill *Meganyctiphanes norvegica*, with respect to reproduction. *Comp. Biochem. Physiol. A* **129,** 373–390.

Alvarino, A. (1957). Zooplancton del Atlantico iberico. Campana del "Xauen" en el Verano del 1954. *Boln. Inst. Esp. Oceanogr.* **82,** 1–51.

Astthorsson, O. S. (1990). Ecology of the Euphausiids *Thysanoessa raschi*, *T. inermis* and *Meganyctiphanes norvegica* in Isafjord-Deep, northwest Iceland. *Mar. Biol.* **107,** 147–157.

Astthorsson, O. S., and Gislason, A. (1997). Biology of euphausiids in the subarctic waters north of Iceland. *Mar. Biol.* **129,** 319–330.

Atkinson, A., Siegel, V., Pakhomov, E. A., Rothery, P., Loeb, V., Ross, R. M., Quetin, L. B., Schmidt, K., Fretwell, P., Murphy, E. J., Tarling, G. A., and Fleming, A. H. (2008). Oceanic circumpolar habitats of Antarctic krill. *Mar. Ecol. Prog. Ser.* **362,** 1–23.

Atkinson, A., Siegel, V., Pakhomov, E. A., Jessopp, M. J., and Loeb, V. (2009). A re-appraisal of the total biomass and annual production of Antarctic krill. *Deep Sea Res. I* **56,** 727–740.

Baker, A. d. C., Clarke, M. R. *et al.* (1973). The N. I. O. Combination net (RMT 1+8) and further developments of rectangular midwater trawls. *J. Mar. Bio. Association, U.K.* **53,** 167–184.

Baker, A. d. C., Boden, B. P., and Brinton, E. (1990). A practical guide to the euphausiids of the world Vol. British Museum (Natural History), London.

Båmstedt, U., and Karlson, K. (1998). Euphausiid predation on copepods in coastal waters of the Northeast Atlantic. *Mar. Ecol. Prog. Ser.* **172,** 149–168.

Bargelloni, L., Zane, L., Derome, N., Lecointre, G., and Patarnello, T. (2000). Molecular zoogeography of Antarctic euphausiids and notothenioids: From species phylogenies to intraspecific patterns of genetic variation. *Antarct. Sci.* **12,** 259–268.

Beyer, F. (1992). *Meganyctiphanes norvegica* (M. Sars) (Euphausiacea) a voracious predator on *Calanus*, other copepods, and ctenophores, in Oslofjorden, southern Norway. *Sarsia* **77,** 189–206.

Bigelow, H. B. (1926). Plankton of the offshore waters of the Gulf of Maine. *Bull. Bur. Fish. Wash.* **40,** 509, (1924).

Bigelow, H. B., and Sears, M. (1939). Studies of the waters of the continental shelf, Cape Cod to Chesapeake Bay. III. A volumetric study of the zooplankton. *Mem. Mus. Comp. Zool. Harv.* **54,** 183–378.

Bjorck, W. (1916). Zoologische Ergebnisse der schwedischen Expedition nach Spitsbergen 1908. Teil II. 6. Die Schizopoden des Eisfjords. *Kungl. Svenska Vetensk. Handl.* **54,** 1–10.

Boldovsky, G. V. (1937). Warm-water Euphausiidae (Crustacea) of the Murman Coast. *Dokl. Akad. Nauk S.S.S.R* **17,** 85–87.

Bouvier, E.-L. (1907). A propos du *Nyctiphanes norvegica. Bull. Soc. ent. Fr.* 183–184.

Boysen, E., and Buchholz, F. (1984). *Meganyctiphanes norvegica* in the Kattegat. *Mar. Biol.* **79,** 195–207.

Brancato, G., Minutoli, R., Granata, A., Sidoti, O., and Guglielmo, L. (2001). Diversity and vertical migration of euphausiids across the Straits of Messina area. *Mediterr. Ecosyst. Struct. Process.* **131,** 141.

Buchholz, F., and Boysen, E. (1988). *Meganyctiphanes norvegica* in the Kattegat: Studies on the horizontal distribution in relation to hydrography and zooplankton. *Ophelia* **29,** 71–82.

Buchholz, F., Buchholz, C., Reppin, J., and Fisher, J. (1995). Diel vertical migrations of *Meganyctiphanes norvegica* in the Kattegat: Comparison of net catches and measurements with Acoustic Doppler Current Profilers. *Helgolander wiss. Meeresunters.* **49,** 849–866.

Buchholz, F., Buchholz, C., and Weslawski, J. C. (2010). Ten years after: Krill as indicator of changes in the macro-zooplankton communities of two Arctic fjords. *Polar Biol.* **33,** 101–113.

Casanova, B. (1984). Phyloenie des euphausiaces (crustaces eucarides). *Bull. Mus. Natn. Hist. Nat.* **6,** 1077–1089.

CCAMLR (2010). Statistical bulletin. Commission for the conservation of Antarctic marine living resources, Hobart, Tasmania.

Clarke, W. D. (1963). Function of bioluminescence in mesopelagic organisms. *Nature* **198,** 1244–1246.

Clarke, A., and Tyler, P. A. (2008). Adult Antarctic krill feeding at abyssal depths. *Curr. Biol.* **18,** 282–285.

Cochrane, N. A., Sameoto, D. D., and Herman, A. W. (2000). Scotian Shelf euphausiid and silver hake population changes during 1984–1996 measured by multi-frequency acoustics. *ICES J. Mar. Sci.* **57,** 122–132.

Colosi, G. (1918). Sul genere *Meganyctiphanes* (Eufausiacei). *Monitore Zool. Ital.* **29,** 178–181.

Conti, S. G., and Demer, D. A. (2006). Improved parameterization of the SDWBA for estimating krill target strength. *ICES J. Mar. Sci.* **63,** 928–935.

Conti, S. G., Demer, D. A., and Brierley, A. S. (2005). Broad-bandwidth, sound scattering, and absorption from krill (*Meganyctiphanes norvegica*), mysids (*Praunus flexuosus* and *Neomysis integer*), and shrimp (*Crangon crangon*). *ICES J. Mar. Sci.* **62,** 956–965.

Cotté, C., and Simard, Y. (2005). Formation of dense krill patches under tidal forcing at whale feeding hot spots in the St. Lawrence Estuary. *Mar. Ecol. Prog. Ser.* **288,** 199–210.

Cottier, F., Tverberg, V., Inall, M., Svendsen, H., Nilsen, F., and Griffiths, C. (2005). Water mass modification in an Arctic fjord through cross-shelf exchange: The seasonal hydrography of Kongsfjorden, Svalbard. *J. Geophys. Res. Oceans* **110**.

Cottier, F. R., Nilsen, F., Inall, M. E., Gerland, S., Tverberg, V., and Svendsen, H. (2007). Wintertime warming of an Arctic shelf in response to large-scale atmospheric circulation. *Geophysical Res. Lett.* **34**.

Cuzin-Roudy, J., and Buchholz, F. (1999). Ovarian development and spawning in relation to the moult cycle in Northern krill *Meganyctiphanes norvegica* (Crustacea: Euphausiacea), along a climatic gradient. *Mar. Biol.* **133**, 267–281.

Cuzin-Roudy, J., Tarling, G. A., and Stromberg, J. O. (2004). Life cycle strategies of Northern krill (*Meganyctiphanes norvegica*) for regulating growth, moult, and reproductive activity in various environments: The case of fjordic populations. *ICES J. Mar. Sci.* **61**, 721–737.

Dalpadado, P. (2006). Distribution and reproduction strategies of krill (Euphausiacea) on the Norwegian shelf. *Polar Biol.* **29**, 849–859.

Dalpadado, P., and Skjoldal, H. R. (1991). Distribution and life history of krill from the Barents Sea. *Polar Res.* **10**, 443–460.

Dalpadado, P., Ellertsen, B., Melle, W., and Skjoldal, H. R. (1998). Summer distribution patterns and biomass estimates of macrozooplankton and micronekton in the Nordic Seas. *Sarsia* **83**, 103–116.

Dalpadado, P., Yamaguchi, A., Ellertsen, B., and Johannessen, S. (2008). Trophic interactions of macro-zooplankton (Krill and amphipods) in the Marginal Ice Zone of the Barents Sea. *Deep Sea Res. II* **55**, 2266–2274.

Damas, D., and Koefoed, E. (1909). Le plankton de la mer du Grönland. Duc d'Orleans: Crois. *Océanogr. "Belgica" Mer Grönl.* **1905**, 109.

D'Amato, M. E., Harkins, G. W., de Oliveira, T., Teske, P. R., and Gibbons, M. J. (2008). Molecular dating and biogeography of the neritic krill *Nyctiphanes. Mar. Biol.* **155**, 243–247.

Dion, Y., and Nouvel, H. (1960). Mysidaces et Euphausiaces recoltes en Mediterranee occidentale par le Navire Oceanographique 'Président-Théodore-Tissier' en 1949. *Bull. Sta. Aquic. Pêche Castiglione* **10**, 9–19.

Drobysheva, S. S. (1957). The effect of some aspects of the biology of Euphausiacea upon the summer feeding conditions of cod in the Barents Sea. *Tr. Polyarn. Nauchno-Issled. Preoktn. Inst. Morsk. Rybn. Khoz. Okeanogr.* **10**, 106–124.

Drobysheva, S. S. (1961). On the extent of isolation of the stock of euphausiid cray fish (Euphausiacea) in the Barents Sea. *Dokl. Akad. Nauk S.S.S.R. Biol. Sci (Transl.)* **133**, 635–638, 1960.

Drobysheva, S. S. (1979). Distribution of the Barents Sea euphausiids (fam. Euphuausiacea). *ICES CM* **L8**, 1–18.

Drobysheva, S. S., and Panasenko, L. D. (1984). On consumption of the Barents Sea euphausiids. *ICES CM* **L7**, 1–12.

Dunbar, M. J. (1942). Marine macroplankton from the Canadian eastern arctic. I. Amphipoda and Schizopoda. *Can. J. Res. D* **20**, 33–46.

Einarsson, H. (1945). Euphausiacea I. Northern Atlantic Species. *Dana Report* **27**, 1–184.

Endo, Y., Hanamura, Y., and Taniguchi, A. (1985). In-situ observations on surface swarming *Euphausia pacifica* in Sendai Bay in early spring with special reference to their biological characteristics. *Bull. Mar. Sci.* **37**, 764.

Everson, I. (2000a). Krill Biology, Ecology and Fisheries Fish and Aquatic Resources Series Vol. 6. p. 372. Blackwell Science, Oxford.

Everson, I. (2000b). Management of krill fisheries in Canadian waters. *In* "Krill: Biology, Ecology and Fisheries" (I. Everson, ed.), Fish and Aquatic Resources Series. Vol. 6, pp. 338–344. Blackwell Science, Oxford.

Everson, I., Tarling, G. A., and Bergström, B. (2007). Improving acoustic estimates of krill: Experience from repeat sampling of northern krill (*Meganyctiphanes norvegica*) in Gullmarsfjord, Sweden. *ICES J. Mar Sci.* **64**, 39–48.

Falk-Petersen, S., and Hopkins, C. C. E. (1981). Ecological investigations on the zooplankton community of Balsfjorden, northern Norway: Population dynamics of the euphausiids *Thysanoessa inermis*, *T. raschii*. *J. Plankton Res.* **3**, 177–192.

Fish, C. J., and Johnson, M. W. (1937). The biology of the zooplankton population in the Bay of Fundy and Gulf of Maine with special reference to production and distribution. *J. Biol. Bd. Can.* **1936–1937**, (3), 189–322.

Fisher, L. R., and Goldie, E. H. (1959). The food of *Meganyctiphanes norvegica* (M. Sars) with an assessment of the contributions of its components to the Vitamin A reserves of the animal. *J. Mar. Biol. Assoc. UK* **38**, 291–312.

Foote, K. G., Everson, I., Watkins, J. L., and Bone, D. G. (1990). Target strength of Antarctic krill (*Euphausia superba*) at 38 and 120 kHz. *J. Acoust. Soc. Am.* **87**, 16–24.

Fowler, S. W., Small, L. F., and Keckes, S. (1971). Effects of temperature and size on molting of euphausiid crustaceans. *Mar. Biol.* **11**, 45–51.

Fregin, T., and Wiese, K. (2002). The photophores of *Meganyctiphanes norvegica* (M. Sars) (Euphausiacea): Mode of operation. *Helgoland Mar. Res.* **56**, 112–124.

Frost, W. E. (1932). Observations on the reproduction of *Nyctiphanes couchii* (Bell) and *Meganyctiphanes norvegica* (Sars) off the south coast of Ireland. *Proc. R. Ir. Acad.* **40**, (Sect, B), 194–232.

Furnestin, M.-L. (1960). Zooplankton du Golfe du Lion et de la côte orientale de Corse. *Rev. Trav. Inst. (Scient. Tech.) Pêch marit.* **24**, 153–252.

Glover, R. S. (1952). Continuous plankton records: The Euphausiacea of the north-eastern Atlantic and North Sea, 1946–1948. *Hull. Bull. Mar. Ecol.* **3**, 185–214.

Gómez-Gutiérrez, J., Tremblay, N., Martínez-Gomez, S., Robinson, C. J., Ángel-Rodríguez, J. D., Rodríguez-Jaramillo, C., and Zavala-Hernández, C. (2010). Biology of the subtropical sac-spawning euphausiid *Nyctiphanes simplex* in the northwestern seas of Mexico: Vertical and horizontal distribution patterns and seasonal variability of brood size. *Deep Sea Res. II* **57**, 606–615.

Gordon, I. (1955). Systematic position of the Euphausiacea. *Nature* **176**, (911), 934–935.

Grinnell, A. D., Narins, P. M., Awbrey, F. T., Hamner, W. M., and Hamner, P. P. (1988). Eye/Photophore coordination and light-following in krill, *Euphausia superba*. *J. Exp. Biol.* **134**, 61–77.

Guglielmo, L. (1979). Observations on euphausiids vertical distribution in the southern Adriatic Sea deep waters July 1974. *Mem. Biol. Mar. di Oceanogr.* **9**, 25–34.

Hamner, W. M. (1984). Aspects of schooling of *Euphausia superba*. *J. Crust. Biol.* **4**, (Special Issue), 67–74.

Hamner, W. M., and Hamner, P. P. (2000). Behavior of Antarctic krill (*Euphausia superba*): Schooling, foraging, and antipredatory behavior. *Can. J. Fish. Aquat. Sci.* **57**, 192–202.

Hampton, I. (1982). Observation and measurement of visible krill swarms during FIBEX. Biomass Sci. Ser. 14:pp. 1-21, SCAR. Cambridge, UK.

Hanamura, Y., Endo, Y., and Taniguchi, A. (1984). Underwater observations on the surface swarm of a euphausiid *Euphausia pacifica* in Sendai Bay, northeastern Japan. *Mer (Tokyo)* **22**, 63–68.

Hansen, H. J. (1908). Crustacea Malacostraca. *Dan. Ingolf. Exped.* **3**, 1–120.

Hansen, H. J. (1915). The Crustacea Euphausiacea of the United Staes National Museum. *Proc. US Natl. Mus.* **48**, 59–114.

Harding, G.C.H., 1996. Ecological factors to be considered in establishing a new krill fishery in the Maritimes Region.

Hardy, A. C., and Kay, R. H. (1964). Experimental studies of plankton luminescence. *J. Mar. Biol. Assoc. UK* **44**, 435–478.

Harvey, M., Galbraith, P. S., and Descroix, A. (2009). Vertical distribution and diel migration of macrozooplankton in the St. Lawrence marine system (Canada) in relation with the cold intermediate layer thermal properties. *Prog. Oceanogr.* **80**, 1–21.

Hassan, H. U. (1999). Bioecology and population dynamics of *Meganyctiphanes norvegica* in mass and Fens Fjords near Bergen, Western Norway. *Pak. J. Zool.* **31**, 93–103.

Heegard, P. (1948). Larval stages of Meganyctiphanes (Euphausiacea) and some general phylogenetic remarks. *Meddr Kommn Danm. Fisk.-og Havunders. (ser. Plankton)* **5**, 25.

Herring, P. J., and Locket, N. A. (1978). The luminescence and photophores of euphausiid crustaceans. *J. Zool. (Lond.)* **186**, 431–462.

Hewitt, R. P., and Demer, D. A. (2000). The use of acoustic sampling to estimate the dispersion and abundance of euphausiids, with an emphasis on Antarctic krill, *Euphausia superba. Fish. Res.* **47**, 215–229.

Hewitt, R. P., Watkins, J., Naganobu, M., Sushin, V., Brierley, A. S., Demer, D., Kasatkina, S., Takao, Y., Goss, C., Malyshko, A., Brandon, M.Kawaguchi, S. *et al.* (2004). Biomass of Antarctic krill in the Scotia Sea in January/February 2000 and its use in revising an estimate of precautionary yield. *Deep Sea Res. II* **51**, 1215–1236.

Hirche, H. J., Hagen, W., Mumm, N., and Richter, C. (1994). The Northeast Water Polynya, Greenland Sea. III Meso- and macrozooplankton distribution and production of dominant herbivorous copepods during spring. *Polar Biol.* **14**, 491–503.

Hjort, J., and Ruud, J. T. (1929). Whaling and fishing in the North Atlantic. *Rapp. P.v. Reun. Cons. Int. Explor. Mer.* **56**, 1–123.

Hollingshead, K. W., and Corey, S. (1974). Aspects of the life history of *Meganyctiphanes norvegica* (M.Sars), Crustacea (Euphausiacea), in Passamaquoddy Bay. *Can. J. Zool.* **52**, 495–505.

Holt, E. W. L., and Tattersall, W. M. (1905). Schizopodous Crustacea from the north-east Atlantic slope. *Scient. Invest. Fish. Brch. Ire. Ann. Rep.* **1902–1903,** (Pt II, App. IV), 99–152.

Hovekamp, S. (1989). Avoidance of nets by *Euphausia pacifica* in Dabob Bay. *J. Plankton Res.* **11**, 907–924.

Hudson, I. R., and Wigham, B. D. (2003). In situ observations of predatory feeding behaviour of the galatheid squat lobster *Munida sarsi* using a remotely operated vehicle. *J. Mar. Biol. Assoc. U.K.* **83**, 463–464.

Isaacs, J. D., and Kidd, L. W. (1953). Isaacs-Kidd mid-water trawl. *Scripps Inst. Oceanogr.* **Ref. 53**, 18.

Jamieson, B. G. M. (1991). Ultrastructure amd phylogeny of crustacean spermatozoa. *Mem. Queenland. Mus.* **31**, 109–142.

Jarman, S. N. (2001). The evolutionary history of krill inferred from nuclear large subunit rDNA sequence analysis. *Biol. J. Linn. Soc. Lond.* **73**, 199–212.

Jarman, S. N., Nicol, S., Elliott, N. G., and McMinn, A. (2000). 28S rDNA evolution in the Eumalacostraca and the phylogenetic position of krill. *Mol. Phylogenet. Evol.* **17**, 26–36.

Jenner, R. A., Dhubhghaill, C. N., Ferla, M. P., and Wills, M. A. (2009). Eumalacostracan phylogeny and total evidence: Limitations of the usual suspects. *BMC Evol. Biol.* **9**, 21,10.1186/1471-2148-9-21.

Jespersen, P. (1923). On the quantity of macroplankton in the Mediterranean and the Atlantic. *Rep. Danish Oceanogr. Expeditions* **1908–1910,** (3), 3.

Jonsson, L. G., Lundalv, T., and Johannesson, K. (2001). Symbiotic associations between anthozoans and crustaceans in a temperate coastal area. *Mar. Ecol. Prog. Ser.* **209**, 189–195.

Kaartvedt, S., Larsen, T., Hjelmseth, K., and Onsrud, M. S. R. (2002). Is the omnivorous krill *Meganyctiphanes norvegica* primarily a selectively feeding carnivore? *Mar. Ecol. Prog. Ser.* **228**, 193–204.

Kaartvedt, S., Rostad, A., Fiksen, O., Melle, W., Torgersen, T., Tiseth-Breien, M., and Klevjer, T. A. (2005). Piscovorous fish patrol krill swarms. *Mar. Ecol. Prog. Ser.* **299**, 1–5.

Kay, K. H. (1962). Bioluminescence of the euphausiid crustacean *Meganyctiphanes norvegica*; the influence of 5-hydroxytryptamine. *Proc. Physiol. Soc.* **165**, 63–64.

Kay, R. H. (1965). Light-stimulated and light inhibited bioluminescence of the euphausiid *Meganyctiphanes norvegica* (G. O. Sars). *Proc. R. Soc. Lond. B Biol.* **162**, 365–386.

Klevjer, T. A., and Kaartvedt, S. (2003). Split-beam target tracking can be used to study the swimming behaviour of deep-living plankton in situ. *Aquat. Living Resour.* **16**, 293–298.

Klevjer, T. A., and Kaartvedt, S. (2006). In situ target strength and behaviour of northern krill (*Meganyctiphanes norvegica*). *ICES J. Mar. Sci.* **63**, 1726–1735.

Korneliussen, R. J., and Ona, E. (2002). An operational system for processing and visualizing multi-frequency acoustic data. *ICES J. Mar. Sci.* **59**, 291–313.

Körte, F. (1964). Luminosity of *Meganyctiphanes norvegica* (M. Sars) in relation to vertical movement. *J. Mar. Biol. Assoc. U.K.* **44**, 479–484.

Kramp, P. L. (1913). Schizopoda. *Bull. trimestriel Cons. Perm. pour l.Expl. de la Mer Résumé Planktonique* **3e partie**.

Krönström, J., Dupont, S., Mallefet, J., Thorndyke, M., and Holmgren, S. (2007). Serotonin and nitric oxide interaction in the control of bioluminescence in northern krill, *Meganyctiphanes norvegica* (M. Sars). *J. Exp. Biol.* **210**, 3179–3187.

Krönström, J., Karlsson, W., Johansson, B. R., and Holmgren, S. (2009). Involvement of contractile elements in control of bioluminescence in Northern krill, *Meganyctiphanes norvegica* (M. Sars). *Cell Tissue Res.* **336**, 299–308.

Kulka, D. W., Corey, S., and Iles, T. D. (1982). Community structure and biomass of euphausiids in the Bay of Fundy. *Can. J. Fish. Aquat. Sci.* **39**, 326–334.

Labat, J. P., and Cuzin-Roudy, J. (1996). Population dynamics of the krill *Meganyctiphanes norvegica* (Sars) (Crustacea: Euphausiacea) in the Ligurian Sea (N-W Mediterranean Sea). Size structure, growth and mortality modelling. *J. Plankton Res.* **18**, 2295–2312.

Lacroix, G. (1961). Les migrations verticale journalières des Euphausides à l'entrée de la Baie de Chaleurs. *Contr. Dep. Pêch. Queb No* **83**, 257–316.

Lass, S., Tarling, G. A., Virtue, P., Matthews, J. B. L., Mayzaud, P., and Buchholz, F. (2001). On the food of Northern Krill (*Meganyctiphanes norvegica*) in relation to its vertical distribution. *Mar. Ecol. Prog. Ser.* **214**, 177–200.

Leavitt, B. B. (1938). The quantitative vertical distribution of macrozooplankton in the Atlantic Ocean Basin. *Biol. Bull. Mar. Biol. Lab.* **74**, 376–394, Woods Hole.

Liljebladh, B., and Thomasson, M. A. (2001). Krill behaviour as recorded by acoustic Doppler current profilers in the Gullmarsfjord. *J. Mar. Syst.* **27**, 301–313.

Lindley, J. A. (1982a). Continuous plankton records: Geographical variations in numerical abundance, biomass and production of euphausiids in the North Atlantic Ocean and the North Sea. *Mar. Biol.* **71**, 7–10.

Lindley, J. A. (1982b). Population dynamics and production of euphausiids III. *Meganyctiphanes norvegica* and *Nyctiphanes couchi* in the north Atlantic ocean and the North Sea. *Mar. Biol.* **66**, 37–46.

Lindley, J. A., Robins, D. B., and Williams, R. (1999). Dry weight carbon and nitrogen content of some euphausiids from the north Atlantic Ocean and Celtic Sea. *J. Plankton Res.* **21**, 2053–2066.

Loeng, H. (1989). The Influence of temperature on some fish population parameters in the Barents Sea, Arctic Ocean. *J. Northwest Atlant. Fish. Sci.* **9**, 103–114.

Lucas, C. E., Marshall, N. B., and Rees, C. B. (1942). Continuous plankton records: The Faroe-Shetland channel, 1939. *Hull. Bull. Mar. Ecol.* **2**, 71–94.

Maas, A., and Waloszek, D. (2001). Larval development of *Euphausia superba* Dana, 1852 and a phylogenetic analysis of the Euphausiacea. *Hydrobiol.* 143–169.

Macaulay, M. (2000). Sampling krill—acoustic estimation of krill abundance. *In* "Krill Biology, Ecology and Fisheries" (Everson Ie, ed.), pp. 20–32. Blackwell Science, Oxford.

Macdonald, R. (1927). Food and habits of Meganyctiphanes norvegica. *J. Mar. Biol. Assoc. UK* **14,** 753–784.

Madureira, L. S. P., Everson, I., and Murphy, E. J. (1993a). Interpretation of acoustic data at two frequencies to discriminate between Antarctic krill (*Euphausia superba*) and other scatterers. *J. Plankton Res.* **15,** 787–802.

Madureira, L. S. P., Ward, P., and Atkinson, A. (1993b). Differences in backscattering strength determined at 120 and 38 kHz for three species of Antarctic macroplankton. *Mar. Ecol. Prog. Ser.* **93,** 17–24.

Matthews, J. B. L. (1973). Ecological studies on the deep-water pelagic community of Korsfjorden, western Norway. *Sarsia* **54,** 75–90.

Mauchline, J. (1960). The biology of the euphausiid crustacean, *Meganyctiphanes norvegica* Sars. *Proc. R. Soc. Edin. B* **67,** 141–179.

Mauchline, J. (1980). The biology of mysids and euphausiids. *Adv. Mar. Biol.* **18,** 1–681.

Mauchline, J. (1985). Growth and production of Euphausiacea (Crustacea) in the Rockall Trough. *Mar. Biol.* **90,** 19–26.

Mauchline, J., and Fisher, L. R. (1967). The distribution of the euphausiid crustacean *Meganyctiphanes norvegica. Ser. Atlas Mar. Environ. Folio* **13,** 2.

Mauchline, J., and Fisher, L. R. (1969). The biology of euphausiids. *Adv. Mar. Biol.* **7,** 1–454.

Mayzaud, P., Boutoute, M., Gasparini, S. P., Mousseau, L., and Lefevre, D. (2005). Respiration in marine zooplankton-the other side of the coin: CO_2 production. *Limnol. Oceanogr.* **50,** 291–298.

McLaughlin, P. A. (1980). Comparative morphology of recent crustacea. Vol. Freeman, San Francisco.

Medina, A., Vila, Y., and Santos, A. (1998). The sperm morphology of the euphausiid *Meganyctiphanes norvegica* (Crustacea, Eucarida). *Invertebr. Reprod. Dev.* **34,** 65–68.

Melle, W., Kaartvedt, S., Knutsen, T., Dalpadado, P., and Skjoldal, H. R. (1993). Acoustic visualisation of large scale macroplankton and micronekton distributions across the Norwegian shelf and slopes of the Norwegian Sea. *ICES CM 1993 L* **40,** 25.

Melle, W., Ellertsen, B., and Skjoldal, H. R. (2004). Zooplankton: The link to higher trophic levels. *In* "The Norwegian Sea ecosystem" (H. R. Skjoldal, ed.), pp. 137–202. Tapir Academic, Trondheim, Norway.

Midttun, L. (1989). Climatic fluctuations in the Barents Sea. Rapp. P.v. Reun. Cons. Int. *Explor. Mer.* **188,** 185–205.

Nicol, S. (1986). Shape, size and density of daytime surface swarms of the euphausiid *Meganyctiphanes norvegica* in the Bay of Fundy. *J. Plankton Res.* **8,** 29–39.

Nicol, S., and Brierley, A. S. (2010). Through a glass less darkly-new approaches for studying the distribution, abundance and biology of Euphausiids. *Deep Sea Res. II* **57,** 496–507.

Omori, M. (1965). A 160-cm opening-closing plankton net. *Oceanogr. Soc. Jpn.* **21,** 20–26.

Onsrud, M. S. R., and Kaartvedt, S. (1998). Diel vertical migration of the krill *Meganyctiphanes norvegica* in relation to physical environment, food and predators. *Mar. Ecol. Prog. Ser.* **171,** 209–219.

Onsrud, M. S. R., Kaartvedt, S., and Breien, M. T. (2005). In situ swimming speed and swimming behaviour of fish feeding on the krill *Meganyctiphanes norvegica. Can. J. Fish. Aquat. Sci.* **62,** 1822–1832.

Patarnello, T., Bargelloni, l, Varotto, V., and Battaglia, B. (1996). Krill evolution and the Antarctic ocean currents: Evidence of vicariant speciation as inferred by molecular data. *Mar. Biol.* **126,** 603–608.

Pauly, T., Nicol, S., Higginbottom, I., Hosie, G., and Kitchener, J. (2000). Distribution and abundance of Antarctic krill (*Euphausia superba*) off East Antarctica (80-150 degrees E) during the Austral summer of 1995/1996. *Deep Sea Res. II* **47**, 2465–2488.

Petersson, G. (1968). Studies on photophores in the euphasiacea. *Sarsia* **36**, 1–39.

Préfontaine, G., and Brunel, P. (1962). Liste d'invertébrés marins receuillis dans l'estuaire du Saint-Laurent de 1929 à 1934. *Contr. Dép. Pêch. Québ.* **86**, 237–263.

Ringo, E., Sperstad, S., Myklebust, R., Mayhew, T. M., Mjelde, A., Melle, W., and Olsen, R. E. (2006). The effect of dietary krill supplementation on epithelium-associated bacteria in the hindgut of Atlantic salmon (*Salmo salar* L.): A microbial and electron microscopical study. *Aquac. Res.* **37**, 1644–1653.

Roe, H. S. J., Baker, A. D. C., Carson, R. M., Wild, R., and Shale, D. M. (1980). Behaviour of the Institute of Oceanographic Science's Rectangular Midwater Trawls: Theoretical aspects and experimental observations. *Mar. Biol.* **56**, 247–259.

Rosenberg, R., Dupont, S., Lundalv, T., Skold, H. N., Norkko, A., Roth, J., Stach, T., and Thorndyke, M. (2005). Biology of the basket star *Gorgonocephalus caputmedusae* (L.). *Mar. Biol.* **148**, 43–50.

Ruud, J. T. (1926). Zooplankton of the coast of Møre. *Rapp. P. -V.. Reun. Cons. Int. Explor. Mer* **41**, 120–124.

Ruud, J. T. (1936). Euphausiacea. *Rep. Dan. oceanogr. Exped. Mediterr.* **2**, 1–86.

Saborowski, R., Salomon, M., and Buchholz, F. (2000). The physiological response of krill (*Meganyctiphanes norvegica*) to temperature gradients in the Kattegat. *Hydrobiol.* **426**, 157–160.

Saborowski, R., Brohl, S., Tarling, G. A., and Buchholz, F. (2002). Metabolic properties of Northern krill, *Meganyctiphanes norvegica*, from different climatic zones. I. Respiration and excretion. *Mar. Biol.* **140**, 547–556.

Salomon, M., Mayzaud, P., and Buchholz, F. (2000). Studies on metabolic properties in the Northern Krill, *Meganyctiphanes norvegica* (Crustacea, Euphausiacea): Influence of nutrition and season on pyruvate kinase. *Comp. Biochem. Physiol. A* **127**, 505–514.

Sameoto, D. D. (1976). Respiration rates, energy budgets and molting frequencies of three species of euphausiid found in the Gulf of St. Lawrence. *J. Fish. Res. Bd Can* **33**, 2568–2576.

Sameoto, D. D., Jaroszynski, L. O., and Fraser, W. B. (1980). BIONESS, a new design in multiple net zooplankton samplers. *Can. J. Fish. Aquat. Sci.* **37**, 722–724.

Sameoto, D., Cochrane, N., and Herman, A. (1993). Convergence of Acoustic, Optical and Net-catch estimates of euphausiid abundance: Use of artificial light to reduce net avoidance. *Can. J. Fish. Aquat. Sci.* **50**, 334–346.

Sardou, J., and Andersen, V. (1993). Micronecton et macroplancton en mer Ligure (Mediteranee): Migrations nycthemerales et distributions verticales. *Oceanol. Acta* **16**, 381–392.

Sardou, J., Etienne, M., and Andersen, V. (1996). Seasonal abundance and vertical distributions of macroplankton and micronekton in the northwestern Mediterranean Sea. *Oceanol. Acta* **19**, 645–656.

Sars, M. (1857). Om 3 nye norske Krebsdyr. *Forh. skand. Naturf. Møte 1856* **7**, 160–175.

Sars, G. O. (1883). Preliminary notices on the Schizopoda of H.M.S. "Challenger" Expedition. *Forh. VidenskSelsk. Krist.* **7**, 1–43.

Sars, G. O. (1900). Crustacea. *Scient. Results Norw. N. Polar Exped.* **1893–1896** (1), 1–137.

Saunders, R. A., Ingvarsdottir, A., Rasmussen, J., Hay, S. J., and Brierley, A. S. (2007). Regional variation in distribution pattern, population structure and growth rates of *Meganyctiphanes norvegica* and *Thysanoessa longicaudata* in the Irminger Sea, North Atlantic. *Prog. Oceanogr.* **72**, 313–342.

Schmidt, K., Tarling, G. A., Plathner, N., and Atkinson, A. (2004). Moult cycle related changes in feeding rates of larval krill *Meganyctiphanes norvegica* and *Thysanoessa* spp. *Mar. Ecol. Prog. Ser.* **281**, 131–143.

Schram, F. R. (1986). Belotelsonidea, waterstonellidea, and 'Eocaridacea'. *In* "Crustacea" (FRe Schram, ed.), pp. 101–106. Oxford University Press, Oxford.

Siegel, V. (2000a). Krill (Euphausiacea) demography and variability in abundance and distribution. *Can. J. Fish. Aquat. Sci.* **57,** 151–167.

Siegel, V. (2000b). Krill (Euphausiacea) life history and aspects of population dynamics. *Can. J. Fish. Aquat. Sci.* **57,** 130–150.

Simard, Y., and Lavoie, D. (1999). The rich krill aggregation of the Saguenay—St. Lawrence Marine Park: Hydroacoustic and geostatistical biomass estimates, structure, variability, and significance for whales. *Can. J. Fish. Aquat. Sci.* **56,** 1182–1197.

Simard, Y., and Sourisseau, M. (2009). Diel changes in acoustic and catch estimates of krill biomass. *ICES J. Mar. Sci.* **66,** 1318–1325.

Simard, Y., Rd, Ladurantaye, and Therriault, J. C. (1986). Aggregation of euphausiids along a coastal shelf in an upwelling environment. *Mar. Ecol. Prog. Ser.* **32,** 203–215.

Skjoldal, H. R., and Rey, F. (1989). Pelagic production and variabibility in the Barents Sea ecosystem. *In* "Biomass yields and geography of large marine ecosystems" (K. Sherman and L. M. Alexander, eds.), American Association for the Advancement of Science, Selected Symposium III. pp. 242–286. Westvew Press, Boulder, Colorado, USA.

Soulier, B. (1965). Euphausiaces des bancs de Terre-neuve de Nouvelle-Ecosse et du Golfe du Maine. *Revue des Travaux Inst. des Peches Martines* **29,** 173–190.

Sourisseau, M., Simard, Y., and Saucier, F. J. (2006). Krill aggregation in the St. Lawrence system, and supply of krill to the whale feeding grounds in the estuary from the gulf. *Mar. Ecol. Prog. Ser.* **314,** 257–270.

Sourisseau, M., Simard, Y., and Saucier, F. J. (2008). Krill diel vertical migration fine dynamics, nocturnal overturns, and their roles for aggregation in stratified flows. *Can. J. Fish. Aquat. Sci.* **65,** 574–587.

Spicer, J. I., and Strömberg, J. O. (2002). Diel vertical migration and the haemocyanin of krill *Meganyctiphanes norvegica*. *Mar. Ecol. Prog. Ser.* **238,** 153–162.

Spicer, J. I., Thomasson, M. A., and Strömberg, J.-O. (1999). Possessing a poor anaerobic capacity does not prevent the diel vertical migration of Nordic krill *Meganyctiphanes norvegica* into hypoxic waters. *Mar. Ecol. Prog. Ser.* 181–187.

Strömberg, J. O., Spicer, J. I., Liljebladh, B., and Thomasson, M. A. (2002). Northern krill, *Meganyctiphanes norvegica*, come up to see the last eclipse of the millennium? *J. Mar. Biol. Assoc. UK* **82,** 919–920.

Tarling, G. A. (2003). Sex dependent diel vertical migration in Northern krill and its consequences for population dynamics. *Mar. Ecol. Prog. Ser.* **260,** 173–188.

Tarling, G. A., and Cuzin-Roudy, J. (2003). Synchronization in the moulting and spawning activity of northern krill (*Meganyctiphanes norvegica*) and its effect on recruitment. *Limnol. Oceanogr.* **48,** 2020–2033.

Tarling, G. A., and Johnson, M. L. (2006). Satiation gives krill that sinking feeling. *Curr. Biol.* **16,** 83–84.

Tarling, G. A., Matthews, J. B. L., Sabarowski, R., and Buchholz, F. (1998). Vertical migratory behaviour of the euphausiid *Meganyctiphanes norvegica*, and its dispersion in the Kattegat Channel. *Hydrobiol.* **375/376,** 331–341.

Tarling, G. A., Buchholz, F., and Matthews, J. B. L. (1999a). The effect of a lunar eclipse on the vertical migration of *Meganyctiphanes norvegica* (Crustacea: Euphausiacea) in the Ligurian Sea. *J. Plankton Res.* **21,** 1475–1488.

Tarling, G. A., Cuzin-Roudy, J., and Buchholz, F. (1999b). Vertical migration behaviour in the northern krill *Meganyctiphanes norvegica* is influenced by moult and reproductive processes. *Mar. Ecol. Prog. Ser.* **190,** 253–262.

Tarling, G. A., Matthews, J. B. L., David, P., Guerin, O., and Buchholz, F. (2001). The swarm dynamics of Northern krill (*Meganyctiphanes norvegica*) and pteropods (*Cavolinia inflexa*) during vertical migration in the Ligurian Sea observed by an Acoustic Doppler Current Profiler. *Deep Sea Res. I* **48**, 1671–1686.

Tarling, G. A., Jarvis, T., Emsley, S. M., and Matthews, J. B. L. (2002). Midnight sinking behaviour in *Calanus finmarchicus*: A response to satiation or krill predation? *Mar. Ecol. Prog. Ser.* **240**, 183–194.

Thiriot, A. (1977). Peuplements planctoniques dans les régions de remontées d'eau du littoral atlantique Africain. *Doc. Scient. Centre Rech. Océanogr. Abidjan* Vol. VIII, (n°1), 1–72, Juin 1977.

Thomasson, M. A. (2003). Vertical distribution, behaviour and aspects of the population biology of euphausiids, especially *Meganyctiphanes norvegica* (M. Sars), in the Gullmarsfjord, western Sweden. PhD Thesis, Goteborg University.

Thomasson, M. A., Johnson, M. L., Strömberg, J. O., and Gaten, E. (2003). Swimming capacity and pleopod beat rate as a function of sex, size and moult stage in Northern krill *Meganyctiphanes norvegica*. *Mar. Ecol. Prog. Ser.* **250**, 205–213.

Timofeev, S. F. (2002). Sex ratios in the population of *Thysanoessa raschii* (M. Sars, 1864) (Euphausiacea) in the Barents Sea (with some notes on thesex ratios in the order Euphausiacea). *Crustaceana* **75**, 937–956.

Torgersen, T. (2001). Visual predation by the euphausiid *Meganyctiphanes norvegica*. *Mar. Ecol. Prog. Ser.* **209**, 295–299.

Ussing, H. H. (1938). The biology of some important plankton animals in the fjords at East Greenland. *Meddr. Gronland* **100**, 1–103.

Virtue, P., Mayzaud, P., Albessard, E., and Nichols, P. (2000). Use of fatty acids as dietary indicators in northern krill, *Meganyctiphanes norvegica*, from northeastern Atlantic, Kattegat, and Mediterranean waters. *Can. J. Fish. Aquat. Sci.* **57**, 104–114.

Waagbø, R., Torrissen, O. J., and Austreng, E. (2001). The greatest challenge for continued growth in Norwegian fish farming industry (In Norwegian). Vol. Norwegian Research Council, Oslo, Norway.

Watkins, J. (2000). Sampling krill—direct sampling. *In* "Krill Biology, Ecology and Fisheries" (I. Everson, ed.), pp. 8–19. Blackwell Science, Oxford.

Whiteley, G. C. (1948). The distribution of larger planktonic crustacea on Georges Bank. *Ecol. Monogr.* **18**, 233–264.

Widder, E. A., Greene, C. H., and Youngbluth, M. J. (1992). Bioluminescence of sound-scattering layers in the Gulf of Maine. *J. Plankton Res.* **14**, 1607–1624.

Wiebe, P. H., Morton, A. W., Bradley, A. M., Backus, R. H., Craddock, J. E., Barber, V., Cowles, T. J., and Flierl, G. R. (1985). New developments in the MOCNESS, an apparatus for sampling zooplankton and micronekton. *Mar. Biol.* **87**, 313–323.

Williamson, D. (1956). The plankton of the Irish Sea, 1951 and 1952. *Bull. Mar. Ecol.* **IV**, 87–114.

Yoshitomi, B., Aoki, M., Oshima, S., and Hata, K. (2006). Evaluation of krill (*Euphausia superba*) meal as a partial replacement for fish meal in rainbow trout (*Oncorhynchus mykiss*) diets. *Aquaculture* **261**, 440–446.

Youngbluth, M. J., Bailey, T. G., Davoll, P. J., Jacoby, C. A., Blades-Eckelbarger, P. I., and Griswold, C. A. (1989). Fecal pellet production and diel migratory behaviour of the euphausiid *Meganyctiphanes norvegica* effect benthic-pelagic coupling. *Deep Sea Res.* **36**, 1491–1501.

Zane, L., and Patarnello, T. (2000). Krill: A possible model for investigating the effects of ocean currents on the genetic structure of a pelagic invertebrate. *Can. J. Fish. Aquat. Sci.* **57**, 16–23.

Zelickman, A. E. (1958). Materialy o rasprostraneii i razmnozhenii evfaudsiid v pribrezhnoi zone Murmana (Data about the distribution and breeding of euphausiids in the Murman coastal waters). *Tr. Murm. Biol. Stn.* **4**, 79–118.

Zelickman, E. A. (1961). The behaviour patterns of the Barents Sea euphausiacea and possible causes of seasonal vertical migrations. *Int. Revue. ges. Hydrobiol.* **46,** 276–281.

Zhou, M., Zhu, Y. W., and Tande, K. S. (2005). Circulation and behavior of euphausiids in two Norwegian sub-Arctic fjords. *Mar. Ecol. Prog. Ser.* **300,** 159–178.

Zhukova, N. G., Nesterova, V. N., Prokopchuk, I. P., and Rudneva, G. B. (2009). Winter distribution of euphausiids (Euphausiacea) in the Barents Sea (2000–2005). *Deep Sea Res. II* **56,** 1959–1967.

CHAPTER TWO

GENETICS OF NORTHERN KRILL (*MEGANYCTIPHANES NORVEGICA* SARS)

Tomaso Patarnello,* Chiara Papetti,[†] *and* Lorenzo Zane[†]

Contents

Abstract

Understanding the origin and maintenance of genetic diversity in the oceanic realm is difficult because barriers to gene flow are far less obvious in marine compared to terrestrial species. This is particularly so for planktonic species such as euphausiids, with no fossil record and high rates of dispersal, and in which paleobiology and evolutionary history remains largely obscure. Population genetics may play an important role in this respect, elucidating population connectivity and shedding light on the historical demography of the investigated species. In turn, the relevant factors that can promote speciation over both at short and long evolutionary timescales can be identified. In this chapter, we outline the available approaches for gathering population genetics information on marine organisms, with a particular focus on the recent achievements in the study of *Meganyctiphanes norvegica*. For population structure, we review the data available for mitochondrial DNA (mtDNA) markers that show the presence of four temporally stable and genetically distinct gene pools, one in the Mediterranean samples and three others in the North Atlantic Ocean, potentially associated with the basin-scale pattern of circulation. Unpublished data on some nuclear microsatellite markers adds support to this conclusion. In addition, we apply a newly introduced Bayesian coalescent approach, and demonstrate that previously reported mitochondrial sequence diversity is indicative of a recent expansion at the Northern edge of the species' distribution.

* Department of Public Health, Comparative Pathology, and Veterinary Hygiene, University of Padova, Italy
† Department of Biology, University of Padova, Italy

Advances in Marine Biology, Volume 57
ISSN 0065-2881, DOI: 10.1016/S0065-2881(10)57002-0

This does not hold for the Southern and Mediterranean populations that appear to be stable over time. We also review the literature reporting new advances on the analysis of *M. norvegica* genes and genes' products involved in metabolic pathways that may underline differences at the population level, possibly linked to environmental variation and local adaptations.

1. POPULATION GENETICS IN THE OCEAN

Origin and maintenance of genetic diversity is a central issue in evolutionary biology. Understanding these processes in the ocean realm is particularly difficult because barriers to gene flow are far less obvious in marine compared to terrestrial species. Vicariance, the phenomenon by which biological populations become physically isolated and accumulate genetic differences, is considered the most common process promoting genetic discontinuities across geographic ranges and, ultimately, resulting is allopatric speciation (Coyne and Orr, 2004). Its understanding requires the identification of historical events that produced the initial separation and the recognition of factors that prevented successive homogenisation of genetic pools. This can be a challenging task for marine organisms, and particularly for planktonic species because they (1) frequently occur on a large geographic scale encompassing distinct oceanic basins, (2) have the capability for long-distance dispersal, and (3) achieved their contemporary distributions after a long history of range shifts due to tectonic and paleoclimatic events. When a fossil record is not available for a group, as in the case of euphausiids, its paleobiology and evolutionary history remains largely obscure. Population genetics may play an important role in this respect elucidating how present day conditions, such as oceanographic patterns, influence marine organisms' population connectivity. Shedding light on the historical demography of the investigated species may contribute to understanding the impact of paleoclimatic events on population dynamics. Also interspecific genetic investigations may help to identify relevant factors that, on a longer evolutionary timescale, can promote speciation in the sea. However, any analysis of population structure depends on our ability to properly assess the degree of intraspecific genetic polymorphism combined with a thorough analysis of the partitioning of genetic polymorphism among populations.

2. THE GENETIC TOOLKIT

Mitochondrial DNA (mtDNA) sequencing and the assessment of length polymorphisms of microsatellites are still the two most widely used approaches for gathering population genetics information on marine organisms though several alternatives such as single nucleotide polymorphism are receiving growing attention due to the biotechnological advances. mtDNA is a very flexible tool for population genetics studies since it is useful not only to identify genetic heterogeneity within a species, particularly at the large geographic scale, but also to explore species or population demographic history. mtDNA has long been the choice tool to obtain genetic information because of the availability of universal primers that allow direct amplification of a suite of genes from different taxa, and because of the possibility to select portions of mtDNA with different level of variation, depending on the ecological problem to be addressed. The phylogenetic component inherent in the datasets generated by mtDNA sequencing for population analysis prompted the creation of a new discipline named phylogeography (Avise, 1998). Phylogeography studies the distribution of lineages across distinct geographic areas. The use of mtDNA remains popular despite several limitations, including (1) maternal inheritance, (2) low resolution at the very small geographic scale, and (3) the fact that information obtained for a single locus does not necessarily reflect the overall pattern of genetic differentiation (reviewed in Avise, 2004). This is because of the availability of a well-established procedure of mtDNA data analysis and the recent introduction of powerful analytic tools, such as Bayesian analysis with Markov Chain Monte Carlo (MCMC) samplers (Drummond and Rambaut, 2007; Kuhner, 2006). In fact, the ease of mtDNA sequencing offers a straight forward approach to map lineages according to geography and also allows statistical parsimony analysis aimed at inferring the most probable processes that account for the distribution of a given group of sequences (Templeton, 2004). Additionally, differences between populations can be tested for by comparing the distribution of haplotypic frequencies (Ryman, 2006) or by performing an Analysis of Molecular Variance (AMOVA) that takes into account the partition of molecular variability within and among populations or groups of populations (Excoffier et al., 1992). For historical demography, neutrality tests like Tajima's D statistic (Tajima, 1989) and Fu's Fs test (Fu, 1997) can be used to estimate whether a population/species deviates from the mutation-drift equilibrium that is an indirect estimate of changes in population size (Excoffier et al., 2005). Significant negative D and Fs values are usually interpreted as signatures of population expansion, although we cannot rule out the impact of selection. As a following step, a mismatch analysis of sequences (frequency of pairwise differences between haplotypes) can be used to look for distinctive signatures of historical population growth, because sudden growth is expected to generate a unimodal distribution of pairwise differences between sequences (Rogers and Harpending, 1992). Recently, a new Bayesian approach implemented in BEAST (Drummond and Rambaut, 2007) has added considerable power to the analysis of historical demography by allowing tests for past population expansions or declines using a variety of models, including not only exponential or logistic growth but also flexible models such as Bayesian Skyline Plots (Drummond *et al.*, 2005). These are based on the idea that the number of lineages sampled at any time in a phylogenic tree is linked to the effective dimension of the population at that time, so that it is possible to track population size changes along any intraspecific phylogeny. In addition, this Bayesian approach can incorporate different models of sequence evolution (Drummond *et al.*, 2006) and allow direct rate calibration based on fossil records or biogeographic evidences (Ho and Phillips, 2009).

In the last few years, microsatellites or SSR (simple sequence repeats) have become one of the most popular molecular markers with applications in many different fields. High polymorphism and the relative ease of scoring represent the two major features that make microsatellites of large interest for many genetic studies including population genetics (Zane *et al.*, 2002). A combination of mitochondrial and nuclear markers is desirable when studying the population structure of a species, however, whereas sequence information is suitable for inferring the history of a species or a population, microsatellites are the primary markers used for inferring population connectivity and isolation at ecological timescales.

Like for mtDNA, exact test for differentiation or Wright's F_{ST} (Wright, 1965), largely equivalent to AMOVA in this context, is the most common way to summarise the degree of population differentiation from any genetic marker. F_{ST} potentially provides a common way for comparing the degree of differentiation found in different studies, but further offers a simple way to estimate gene flow at least under several simplifying assumptions (Whitlock and McCauley, 1999). F_{ST} varies between 0 and 1. In calculating F_{ST}, total variation across populations is subdivided into variation arising within and between populations. As variation within populations increases, the remainder left to be credited to differences among populations necessarily becomes lower and F_{ST} will tend towards zero. On the other hand, the more genetic variation that is assigned to differences between populations, the more F_{ST} will increase. In this respect, it is essential to select molecular marker/s with enough, but not excessive, polymorphism to be genetically informative. Typically, highly variable microsatellites are useful at small to moderate levels of genetic differentiation.

Advances in microsatellite data analysis, however, have introduced new analytical tools especially useful for marine organisms. These include the reconstruction of single locus confidence intervals of genetic variability (Kauer *et al.*, 2003) and population divergence (Beaumont and Balding, 2004) to identify outlier loci potentially under the effect of natural selection. In addition, specific tests can be used to assign individuals to its population of origin. By means of a MCMC approach implemented in the software STRUCTURE, individual genotypes can also be assigned to one or more genetically distinct groups without any *a priori* assumption on its origin

(Falush *et al.*, 2003). Additionally, softwares like MIGRATE (Beerli and Felsenstein, 2001) and IMa (Hey and Nielsen, 2004) can be used to estimate the extent of gene flow between populations, taking into consideration realistic models of dispersion between populations, oceanographic models and incorporating the existing knowledge of historical processes. With regard to inference on the demography of a species, microsatellite data can be used to infer effective population size and its changes over both short and long timescales (Cornuet and Luikart, 1996; Garza and Williamson, 2001; Kuhner, 2006).

3. THE SPECIES

Many aspects of *Meganyctiphanes norvegica* biology have been addressed in other chapters of this volume. However, species distribution and dispersal ability are relevant to the population structure issue, therefore, they will be addressed hereafter. *M. norvegica* is distributed in the northern hemisphere, it is present in the North Atlantic between 30° and 80° N and extends its geographic range into the western Mediterranean. The distribution encompasses water masses and oceanographic conditions as varied as open oceans, marginal seas and fjords. The species seems constrained to water temperatures which range between 2 and 15 °C (Einarsson, 1945). Krill of all sizes can potentially be carried away by ocean currents, but only larval and juvenile organisms can be easily transported. Older and larger individuals can achieve swimming speeds similar to average current speeds, thus *M. norvegica* has a certain capacity to determine its distribution on small horizontal (geographic) and vertical (in the water column) scales. The active movements of adults, both vertical and horizontal, may represent an adaptive behaviour, as they allow the species to maintain a stable population structure, while continuously being exposed to heterogeneous oceanic conditions (Buchholz and Boysen-Ennen, 1988; Zane and Patarnello, 2000). This aspect should to be taken into account when dealing with dispersal and gene flow of the northern krill, because active swimming could have an effect on the stability of the population's genetic structure at particular geographic locations, especially where self-sustaining populations of *M. norvegica* are known to occur (Mauchline, 1980; Mauchline and Fisher, 1969). While adults are capable of retaining location, larvae are dispersive agents. The fact that larvae are released from restricted geographic locations implies, however, that there are limits to the maximum geographic range of dispersion linked to prevailing oceanographic currents as well as to larval duration. In the NW Atlantic, *M. norvegica* is known to reproduce in the Gulf of St. Lawrence, in the Gulf of Maine and on the Scotian Shelf. In the Ligurian Sea, krill are consistently associated with a stable oceanographic front occurring off Nice (Cuzin-Roudy, 2000). At least two other self-maintaining krill populations are well documented in the Baltic Sea: one in the Skagerrak Sea and a second one in the Kattegat Sea (Buchholz and Boysen-Ennen, 1988). Despite the strong Atlantic currents, *M. norvegica* also likely consists of a self-sustaining populations in the Clyde Sea and in the Cadiz Bay, where southern currents may play a role in the transport of individuals from the Morocco coast (Mauchline, 1980; Zane *et al.*, 2000).

4. POPULATION GENETICS

Genetic investigations at the intraspecific level need to employ fast-evolving genes suitable to detect the sometimes subtle differences between populations. mtDNA is maternally inherited and has a faster mutation rate than the largest part of the nuclear genome. For this reason, it has been used for studying the population structure of many marine organisms including *M. norvegica*. In this species, cytochrome *b* (Cyt*b*), cytochrome oxidase Subunit I (COI) and NADH dehydrogenase Subunit 1 (ND1) mitochondrial genes were successfully used as molecular markers revealing an emblematic pattern of differentiation within and between oceanic basins. In the study by Bucklin *et al.* (1997), Cyt*b* and COI sequence variation revealed significant structuring of populations at trans-ocean basin scales pointing out that gene flow of *M. norvegica* within the NW Atlantic was sufficient to prevent the formation of distinctive geographic populations in that region whereas genetic differentiation of the NW Atlantic and Norwegian fjord populations indicated restricted gene flow across the North Atlantic Ocean. Zane *et al.* (2000) also found significant genetic differentiation among populations investigated across the NE Atlantic and the Mediterranean Sea, at least three distinct gene pools were identified: (1) in the area of Cadiz Bay population, (2) in the Ligurian Sea population, and (3) in NE Atlantic. A more recent investigation (Papetti *et al.*, 2005) has built upon the findings of the two previously mentioned studies by extending temporal and spatial sampling in order to assess genetic temporal stability of populations and to include the largest possible portion of the species' natural range. ND1 sequence analysis was carried out for 23 population samples representing 13 geographic locations (Fig. 2.1) ranging from NW Atlantic, Greenland, NE Atlantic and Mediterranean (samples analysed in Bucklin *et al.*, 1997 and Zane *et al.*, 2000 were also included in the survey). Five out of 13 locations were sampled for temporal replicates as summarised in Fig. 2.1. AMOVA showed two major results: (1) genetic homogeneity of temporal replicates, (2) hypothesis of panmixia was significantly rejected and four differentiated gene pools were identified. As to the temporal replicates (Fig. 2.1), popula-tions in the Clyde Sea (2 samples), Kattegat Sea (4 samples), Ligurian Sea

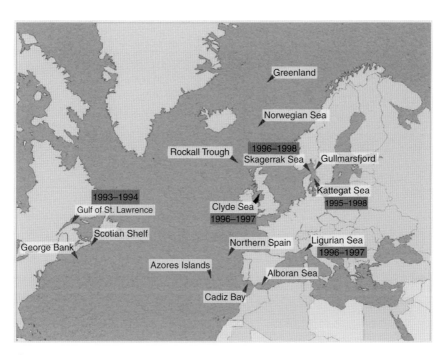

Figure 2.1 Sampling sites. Sites where populations of *Meganyctiphanes norvegica* were collected for population genetics analysis. In red are indicated the populations for which temporal replicates were available (years of collection are also indicated) (from Papetti *et al.*, 2005; modified). (For interpretation of the references to color in this figure legend, the reader is referred to the Web version of this chapter.)

(2 samples), Skagerrak Sea (3 samples) and NW Atlantic (4 samples) were found to be genetically homogeneous over time. AMOVA on 13 geographic populations which included Alboran Sea, NW Atlantic, Azores Islands, Cadiz Bay, Clyde Sea, Greenland, Gullmarsfjord, Kattegat Sea, Ligurian Sea, Norwegian Sea, Rockall Trough, Skagerrak Sea, and NW Spain (Fig. 2.1) clearly rejected the hypothesis of panmixia indicating that a substantial percentage (11.82%) of genetic variance was due to population subdivision. Four distinct gene pools were identified, two of these are on the eastern side of the North Atlantic and are referred as 'northern' NE Atlantic (above 55°N) and 'southern' NE Atlantic (below 45° N). A third group was found in the Ligurian Sea and possibly represents a distinct Mediterranean gene pool (Figs. 2.2 and 2.3). The fourth group, comprising individuals from NW Atlantic populations, was less clearly identified due to the smaller samples size and potential homoplasy resulting in convergence with few haplotypes from the Ligurian Sea; after removal of these convergent haplotypes the differentiation of samples from both side of the Atlantic was strongly supported (Papetti *et al.*, 2005). Further support to NW and NE

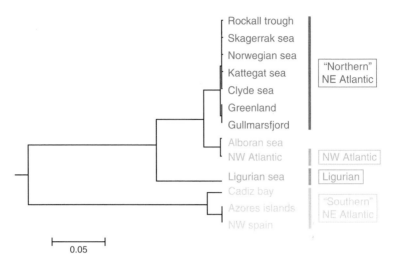

Figure 2.2 Unweighted pair-group method with arithmetic averages (UPGMA) dendrogram, which describes the relationships among populations, tree topology was based on the conventional F_{ST} pairwise matrix as from Papetti *et al.*, 2005 (modified).

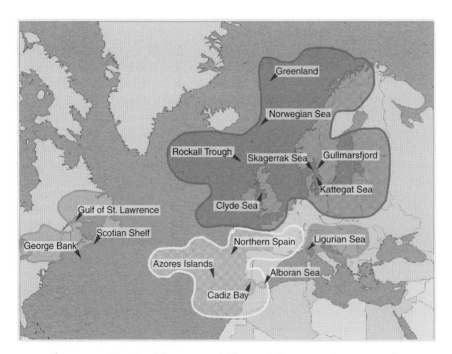

Figure 2.3 Sketch of the proposed *Meganyctiphanes norvegica* gene pools.

Atlantic differentiation was provided by microsatellite analysis (Bortolotto E. and Patarnello T., unpublished, see below).

Surprisingly, the 'southern' NE Atlantic (Azores Islands, Cadiz Bay, and NW Spain that could be defined as 'Iberian' gene pool; Fig. 2.2) was the most genetically divergent group. This is difficult to explain considering the relatively short geographic distance between these populations and the other sampling sites in NE Atlantic. The origin of this southern gene pool is unclear. However, the occurrence of a source population off the Morocco coast was hypothesised which could be advected by ocean currents flowing northward at depths of 200–400 m (Thiriot, 1997; Zane *et al.*, 2000). Also the Ligurian population was well differentiated from all the others supporting the possibility of a Mediterranean gene pool with very limited exchange with the Atlantic as also reported for other marine species (Patarnello *et al.*, 2007). Independent physiological and ecological observations support the distinctiveness of the Ligurian population (Cuzin-Roudy, 2000). A large gene pool, homogeneous over time and space, appears to be composed of the 'northern' NE Atlantic populations to which belong also the Gullmarsfjord, Greenland, and Norwegian samples. Despite some marginal differentiation between populations of this area, it appears that all samples are part of a stable breeding unit with high levels of gene flow, possibly promoted by the oceanographic features of the area, where mixing of water masses favours an extensive larval exchange, thus preventing population differentiation. There is a close relationship between NW and NE Atlantic populations though the two groups are clearly distinct (Bowden, 1975; Papetti *et al.*, 2005). Interestingly, the Alboran sea population is more closely related to the NW Atlantic sample than to the Ligurian one. However, the hydrographic and ecological features of the Alboran Sea are indeed more similar to the Atlantic Ocean than to the Mediterranean Sea (Estrada *et al.*, 1985; Kiortsis, 1985).

The same group of 13 populations analysed for ND1 was recently scored for three microsatellite loci (Bortolotto E. and Patarnello T., unpublished). Two loci were from a previous isolation report (MnC7 and MnO8; Ostellari *et al.*, 2000). It should be noted here that such a limited number of microsatellite markers do not allow any firm conclusion. However, it is worth reporting at least the most statistically supported result of the AMOVA pairwise analysis. A clear genetic differentiation emerged at the nuclear DNA level for the 'Iberian' gene pool and for the Ligurian population (the latter statistically differentiated from the NE Atlantic gene pool). Both results are in keeping with the mitochondrial data supporting the two major findings obtained by the ND1 sequence analysis. In addition, microsatellites also revealed a certain degree of differentiation between the NE Atlantic and NW Atlantic populations that the mtDNA failed to detect.

In general terms, we can conclude that the genetically distinct groups appear to correspond to ocean basin-scale patterns of circulation, which partition the North Atlantic into 3 gyres, and may effectively isolate the entrained ecosystems and populations. A 4th gene pool is clearly associated with the Ligurian samples (Fig. 2.3), corroborating the view of a separate Mediterranean population similar to other marine species (Patarnello *et al.*, 2007). Therefore, major ocean currents and gyres seem to play a relevant role in limiting to the gene flow across the North Atlantic Ocean thus favouring population structuring at large geographic scales (Aksnes and Blindheim, 1996; Bucklin *et al.*, 2000).

However, the present-day population structure of Northern krill could have been established over a very long time scale implying that the limitation to gene flow (by barrier such as oceanic gyres) must have been effective over evolutionary timescales, possibly millions of years. In fact, the divergence among *M. norvegica* gene pools has been dated from a minimum of 0.6 to a maximum of 14 million years ago (Papetti *et al.*, 2005) by using a simple approach based on the average number of pairwise differences (Gaggiotti and Excoffier, 2000) and the use of 'universal' rates of nucleotide substitution. The approach seems now questionable, both because the specific method used has been showed to produce aberrant results in some cases and because universal rates, obtained in interspecific phylogenetic studies, when used to date intraspecific events, can produce a strong overestimation of divergence times (Ho *et al.*, 2005).

To overcome these limitations and thanks to the new methodologies available, we report an updated analysis of *M. norvegica* historical demography and we critically assess the limits resulting from rate uncertainty and divergence overestimation in order to provide a more reasonable (if tentative) timescale for the evolutionary processes affecting *M. norvegica* genetic diversity. To this end, we analysed our previously reported ND1 dataset (936 sequences of 155 bp each, after excluding the potentially admixed Alboran sample; Papetti *et al.*, 2005) using the Bayesian coalescent method implemented in BEAST software (Drummond and Rambaut, 2007). Analyses have been performed separately for each of the four genetically distinct genetic pools identified before (Table 2.1, Figs. 2.2 and 2.3) and, in order to test for the non-stationarity of each gene pool, an exponential coalescent prior was first assumed (Ho *et al.*, 2008). For the Northern NE Atlantic group only, this approach estimated a significant positive value of the rate of exponential growth ($g = 1.01 \times 10^{-4}$ yr^{-1}; 95% highest posterior density, HPD: 9.57×10^{-6}–2.31×10^{-4} yr^{-1}) thus providing compelling evidence that this group of populations has been growing in the past. No evidence for population growth was found for the other three groups, where 95% HPD was centred on 0. Following these results, a second set of BEAST runs was conducted, applying a constant demographic prior for the three samples at equilibrium to allow correct estimation of the time of Most Recent Common Ancestor (tMRCA) for all the four groups. Results, reported in Table 2.1, show a much more recent origin of the distinct genetic pools than previously reported (Papetti *et al.*, 2005) with common

Table 2.1 Time of most recent common ancestor (tMRCA) for different groups of populations. Estimates have been obtained using BEAST ver. 1.5.4, with a HKY model of sequence evolution and strict clock assumed on the basis of preliminary runs. Estimates were obtained by assuming a coalescent prior of constant size or exponential growth and by including all the sequences of each given group. Names are the same as in Fig. 2.2. Markov Chain Monte Carlo (MCMC) was run for 100 million generations, with results logged every 10,000. Convergence was checked by monitoring traces of sampled parameters and effective sample size, after burning 10% of each chain. For each run, the time of the most recent common ancestor (tMRCA) of selected groups was determined assuming a sequence substitution rate for crustaceans of 1% per million years (Goodall-Copestake *et al.*, 2010; Lessios, 2008).

Geographic group	Tree prior	tMRCA (years)	95% highest posterior density interval
Southern NE Atlantic Ocean	Constant	1,497,800	2,648,000–486,150
Mediterranean Sea	Constant	795,140	1,522,900–201,800
Northern NE Atlantic Ocean	Exponential[a]	113,820	246,490–35,200
NW Atlantic Ocean	Constant	891,800	1,741,000–196,160

[a] Results based on an exponential prior are reported because the rate of exponential growth (*g*) was significantly higher than 0 for this group. Convergence was achieved after 5 runs of 100 million steps each.

ancestry dated to about 1.5–0.8 million years ago for the stable Mediterranean, Southern NE Atlantic, and NW Atlantic populations. For the expanding Northern NE Atlantic gene pool, a very recent origin (approximately 110,000 years ago) is suggested. These dates have been obtained by applying a substitution rate of 1% per million year obtained from the mid-point of phylogenetic calibrations available for crustaceans (Goodall-Copestake *et al.*, 2010; Lessios, 2008) and may be upwardly biased (Ho *et al.*, 2005). We cannot avoid this limitation due to the lack of specific rates and recent calibration points for this group, it is conservative to consider our date estimations as the maximum age for each given group, bearing in mind that they could be close to the true value in the case of no, or little, time-dependency of molecular clock calibrations (Burridge *et al.*, 2008; Millar *et al.*, 2008). As a further investigation of the historical demography of the Northern NE Atlantic samples, we used Bayesian Skyline Plots (Drummond *et al.*, 2005) to examine population size changes through time and infer a tentative date for the observed expansion. Relaxation of the assumption of simple exponential growth showed indeed that the Northern NE Atlantic samples could have been constant in size until approximately 25,000 years ago, after which an explosive expansion occurred (Fig. 2.4).

Taken together, *M. norvegica* genetic data provides evidence for a well-established intraspecific genetic structure with three distinct gene pools within the Atlantic Ocean that have possibly experiences different demographic histories. The new tentative timing provided in the present work by

A

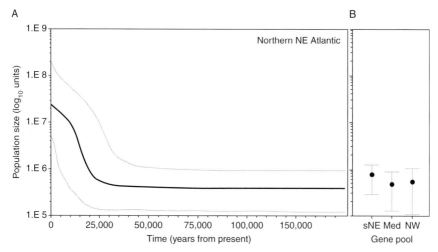

B

Figure 2.4 *Meganyctiphanes norvegica* female effective population size (N_e). Estimates have been obtained by analysis of ND1 haplotypes using BEAST ver. 1.5.4 using HKY model of sequence evolution. (A) Bayesian Skyline Plot showing the recent population increase detected in Northern NE Atlantic *M. norvegica* samples. Reported is a plot of N_e against the time in years from present. The mean N_e is shown as a solid line and light blue lines enclose the 95% highest posterior density interval (HPD). Ten MCMC runs of 100 million steps were performed and 5 groups of constant population size assumed. Time scale was obtained assuming a sequence substitution rate for crustaceans of 1% per million years (Goodall-Copestake *et al.*, 2010; Lessios, 2008) and N_e was scaled assuming a generation time of one year (Mauchline and Fisher, 1969). (B) For the sake of comparison, mean N_e (dots) and HPD (light blue bars) for the non-expanding Southern NE Atlantic Ocean, Mediterranean, and North West Atlantic Ocean gene pools are reported; estimates are based on coalescent prior of constant population size and scaled as before. (For interpretation of the references to color in this figure legend, the reader is referred to the Web version of this chapter.)

re-analysing data by Papetti *et al.* (2005) sets the origin of these difference within a much more recent timeframe, however, persistence over time of limitations to the gene flow by oceanographic conditions have to be postulated. The finding of ancient and non-expanding populations at the southern edge of distribution and in the warmer Mediterranean waters suggests that *M. norvegica* demography is weakly affected by climatic events in this area. On the other hand, keeping in mind the caution needed when dealing with date estimates based on genetic data, it worth to underline that the signal of population expansion detected for the Northern NE Atlantic genetic pool is compatible with a recent northward expansion of the species possibly in consequence of new colonisation areas available in the Northern edge of species' distribution after the last glaciations maxima.

This observation, if validated by more extensive sequence data and by more careful rate calibration, could provide a mechanistic link between climatic events and demography in this species and may help more generally to understand how the generation and maintenance of genetic variation is achieved in marine species as a whole.

5. GENES, ENZYMES, AND ADAPTATION

In addition to nuclear microsatellites loci and mitochondrial genes employed for population genetics studies, other genes were investigated in *M. norvegica* aimed at characterising metabolic pathways that may have a role in the adaptation to local environmental condition. This was the case of key metabolic genes, citrate synthase (CS) and pyruvate kinase (PK) that were investigated in *M. norvegica* populations of warm and cold waters. Biochemical properties of CS gene products (*CS* enzymes; Saborowski and Buchholz, 2002) differed in the Mediterranean (Ligurian) population as compared to Northern Atlantic samples suggesting that CS may be one of the key response in krill energetic. It was hypothesised that the short pulse of phytoplankton production in the Ligurian Sea, coupled with low year-round food availability, have driven expressional differences in metabolic regulation to save energy (Saborowski and Buchholz, 2002). Several other genes, which have been suggested to show environmentally related variation in other organisms, have been characterised for *M. norvegica*. Oxygen regulation pathways (i.e. hypoxia tolerance) and energy-production pathways (e.g. lactate dehydrogenase, LDH) have been linked to physiological response and environmental variation. Functional responses to oxygen stress (Spicer and Strömberg, 2003) have been shown to vary with developmental stage in *M. norvegica*, affecting onset of diel vertical migration and changing the organism's ability to deal with hypoxia at different stages. LDH has been characterised in *M. norvegica*, (Mulkiewicz *et al.*, 2001), and may also be involved in physiological adaptation of the warmer Mediterranean versus the colder northern Atlantic. The glycolytic enzyme phosphoglucose isomerase (PGI), which is involved in the same pathway as LDH, was sequenced in full by T. Patarnello (unpublished data). These genes provide additional targets to investigate environmentally meaningful genetic variation. In another krill species, such as *Euphausia superba* and *Nematoscelis difficilis*, LDH, PGI, and CS showed geographic differences in concentrations in association with environmental variation (Bucklin *et al.*, 2002; Fevolden and Schneppenheim, 1989). Genes controlling reproductive processes were also investigated in *M. norvegica* in relation to species' timing of reproduction. In general, individuals are sexually mature at one year, and the majority of the population survives to reproduce again the following year (Mauchline, 1980). However, Ligurian Sea populations tend to reproduce in early spring, whereas northern populations reproduce in late spring, and western Atlantic populations in the summer (Cuzin-Roudy, 1993). These differences are apparently not linked to latitudinal variation in temperature, but instead are postulated to be driven by differences in patterns of lipid storage and metabolism (see below). It appears that different geographical populations of *M. norvegica* vary the phases and timing of reproduction in response to some environmental parameter, for example food availability. Further, developmental pathways are also variable (i.e. not all larval stages occur), as is the sex ratio of the adult population as a result of differential mortality patterns between sexes, probably linked to reproductive activities (Cuzin-Roudy and Buchholz, 1999; Mauchline, 1980). Differential timing of reproduction of Mediterranean and North Atlantic populations of *M. norvegica* is supported by analyses of vitellogenesis, lipid allocation, and metabolic processes. The timing of vitellin deposition in Northern krill follows a south-to-north gradient, first appearing in Ligurian Sea populations and then in the Northeast Atlantic (Cuzin-Roudy and Buchholz, 1999). This pattern is mirrored in lipid allocation, with increased lipid deposition to mature ovaries occurring first in Mediterranean and later North Atlantic populations (Albessard and Mayzaud, 2003; Albessard *et al.*, 2001). In addition, populations in North Atlantic regions allocate lipids (which are linked to the vitellogenin pathway) to pre- and post-reproductive females in a different manner than do populations in the Mediterranean. These findings suggest that *M. norvegica* does not respond homogeneously to environmental variation. Rather, populations in different regions of the species' range have developed different adaptations in response to fluctuating seasonal conditions. Albessard and Mayzaud (2003) hypothesised that these differential responses in lipid allocation are linked to proper functioning of the main energy pathways in krill mitochondria and reflect a response to changes in salinity and perhaps temperature. Their hypothesis is supported by the fact that males in these studied populations also exhibit allocation differences (i.e. which types of lipids are stored, in which organs they are stored, and when storage and depletion occur). Vitellogenin and lipid allocation have both been used as determinants of population health and physiological state in Northern krill (Mayzaud *et al.*, 1999).

ACKNOWLEDGEMENTS

We thank Dr. Erica Bortolotto for her significant contribution in elucidating the population structure of *Meganyctiphanes norvegica*. A special acknowledgement is due to Dr. Ann Bucklin for her long standing and precious collaboration with the authors of this chapter, without Dr. Bucklin's work on krill a large part of this review would have not been possible.

REFERENCES

Aksnes, D. L., and Blindheim, J. (1996). Circulation patterns in the North Atlantic and possible impact on the population of *Calanus finmarchicus*. *Ophelia* **44**, 7–28.

Albessard, E., and Mayzaud, P. (2003). Influence of tropho-climatic environment and reproduction on lipid composition of the euphausiid *Meganyctiphanes norvegica* in the Ligurian Sea, the Clyde Sea, and the Kattegat. *Mar. Ecol. Prog. Ser.* **253**, 217–232.

Albessard, E., Mayzaud, P., and Cuzin-Roudy, J. (2001). Variation of lipid classes among organs of the Northern krill *Meganyctiphanes norvegica*, with respect to reproduction. *Comp. Biochem. Physiol. A Mol. Integr. Physiol.* **129**, 373–390.

Avise, J. C. (1998). The history and purview of phylogeography: A personal reflection. *Mol. Ecol.* **7**, 371–379.

Avise, J. C. (2004). Molecular markers, Natural History, and Evolution (Second edition) Sinauer, Sanderland MS (USA).

Beaumont, M. A., and Balding, D. J. (2004). Identifying adaptive genetic divergence among populations from genome scans. *Mol. Ecol.* **13**, 969–980.

Beerli, P., and Felsenstein, J. (2001). Maximum likelihood estimation of a migration matrix and effective population sizes in n subpopulations by using a coalescent approach. *Proc. Nat. Acad. Sci. USA* **98**, 4563–4568.

Bowden, K. F. (1975). Oceanic and estuarine mixing processes. *In* "Chemical oceanography" (J. P. Riley and G. Skirrow, eds.), Vol. 1, pp. 1–41. Academy Press, New York.

Buchholz, F., and Boysen-Ennen, E. (1988). *Meganyctiphanes norvegica* in the Kattegat: Studies on the horizontal distribution in relation to hydrography and zooplankton. *Ophelia* **29**, 71–82.

Bucklin, A., Smolenack, S. B., Bentley, A. M., and Wiebe, P. H. (1997). Gene flow patterns of the euphausiid, *Meganyctiphanes norvegica*, in the N Atlantic based on DNA sequences for mitochondrial cytochrome oxidase I and cytochrome b. *J. Plankton Res.* **19**, 1763–1781.

Bucklin, A., Wiebe, P. H., Astthorsson, O. S., Gislason, A., Allen, L. D., and Smolenack, S. B. (2000). Population genetic variation of *Calanus finmarchicus* in Icelandic waters: Preliminary evidence of genetic differences between Atlantic and Polar populations. *ICES J. Mar. Res.* **57**, 1592–1604.

Bucklin, A., Wiebe, P. H., Smolenack, S. B., Copley, N. J., and Clarke, M. E. (2002). Integrated biochemical, molecular genetic, and bioacoustical analysis of mesoscale variability of the euphausiid *Nematoscelis difficilis* in the California Current. *Deep. Sea. Res. I* **49**, 437–462.

Burridge, C. P., Craw, D., Fletcher, D., and Waters, J. M. (2008). Geological dates and molecular rates: Fish DNA sheds light on time dependency. *Mol. Biol. Evol.* **25**, 624–633.

Cornuet, J. M., and Luikart, G. (1996). Description and power analysis of two tests for detecting recent population bottlenecks from allele frequency data. *Genetics* **144**, 2001–2014.

Coyne, J. A., and Orr, H. A. (2004). "Speciation" Sinauer, 545.

Cuzin-Roudy, J. (1993). Reproductive strategies of the Mediterranean krill, *Meganyctiphanes norvegica* and the Antarctic krill, *Euphausia superba* (Crustacea: Euphausiacea). *Invertebr. Reprod. Dev.* **23**, 105–114.

Cuzin-Roudy, J. (2000). Seasonal reproduction, multiple spawning, and fecundity in northern krill, *Meganyctiphanes norvegica*, and Antarctic krill, *Euphausia superba*. *Can. J. Fish. Aquat. Sci.* **57**, 6–15.

Cuzin-Roudy, J., and Buchholz, F. (1999). Ovarian development and spawning in relation to the moult cycle in Northern krill, *Meganyctiphanes norvegica* (Crustacea: Euphausiacea), along a climatic gradient. *Mar. Biol.* **113**, 267–281.

Drummond, A. J., and Rambaut, A. (2007). BEAST: Bayesian evolutionary analysis by sampling trees. *BMC Evol. Biol.* **7**, 214.

Drummond, A. J., Rambaut, A., Shapiro, B., and Pybus, O. G. (2005). Bayesian coalescent inference of past population dynamics from molecular sequences. *Mol. Biol. Evol.* **22,** 1185–1192.

Drummond, A. J., Ho, S. Y. W., Phillips, M. J., and Rambaut, A. (2006). Bayesian coalescent inference of past population dynamics from molecular sequences. *PLoS Biol.* **4,** 88.

Einarsson, H. (1945). Euphausiacea I. Northern Atlantic Species. *Dana Report* **27,** 1–184.

Estrada, M., Vives, F., and Alcaraz, M. (1985). Life and productivity of the open sea. *In* "Western Mediterranean" (R. Margalef, ed.), pp. 148–197, 363. Pergamon Press, Oxford.

Excoffier, L., Smouse, P. E., and Quattro, J. M. (1992). Analysis of molecular variance inferred from metric distances among DNA haplotypes: Application to human mitochondrial DNA restriction data. *Genetics* **131,** 479–491.

Excoffier, L., Laval, G., and Schneider, S. (2005). Arlequin ver. 3.0: An integrated software package for population genetics data analysis. *Evol. Bioinform. Online* **1,** 47–50.

Falush, D., Stephens, M., and Pritchard, J. K. (2003). Inference of population structure using multilocus genotype data: Linked loci and correlated allele frequencies. *Genetics* **164,** 1567–1587.

Fevolden, S. E., and Schneppenheim, R. (1989). Genetic homogeneity of krill (*Euphausia superba* Dana) in the Southern Ocean. *Polar Biol.* **9,** 533–539.

Fu, Y. X. (1997). Statistical tests of neutrality of mutations against population growth, hitchhiking and background selection. *Genetics* **147,** 915–925.

Gaggiotti, O. E., and Excoffier, L. (2000). A simple method of removing the effect of a bottleneck and unequal population sizes on pairwise genetic distances. *Proc. R. Soc. Lond. B* **267,** 81–87.

Garza, J. C., and Williamson, E. G. (2001). Detection of reduction in population size using data from microsatellite loci. *Mol. Ecol.* **10,** 305–318.

Goodall-Copestake, W. P., Perez-Espona, S., Clark, M. S., Murphy, E. J., Seear, P. J., and Tarling, G. A. (2010). Swarms of diversity at the gene cox1 in the Antarctic krill. *Heredity* 1–6.

Hey, J., and Nielsen, R. (2004). Multilocus methods for estimating population sizes, migration rates and divergence time, with applications to the divergence of *Drosophila pseudoobscura* and *D. persimilis*. *Genetics* **167,** (2), 747–760.

Ho, S. Y. W., and Phillips, M. J. (2009). Accounting for calibration uncertainty in phylogenetic estimation of evolutionary divergence times. *Syst. Biol.* **58,** 367–380.

Ho, S. Y. W., Phillips, M. J., Cooper, A., and Drummont, A. J. (2005). Time dependency of molecular rate estimates and systematic overestimation of recent divergence times. *Mol. Biol. Evol.* **22,** 1561–1568.

Ho, S. Y. W., Larson, G., Edwards, C. J., Heupink, T. H., Lakin, K. E., Holland, P. W. H., and Shapiro, B. (2008). Correlating Bayesian date estimates with climatic events and domestication using a bovine case study. *Biol. Lett.* **4,** 370–374.

Kauer, M. O., Dieringer, D., and Schlotterer, C. (2003). A microsatellite variability screen for positive selection associated with the "Out of Africa" habitat expansion of *Drosophila melanogaster*. *Genetics* **165,** 1137–1148.

Kiortsis, V. (1985). Mediterranean marine ecosystems: Establishment of zooplanktonic communities in transitional and partly isolated areas. *In* "Mediterranean marine ecosystems" (M. Moroitou-Apostolopoulou and V. Kiortsis, eds.), pp. 377–385. Plenum Press, New York. NATO Conference series I, Ecology 8.

Kuhner, M. K. (2006). LAMARC 2.0: Maximum likelihood and Bayesian estimation of population parameters. *Bioinformatics* **22,** 768–770.

Lessios, H. A. (2008). The great American schism: Divergence of marine organisms after the rise of the Central American Isthmus. *Annu. Rev. Ecol. Syst.* **39,** 63–91.

Mauchline, J. (1980). The biology of mysids and euphausiids. Part 2. The biology of euphausiids. *Adv. Mar. Biol.* **18**, 1–681.

Mauchline, J., and Fisher, L. R. (1969). The biology of euphausiids. *Adv. Mar. Biol.* **7**, 1–454.

Mayzaud, P., Virtue, P., and Albessard, E. (1999). Seasonal variations in the lipid and fatty acid composition of the euphausiid *Meganyctiphanes norvegica* from the Ligurian Sea. *Mar. Ecol. Prog. Ser.* **186**, 199–210.

Millar, C. D., Dodd, A., Anderson, J., Gibb, G. C., Ritchie, P. A.Baroni, C. *et al.* (2008). Mutation and evolutionary rates in Adélie Penguins from the Antarctic. *PLoS Genet.* **4**, e1000209.

Mulkiewicz, E., Ziętara, M. S., Strömberg, J. O., and Skorkowski, E. F. (2001). Lactate dehydrogenase from the northern krill *Meganyctiphanes norvegica*: Comparison with LDH from the Antarctic krill *Euphausia superba*. *Comp. Biochem. Physiol. B, Biochem. Mol. Biol.* **128**, 233–245.

Ostellari, L., Zane, L., Maccatrozzo, L., Bargelloni, L., and Patarnello, T. (2000). Novel microsatellite loci isolated from the Northern krill, *Meganyctiphanes norvegica* (Crustacea, Euphausiacea). *Mol. Ecol.* **9**, 377–378.

Papetti, C., Zane, L., Bortolotto, E., Bucklin, A., and Patarnello, T. (2005). Genetic differentiation and local temporal stability of population structure in the euphausiid *Meganyctiphanes norvegica*. *Mar. Ecol. Prog. Ser.* **289**, 225–235.

Patarnello, T., Volckaert, F. A., and Castilho, R. (2007). Pillars of hercules: Is the atlantic-mediterranean transition a phylogeographical break? *Mol. Ecol.* **16**, 4426–4444.

Rogers, A. R., and Harpending, H. (1992). Population growth makes waves in the distribution of pairwise genetic differences. *Mol. Biol. Evol.* **9**, 552–569.

Ryman, N. (2006). Chifish: A computer program testing for genetic heterogeneity at multiple loci using chi-square and Fisher's exact test. *Mol. Ecol. Notes* **6**, 285–287.

Saborowski, R., and Buchholz, F. (2002). Metabolic properties of Northern krill, *Meganyctiphanes norvegica*, from different climatic zonesII. Enzyme characteristics and activities. *Mar. Biol.* **140**, 557–565.

Spicer, J. I., and Strömberg, J. O. (2003). Developmental changes in the responses of O_2 uptake and ventilation to acutely declining O_2 tensions in larval krill *Meganyctiphanes norvegica*. *J. Exp. Mar. Biol. Ecol.* **295**, 207–218.

Tajima, F. (1989). The effect of change in population size on DNA polymorphism. *Genetics* **123**, 597–601.

Templeton, A. R. (2004). Statistical phylogeography: Methods of evaluating and minimizing inference errors. *Mol. Ecol.* **13**, 789–809.

Thiriot, A. (1997). Peuplements zooplanctoniques dans les regions de remontees d'eau du littoral Atlantique africain. Doc Sci Cent Rech Oceanogr Abidjan 8:1–72. Western Mediterranean, Pergamon Press, Oxford, pp. 148–197.

Whitlock, M. C., and McCauley, D. E. (1999). Indirect measures of gene flow and migration: F_{ST} not equal $1/(4Nm+1)$. *Heredity* **82**, 117–125.

Wright, S. (1965). The interpretation of population structure by F-statistics with special regard to systems of mating. *Evolution* **19**, 395–420.

Zane, L., and Patarnello, T. (2000). Krill: A possible model for investigating the effects of ocean currents on the genetic structure of a pelagic invertebrate. *Can. J. Fish. Aquat. Sci.* **57**, 1–8.

Zane, L., Ostellari, L., Maccatrozzo, L., Bargelloni, L., Cuzin-Roudy, J., Buchholz, F., and Patarnello, T. (2000). Genetic differentiation in a pelagic crustacean (*Meganyctiphanes norvegica* Euphausiacea) from the North east Atlantic and the Mediterranean Sea. *Mar. Biol.* **136**, 191–199.

Zane, L., Bargelloni, L., and Patarnello, T. (2002). Strategies for microsatellite isolation: A review. *Mol. Ecol.* **11**, 1–16.

CHAPTER THREE

POPULATION DYNAMICS OF NORTHERN KRILL (*MEGANYCTIPHANES NORVEGICA* SARS)

Geraint A. Tarling

Contents

Abstract

This chapter reviews the short- and long-term changes in the size and age composition of Northern krill (*Meganyctiphanes norvegica*) populations and the environmental processes influencing those changes. It examines how populations of this species are affected by rates of reproductive output and mortality, and the effects of development, immigration and dispersion on population structure. This review also takes account of the many behavioural features

British Antarctic Survey, Natural Environment Research Council, High Cross, Cambridge, United Kingdom

Advances in Marine Biology, Volume 57
ISSN 0065-2881, DOI: 10.1016/S0065-2881(10)57003-2

that directly or indirectly influence *M. norvegica* population dynamics, such as swarming behaviour, diel vertical migration and diverse feeding strategies. What becomes evident is that *M. norvegica* shows a wide variability in population size-structure over the species extensive distributional range. Nevertheless, there are limits to this variability, as a result of the common life-cycle pattern.

1. INTRODUCTION

The life-cycle of *Meganyctiphanes norvegica* was first described by Einarsson (1945) based on samples from a number of different locations. It still stands as a good rule of thumb by which to compare other observations. Einarsson (1945) states that *M. norvegica* becomes mature at the age of 1 year throughout its area of reproduction. In the southern part of the reproductive area, the species spawns in the first months (February–April) and becomes mature after 1 year at 27–35 mm total body length. Part of the stock probably lives longer. In more northern waters, *M. norvegica* spawns a little later in the year (March–July) and the first time spawners are, as a rule, somewhat smaller than those found in the southern part of the distributional area ($\sim$25 mm). At least some portion of this stock grow further in their second summer season to a length above 30 mm and spawn again at the end of their second year.

I will make much reference to the different generations of krill that co-exist in a standard *M. norvegica* population. These have been given various nomenclatures in the literature so the following system has been adopted to provide a common framework with which to make comparisons. The youngest generation is the '0-group', which refers to individuals from the point of hatching to 31st March the following year. The next oldest generation is the 'I-group', covering individuals in their second year of life, starting 1st April and ending 31st March. Beyond that, the generations are referred to as the II-group and the III-group, although the latter is rarely observed.

2. BREEDING SEASON AND SPAWNING

Overall, the breeding season in *M. norvegica* is longer and less defined than in other co-occurring species, such as *Thysanoessa inermis* and *Thysanoessa raschii*, with spawning occurring in spring as well as in summer (Mauchline and Fisher, 1969). It may also vary interannually in length and timing; the onset and completion of the reproductive period can change by approximately ± 1 month over its distributional range (Cuzin-Roudy and Buchholz, 1999).

As a typical example of the timing of events in the reproductive cycle, Mauchline and Fisher (1969) describe that, in the Clyde, the male genital system starts developing about the same time as the ovary, in November, with most males having spermatozoa present in the vasa deferentia by December. By the beginning of February, many males have fully formed spermatophores present in their ejaculatory ducts and mating usually takes place with nearly every female bearing spermatozoa in their thelycums by the end of February. The eggs are not usually released until the beginning of April. The exact timings of each of the maturation steps nevertheless vary between locations, as detailed in Table 3.1, which also illustrates the range of indicators used by various authors to signify the period of breeding.

Many authors agree that *M. norvegica* becomes sexually mature at an age of 9–12 months across its distributional range (Boysen and Buchholz, 1984), although this may translate into different body lengths according to location. According to Wiborg (1971), large females mate earlier than small ones. Boysen and Buchholz (1984) posited that the spawning of older females of the II-group accounts for the first maximum of egg occurrence and the spawning of the younger I-group for the second one. In April and May, 10% of the females had spawned while from July to October, 90% had done so. A pattern of I-group females maturing after II-group females, with similar offsets in fertilisation and egg laying was also reported by Hollingshead and Corey (1974) in the Passamaquoddy Bay area. Similarly, Ruud (1936) considered that older euphausiids in the Mediterranean were fertilised before the younger ones. However, Mauchline (1960) considered that I-group and II-group females reproduced almost simultaneously in the Clyde. Overall, there appear to be regional differences in the relative timing of spawning between the two age groups.

The actual start of spawning has been considered by some to be a function of latitude (Boysen and Buchholz, 1984), occurring progressively later from south to north. Ruud (1936), Labat and Cuzin-Roudy (1996) and Cuzin-Roudy (2000) report spawning in the months of February–April for the Mediterranean. In various areas of the North Atlantic, a number of authors describe a later and less defined spawning time, from the end of March to the end of July (Berkes, 1976; Einarsson, 1945; Falk-Petersen and Hopkins, 1981; Mauchline, 1960). Lucas *et al.* (1957) and Mauchline (1960) noticed two maxima of egg occurrence in the channel between the Faroes and Shetland Islands. In the Clyde, the first maximum occurs in the first days of April and the second and more pronounced one, in the latter part of June.

With respect to the initiation of spawning, there is growing support for the strongest link being with the onset of the spring phytoplankton bloom (Astthorsson, 1990; Berkes, 1976; Dalpadado, 2006; Dalpadado and Skjoldal, 1991; Einarsson, 1945; Falk-Petersen and Hopkins, 1981; Gislason and Astthorsson, 1995; Hopkins, 1981). This may take precedence over other factors such as daylength and temperature as spawning initiation

Table 3.1 Indicators of the extent of the breeding season of *Meganyctiphanes norvegica* over its distributional range

Location	Earliest maturity size (total body length, mm)	Females with implanted spermatophores	Female with mature oocytes/ swollen thorax	Observed spawning	Notes	Authority
Clyde, Scotland	25	Mid-January to mid-September		March to July	Eggs in plankton beginning of April to beginning of July	Mauchline (1960)
Ligurian Sea, Mediterranean	22–25		February to May			Cuzin-Roudy (1993); Labat and Cuzin-Roudy (1996)
Gullmarsfjorden, Sweden			March to September			Thomasson (2003)
Kattegat, Denmark	21–25	Starts in January	March to September	April to October	A gap of spawning in June	Boysen and Buchholz (1984)
Rockall Trough, North Atlantic	15–20	March to September				Mauchline (1985)
Passamaquoddy Bay, N Amer. coast				July to early September		Hollingshead and Corey (1974)
Isafjord-deep, north-west Iceland	18*	February to end of May			*Converted from carapace length	Astthorsson (1990)
Balsfjorden, sub-Arctic Norway		February to June			No eggs were seen in the plankton	Falk-Petersen and Hopkins (1981)

*Indicates where total body length has been estimated from carapace length.

factors. For example, Astthorsson and Gislason (1997) found a relatively early onset of the spawning season in the waters north of Iceland, with males possessing spermatophores from February to April and euphausiid eggs appearing in the latter part of April. Despite the low temperatures there, phytoplankton spring growth begins earlier than in the Atlantic water of the south coast, possibly due to an earlier stratification of the surface waters in spring, mainly arising from lower salinities and faster warming.

Both Astthorsson and Gislason (1997) and Tarling and Cuzin-Roudy (2003) were also able to link the timing of the spring phytoplankton bloom with the appearance of eggs and early-stage larvae in the water column approximately 1 month later. Tarling and Cuzin-Roudy (2003) explain that, once ovarian maturation has been initiated by the spring bloom, females must go through a full reproductive cycle before spawning (see Chapter 7).

3. DEVELOPMENT OF EGGS AND LARVAE

3.1. Egg development

After spawning, euphausiid eggs are fertilised on departing the oviduct and, depending on the genus, are liberated into the water or are carried attached to the underside of the female. The attachment method offers protection but does limit egg numbers. *M. norvegica* adopts the alternative free-release method, which allows the production of high numbers of eggs but includes the risk of the eggs sinking to the bottom if they are too heavy or of being consumed by pelagic predators if they are neutrally buoyant.

Like other eggs that are laid freely into the water column, the eggs of *M. norvegica* have a vitelline membrane around the embryo, a perivitelline space and an outer membrane which is possibly two membranes stuck together. There is considerable variation in both the diameter of the embryo and the size of the perivitelline space, even in the eggs taken in the same time and place. Differences in the sizes of the eggs are not directly related to the differences in the size of the females producing them (Mauchline, 1968). Reported ranges in sizes of the eggs are 0.67–0.75 mm (Lebour, 1924), 0.59–0.85 mm (Macdonald, 1927), and 0.36–0.41 mm (Ruud, 1932).

The sinking speed of *M. norvegica* eggs have been estimated by Mauchline (unpublished and reported in Mauchline and Fisher, 1969) and by Marschall (1983). Mauchline found experimentally that eggs of *T. raschii* and *M. norvegica*, no matter at what stage of development, sink at comparable rates, namely 132–180 m d^{-1} in water of salinity 33 units and 15 °C. At the same salinity but at 0 °C, the eggs sank at 96–120 m d^{-1}. Marschall (1983) measured sinking speeds of between 61 and 104 m d^{-1}. His calculations, using a Navier–Stokes' equation, showed that egg density is always higher than seawater density found in their natural environment, meaning that the eggs will never float. *M. norvegica* eggs sink at one of the fastest rates amongst euphausiid species (Ross and Quetin, 2000). The fastest rate is in *Euphausia superba* at between 203 and 231 m d^{-1} (Marschall, 1983). *Euphausia crystallorophias*, by contrast, produces eggs of neutral density (Harrington and Thomas, 1987). Bollens *et al.* (1992) proposed that although the eggs of *Euphausia pacifica* sink at relatively high rates, their fast development time means that they do not experience much of a vertical descent before hatching.

The main significance of sinking speed is the determination of whether the eggs will reach the benthos before hatching. Marschall (1983) estimated a developmental time to hatching of 4.5 days at 5 °C and 2.5 days at 10 °C. Assuming the eggs are spawned at the surface, and in the absence of mixing processes, the hatching depth is between 275 and 515 m at 5 °C and between 150 and 290 m at 10 °C. What is striking about these estimates is that, even at a relatively fast development rate, there is a reasonable chance that the eggs will encounter the bottom before hatching in many environments believed to contain self-sustaining populations of *M. norvegica*, such as fjords and certain shallow seas, where water depths are often no deeper than 200 m. It suggests that either water turbulence has an important role in slowing the sinking speed (Marschall, 1983) or that contact with the benthos is not necessarily terminal for the eggs.

3.2. Larval developmental stages

Once hatched, the larvae develop through a series of stages, punctuated by moults, starting with the nauplius, which is the first stage after hatching and swims freely by means of its antennae and mandibles (Fig. 3.1; Mauchline and Fisher, 1969). It does not feed but obtains nourishment from the remaining yolk. There is a first and second stage generally separated according to size (Table 3.2). The metanauplius has two pairs of limbs, uniramous antennules and biramous antennae. It is also a non-feeding stage. The metanauplius moults to the first calyptopis, the first of the three larvae in the next phase of development, during which the abdomen and the eyes develop. The first calyptopis has functional mouthparts and filter-feeds. It swims by means of biramous segmented antennae. The different calyptopis stages differ according to size and degree of abdominal segmentation. The first calyptopis has an unsegmented abdomen and eyes that are just beginning to differentiate. By the second stage, the abdomen has five segments and the uropods can be seen developing under the cuticle. The eyes are better defined with the first traces of pigment. In the third stage, the abdomen has six segments and exceeds the carapace in length. The carapace still covers the eyes, which are becoming globular in form, pigmented and with visual elements discernable.

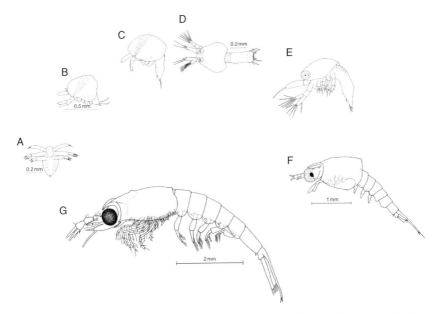

Figure 3.1 Larval stages of *Meganyctiphanes norvegica*. A, nauplius—ventral view; B, metanauplius—lateral view; C, calyptopis I—lateral view; D, calyptopis I—dorsal view; E, calyptopis II—lateral view; F, furcilia I; G, furcilia V. Scales given apart from calyptopis II where average total body length is 1.6 mm. A, D and E are redrawn from Heegard (1948), all others redrawn from Einarsson (1945).

The next developmental stages are the furcilia in which there are a number of stages generally signified by the incremental increase in setose pleopods and decrease in the number of telson spines (Table 3.2). Although there is a dominant pathway through these stages, a number of variants have been reported. For this reason, one of the most useful schemes was that described by Mauchline (1959) in which this stage of development was split into four main phases. During the first two, the abdominal pleopods become functional while, in the latter two, development of the thoracic limbs, abdominal photophores and telson takes place. Phase I (furcilia I): The third calyptopis stage moults to a furcilia which either has or acquires at the moult non-setose pleopods. There are either one or two moults depending on whether there is an instar without pleopods. Phase II (furcilia II): In the next moult, some pleopods become setose. There may be several moults in this phase until all pleopods eventually become setose. Phase III (furcilia III): The furcilia all have five pairs of setose pleopods and the terminal spines on the telson are not reduced in number from the original number of seven. The number of moults within this stage varies from one to four. Phase IV (furcilia IV–VI): The terminal spines on the telson are reduced in number, normally in the sequence 5 (FIV), 3 (FV), and 1 (FVI).

Table 3.2 Stages in the sequence of larval development in *Meganyctiphanes norvegica* from the point of hatching to the last stage before adolescence, based on a combination of Mauchline (1959) and Mauchline (1977)

Developmental stage	Pleopods	Telson spines	Total length (mm)
NI			0.48
NII			0.48
MN			0.51
CI	**0**		1.03
CII	**0**		1.59
CIII	**0**		2.40
FI	0		2.76
FI	2N		to
FI	**3N**		3.35
FI	4N		
FII	2S 3N	7	3.65
FII	**3S 2N**		to
FII	4S 1N		4.50
FII	5S		
FIII	**5S**	7	4.60
			to
			5.70
FIV	5S	6	5.56
FIV	**5S**	**5**	to
FIV	5S	4	7.70
FV	**5S**	**3**	6.64
			to
			7.50
FVI	**5S**	**1**	7.84
			to
			9.16

N, nauplius; MN, metanauplius; C, calyptopes; F, furcilia; S, setose; N, non-setose. The main developmental sequence is indicated in bold, with non-bold representing observed variants to these forms.

The reason for the variation in development during these stages is unclear. MacDonald (1927) reported that shortening of the larval period in the Clyde which involved a reduction in the number of instars and an accelerated development of the appendages was promoted by 'favourable conditions'. Suh *et al.* (1993) proposed that both temperature and the quantity and quality of food affect larval developmental times and dominant pathways. The fact that the breeding season can be so extended means that different cohorts may experience very different environmental conditions, a fact noted by both Zelickman (1968) and Makarov (1971, 1974). More mechanistic studies of the factors affecting early development are still warranted.

A larva is usually designated 'adolescent' once it has five pairs of setose pleopods and all luminescent organs, and the numbers of lateral and terminal spines on the telson are reduced to adult numbers (Mauchline and Fisher, 1969). By that stage, the thoracic appendages are very nearly, if not already, adult in form, but the maxillules and maxillae may still be more similar to those of larvae. Most of the development in adolescent animals involves the reproductive organs and secondary sexual characteristics.

3.3. Larval developmental rates

An initial attempt at estimating the rate of development of the larval stages of *M. norvegica* was made by Mauchline (1977, table 3) in the Clyde (Loch Fyne) during spring. The development of the first larval stages (egg to furcilia I) was estimated to take 29 days. These calculations assumed that spawning took place on 1st April and development occurred over the late spring and early summer. Further development to adolescence took a further 39 days, making total development time to adolescence, 68 days.

Developmental rates for cohorts spawned at different times of year or at different locations are likely to be altered by the ambient temperature (Hirst *et al.*, 2003). Tarling and Cuzin-Roudy (2003) sought to develop a method of estimating *M. norvegica* larval development times as a function of ambient temperature. The effect of temperature on euphausiid larval development was previously carried out on *E. pacifica* by Ross (1981), in which a developmental Q_{10} value of 2.81 was derived through incubations at 8 and 12 °C. Tarling and Cuzin-Roudy (2003) applied this Q_{10} value to the larval development rates of Mauchline (1977) to estimate developmental times from nauplius I to adolescence at other times of the year, where ambient temperatures were different. They estimated that, for cohorts spawned later in the summer (June and July), where ambient temperatures were around 2.5 °C warmer than in March–April, total larval development time would be 57 days, 11 days shorter than the spring cohort. Parameterisation of the effect of food on developmental rates is the next major step in assessing the sensitivity of this life-cycle phase to environmental conditions.

3.4. Larval behaviour and dispersion

The pattern of sinking and hatching generates a vertical gradient in larval stages. Mauchline and Fisher (1969) report that, in the Clyde, the earliest embryo stages were found in surface waters, prior to descent, with more developed embryos occurring deeper and nauplii deeper still. Then there is an upward movement of each subsequent larval stage with maximum abundance of furcilia stages being once again close to the surface.

As the larvae develop, they show increasingly pronounced diel vertical migrations (DVMs; Mauchline, 1959, 1965; Lacroix, 1961). Nauplii and metanauplii show no obvious DVM but a migration range of up to 50 m is apparent in the calyptopes stages and up to 100 m in the early furcilia stages. Nevertheless, vertical dispersion of these stages through much of the water column, particularly in daytime, is also reported (Mauchline and Fisher, 1969). The last one or two furciliae and the adolescents tend to live at the same depths as the adults and are absent from the surface layers during the hours of daylight.

The differing vertical distributions of the different larval stages will no doubt have an influence on their diet. However, there has been little consideration of the diet of *M. norvegica* larvae since that reported by Mauchline and Fisher (1969). They considered that the earlier larval stages mainly fed through filter feeding but that older stages had a more omnivorous diet. This was based on their observation that later furcilia have remains of crustaceans, probably copepodite stages, in their stomach contents. A detailed re-examination of this aspect of *M. norvegica* biology is very much needed, particularly when considering what environmental conditions are viable for successful larval development.

The differences in vertical distribution and behaviour of the larvae relative to the adults also have consequences to their relative spatial distributions. Einarsson (1945) mentioned that the drift of early developmental stages resulted in a much wider dispersion from the more restricted spawning regions, such that populations are maintained in areas beyond the main spawning grounds. For instance, he posited that specimens in west and east Greenland and the Barents Sea originated from the drifting of new recruits from spawning grounds further south, close to Iceland. Similarly, Skjoldal *et al.* (2004) and Melle *et al.* (2004) describe how the spawning products of krill on the Norwegian shelf are advected northwards along the Norwegian continental slope to the frontal region in the west Norwegian Sea. On a more local scale, Kulka *et al.* (1982) describe how, in the Bay of Fundy, the reduced vertical migration of larvae is such that they remain within a surface eddy away from the adults and are consequently not washed out of the Bay in the deeper currents. As they transform into adults, they complete a horizontal migration back into the area of highest adult concentration. In Passamaquoddy Bay, Hollingshead and Corey (1974) describe how stages beyond furcilia III were not caught because of dispersal by strong tidal currents in the Western Passage area. Such local-scale observations are now being simulated in 3D advections models (Simard *et al.*, 2003; Sourisseau *et al.*, 2006; Zhou *et al.*, 2005).

4. GROWTH AND MATURATION

4.1. Population size structure

The first description of the growth and maturation pattern of a local resident population of *M. norvegica* (i.e. where immigration and emigration can be ruled out as a confounding factor) was provided by Mauchline (1960) for the population in the Clyde. The description still stands as a good baseline with which to compare the structure of other populations. Mauchline (1960) states that 'the eggs are laid from March to July, the 0-group attaining an average length of about 25 mm by late autumn'. Little growth takes place throughout the winter, but by January, the males and by March to April, the females become mature. Spermatophores are transferred in January and February in the first instance, the eggs laid in March–April. This I-group is now 1 year old and grows throughout the spring and summer, reaching a mean length of 36–37 mm by the autumn. A high percentage, in some years, as much as 80% do not survive the second summer. Those living to the autumn are slowly reduced in numbers throughout the winter, a variable percentage (10–40%) taking part in a second breeding season (II-group). A very small percentage of the original population survive until the following winter when they are about 2.5–2.75 years old. A still smaller percentage seem to be able to survive to a third breeding season (III-group), when they reach a size of almost 46 mm.

Figure 3.2 presents two snapshots of the Clyde population, from mid-summer (July 1999) and December (1999), which illustrate how the various cohorts appear within the size-distribution pattern of the population. In summer, the youngest cohort (0-group) consists of juveniles with a modal peak around 15 mm total body length and a range of between 7 and 23 mm. Two other cohorts are evident in the adult size classes, one with a modal peak around 34 mm (range 29–36 mm) and another with a peak at 38 mm (range 35–42 mm), the former representing the I-group and the latter principally the II-group. By winter, the 0-group cohort has developed into sub-adults and now has a modal peak around 22 mm with a relatively large range stretching from around 13 mm to 17 mm. The I-group has a peak around 36 mm (range 32 mm–37 mm). The II-group, consisting of individuals generally larger than 40 mm (Mauchline, 1960) has all but disappeared in this particular sample set.

Compared to this baseline, examination of the size-structure patterns from different parts of the distribution of *M. norvegica* at (or close to) mid-summer and mid-winter provides a useful insight into the dynamics of the respective populations (Fig. 3.2). Note that each represents the situation in only one year, so the further influence of inter-annual variability in the described patterns must also be kept close to mind. For this reason, I have mainly focused body-size ranges and peaks of cohorts rather than relative cohort size, given the inter-annual variability in year class strength.

4.1.1. Mediterranean
In the Ligurian region of the Mediterranean (raw data from Labat and Cuzin-Roudy, 1996), the summertime size structure appears relatively similar to that described in the Clyde, although the 0-group is relatively more advanced with a modal peak close to 20 mm. This reflects the earlier

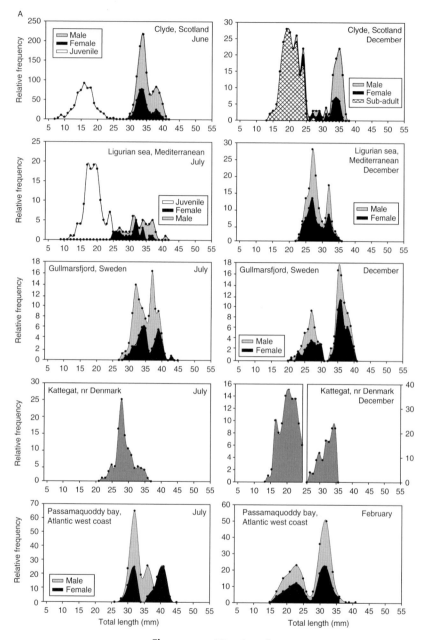

Figure 3.2   (Continued)

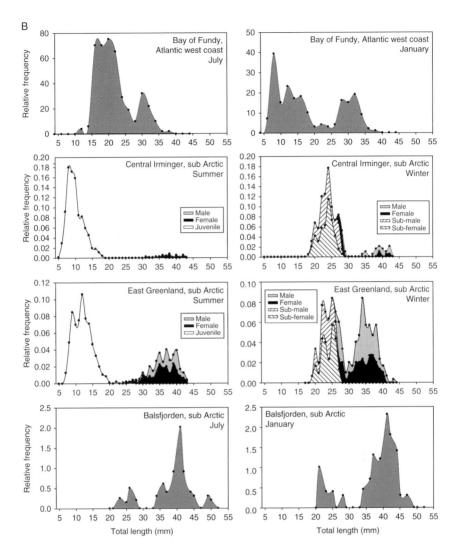

Figure 3.2 Patterns of size structure (in terms of total body length) from different locations within the distributional range of *Meganyctiphanes norvegica*. Plots on the left-hand side represent the size structure at (or close to) mid-summer while those on the right-hand side are from approximately mid-winter. Each plot has been redrawn from the original raw data or by extraction from published plots. The scale of the *y*-axis varies according to the reported values within the dataset and is adjusted to show the main features of the size-structure pattern. In the case of the Kattegat (nr Denmark) two different scales are used for the juvenile and adult fractions of the population since these comprise separate datasets with the source publication (Boysen and Buchholz, 1984).

onset of reproduction in this region (February–March, Cuzin-Roudy, 1993). There appears to be evidence of a I-group and II-group present in the population at this time although the cohorts are not easily distinguishable. By winter, the 0-group mode has already reached around 26 mm and individuals have matured sexually. A I-group cohort is more clearly discernable, with a peak of around 32 mm, 4 mm shorter than Clyde specimens of the same age. There is little evidence that any individuals of the II-group having survived from the summer.

4.1.2. Skagerrak and Kattegat

Although in close geographic proximity, the respective size structure of populations in the Kattegat, Denmark (data extracted from Boysen and Buchholz, 1984) and Gullmarsfjord (Sweden, data extracted from Thomasson, 2003) shows some clear differences. In summertime, both show the presence of I-group and II-group cohorts but the lack of a 0-group cohort. This may reflect the fact that populations at both sites are in relatively constrained basins where smaller individuals may be displaced from the adult population and hence missed by the sampling programme. What is quite marked is that the average length of Gullmarsfjord individuals is around 3 mm larger in the I-group and almost 5 mm larger in the II-group compared to specimens in the Kattegat. The body lengths of the Gullmarsfjord population are similar to those in the Clyde, while Kattegat individuals are comparatively small. It is also interesting to note that females in Gullmarsfjord appear to be around 1–2 mm larger than the males in both the I-group and II-group during summer. During the winter, the 0-group in the Kattegat shows a similar size-distribution pattern to the Clyde, with the bulk of individuals between 12 and 23 mm. In Gullmarsfjord, by contrast, the 0-group appears to have accelerated growth and has reached an average length of 26 mm by mid-winter. The I-group shows a similar offset between the 2 regions with Gullmarsfjord specimens around 2–3 mm larger. There is an indication of the persistence of II-group individuals into winter in Gullmarsfjord but not in the Kattegat.

4.1.3. Atlantic west-coast

In Passamaquoddy Bay, on the North American coast, there is again an absence of 0-group specimens in population samples in mid-summer (data extracted from Hollingshead and Corey, 1974). Hollingshead and Corey (1974) postulate that the absence of the recruiting cohort may be the result of dispersion by strong tidal currents from Passamaquoddy Bay into the Western passage area. As in the Clyde, the modal peak of the I-group occurs around 33–34 mm. However, in the II-group, females appear larger than males by around 4 mm, the former showing a modal peak around 40 mm, the latter around 36 mm. By winter, the 0-group has reappeared, presumably through immigration back into the Bay, with a modal peak around

23 mm. The modal peak of the I-group has remained relatively static around 33 mm. There are relatively few II-group individuals still around at this time.

Further up the coast, in the Bay of Fundy, the recruitment pattern appears to be quite different (data extracted from Berkes, 1976). In mid-summer, there is a large cohort with a modal peak around 19 mm and range from 10 to 25 mm. Superficially, this appears to mirror that found in the Mediterranean, which stems from an early reproductive season in February–March. However, Berkes (1976) states that the reproductive season in the Bay of Fundy is relatively late in the year, between July and October, which means that the cohort represents the I-group. There is a further distinct cohort with a modal peak around 31 mm, which is most likely the II-group. There is also a small residual cohort with a modal peak around 39 mm which could contain III-group specimens. In the winter, the product of the late-season spawning episode is clearly evident, with the presence of the 0-group cohort, which has a modal peak around 12 mm and a range of between 5 and 20 mm. The I-group has appeared to have grown rapidly during the latter half of the year and has advanced almost 10 mm to a modal peak of around 30 mm. It is likely that a number of II-group specimens still persist at this time, although not as a clearly discernable cohort.

4.1.4. Sub-Arctic

Sub-Arctic environments represent the northern boundary of successful spawning in *M. norvegica* (Einarsson, 1945) so I have not considered environments such as the Barents Sea and sub-Arctic waters north of Iceland, where spawning is sporadic at best and recruitment is most likely to have been via immigration from population centres further to the south (Astthorsson and Gislason, 1997; Dalpadado and Skjoldal, 1991, 1996). The three examples I have considered: Central Irminger Sea, East Greenland and Balsfjorden (Northern Norway), are themselves subject to variance in recruitment levels. However, they are each interesting case studies to compare with patterns from more central parts of the distributional range.

In the central Irminger Sea (raw data from Saunders *et al.*, 2007), there was a predominance of 0-group juveniles with a modal peak at 8 mm body length (range 5–18 mm) during summer. The relative number of adult specimens was low but contained a very large range of body lengths, from 31 to 43 mm, indicating a mixture of I-group and II-group individuals. By winter-time, the 0-group cohort had progressed to a model peak of 24 mm and matured into sub-adults. The body-length distribution of the adult population remained approximately the same, between 33 and 43 mm, indicating that some II-group specimens had persisted from the summer period.

The 0-group was comparatively more advanced during summer in East Greenland (raw data from Saunders *et al.*, 2007), with a modal peak around 13 mm, suggesting that the reproductive season and recruitment occurred earlier than in the neighbouring Irminger Sea. The range of sizes in the adult population was similar to the Irminger Sea, with both I- and II-group specimens being present. There was little difference in the range and modal peak of the 0-group populations in East Greenland and the Irminger Sea in winter, which indicates that the Irminger population had caught up despite their later recruitment. Both populations achieved body lengths similar to those observed in the Clyde. The adult population contained a good representation of II-group as well as I-group specimens, with maximum body lengths reaching up to 44 mm. This suggests that some may continue to reproduce for a third year (III-group).

The longevity of Northern krill observed in East Greenland is even more evident in the population found in Balsfjorden, Northern Norway (data extracted from Falk-Petersen and Hopkins, 1981). During summer, three distinct cohorts were apparent which Falk-Petersen and Hopkins (1981) determine to be three different generations equivalent to I-group, II-group and III-group. The latter group contained the largest specimens reported for *M. norvegica*, ranging between 47 and 50 mm total body length (although note that these values are converted from carapace lengths). No 0-group specimens were found in summer, but were present in winter, and were clearly distinct from the adult population, in which the generations had become merged and spanned body lengths of between 34 and 47 mm.

4.1.5. Perspective on size-structure patterns

Towards the northern limits of their range, it appears that there is a tendency for Northern krill to live longer and to grow to larger sizes. It would seem that the limitation to their further expansion northwards lies not in their ability to survive and grow, but to reproduce. Falk-Petersen and Hopkins (1981) report that although male and female specimens in Balsfjorden had spermatophores, no eggs or larvae could be seen in the plankton. In sub-Arctic waters, north of Iceland, Astthorsson and Gislason (1997) found only a single female bearing spermatophores. Dalpadado and Skjoldal (1991) reported that adult specimens in the Barents Sea had not spawned or reached spawning condition by mid-summer. A focus on the factors affecting sexual maturation and spawning in *M. norvegica* will be insightful in determining how their biogeographic range is ultimately limited.

4.2. Growth curves

An alternative means of comparing life-cycle patterns between environments is to examine the growth curves (Fig. 3.3). These are based on information taken from time-series size-structure data to chart the change in mean or modal body length with age over the lifespan of the organism. This process is difficult in open-ocean populations because of the problems of tracking cohorts that are being continuously mixed through immigration

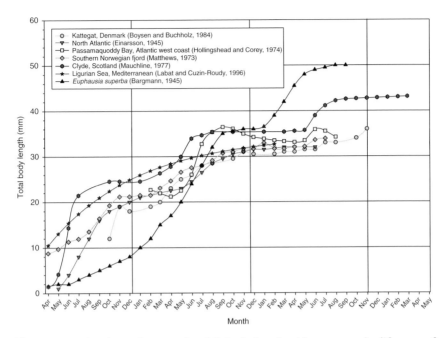

Figure 3.3 The change in mean (modal) body length with age over the life-span of *Meganyctiphanes norvegica* at different locations within its distributional range. The curves are constructed from time-series data on population total body length. Most curves have been constructed through the empirical fitting of spline curves between the approximate monthly data points. In the case of Labat and Cuzin-Roudy (1996), the curve represents the best fit of a Von-Bertalanffy growth function to the time-series data.

and emigration, so the best descriptions are generally derived from enclosed populations in fjords and embayments. Referring back to the life-cycle description of Mauchline (1960) in Section 4.1, in the Clyde it is apparent that individuals oscillate between periods of high growth in the productive months and of little or no growth during the winter. The most rapid growth (in terms of body length) is observed in the first year (0-group individuals) where individuals in the Clyde grow to an average body length of 24 mm. The most rapid phase of growth is between June and September, where individuals grow from 1 to 24 mm, at an average rate of 0.22 mm d^{-1}. The Clyde represents an upper limit to the amount of growth in length that could be attained in the first year. By comparison, individuals in the North Atlantic and in Kattegat attain a size of only 19 mm after the first year of growth (Boysen and Buchholz, 1984; Einarsson, 1945). Lindley (1982) also made estimates of 0-group growth rates in various sectors on the North Atlantic (not plotted in Fig. 3.3). He found that the fastest rates occurred in the oceanic areas west of Ireland where a size of 25 mm was attained by winter. The slowest were found in the Norwegian Sea, where 0-group individuals overwintered at a mean size of 16 mm. Nevertheless, taking this variation in account, it is interesting to note that the size attained by 0-group individuals at the end of the first season of growth is approximately the same as that achieved by *E. superba* (Fig. 3.3; Bargmann, 1945; Mauchline and Fisher, 1969; Mackintosh, 1972) despite the 4–12 °C difference in average temperature experienced by the respective species.

In winter, there is a stasis in growth which generally lasts between November and March, after which growth resumes in what has now become the I-group. These are generally sexually mature animals in which rates of growth in body length are slower than 0-group individuals, probably because a certain fraction of resources are now dedicated to the development of sexual organs. In this second year of growth, individuals can expect to grow an average of 10 mm in body length. Over a 4 month growing season, this would equate to an average rate of 0.08 mm d^{-1}. This is similar to that measured in *E. superba*, which approximates to 0.1 mm d^{-1} during the productive months (Atkinson *et al.*, 2006; Kawaguchi *et al.*, 2006).

For those surviving another year to become II-group individuals, a growth in length of up to 5 mm can be expected at an average rate of around 0.04 mm d^{-1} (assuming a 4-month growing season). Such a difference in growth rate between I-group and II-group individuals has been directly observed by Thomasson (2003) who reported growth to be 0.07 mm d^{-1} between March and April in I-group individuals but only 0.04 mm d^{-1} over the same period in II-group individuals in Gullmarsfjord, which contains semi-enclosed population of krill. The difficulty in making such measurements in open ocean populations is illustrated by Saunders *et al.* (2007) who reported rates of growth in length in summer that varied between -0.0394 and 0.2038 mm d^{-1} for the 0-group population and between -0.1735 and 0.1245 mm d^{-1} in the I-group in the Irminger Sea population.

The majority of growth trajectories in Fig. 3.3 are fitted empirically with spline curves. However, that of Labat and Cuzin-Roudy (1996) represents the best fit of a Von-Bertalanffy (VB) growth function, as follows:

$$BL_t = BL_{inf} - (BL_{inf} - BL_{min}) \exp^{(-K(t-T_0))} \qquad (3.1)$$

where BL_t is total body length at time t, BL_{inf} is the maximal theoretical body length, BL_{min} is the minimal body length, and T_0 is the time at the minimal size. The parameter K or Brody coefficient determines the initial rate of exponential increase in body length and also how fast this rate declines as the individual becomes larger.

Labat and Cuzin-Roudy (1996) derived a K of 0.0037 (other constants being $BL_{inf} = 34.62$, $BL_{min} = 9$ and $T_0 = 64.6$). The VB curve actually did not achieve as good a fit to the data as two linear curves, which better captured the rapid rate of growth in the juvenile stages in year 1 (0-group) and the lower growth in year 2 (I-group). Neither does the standard VB functional form allow for the seasonal oscillation in growth rate between summer and winter. However, the VB growth function has a strong theoretical basis in linking body size, metabolism and growth rate. The relatively good fit of the model in the present instance is a useful demonstration of the action of universal biological constraints on growth in this species.

4.3. Instantaneous growth rate estimates

Growth rates can also be measured directly on individuals using the Instantaneous Growth Rate (IGR) method, adapted for use in krill by Quetin and Ross (1991) and Nicol et al. (1992). The method involves taking around 100 krill (minimum) and incubating them for 2–3 days in individual containers. The animals must be checked daily and any moulted individuals extracted from the experiment with the cast-exoskeleton and the newly moulted animal measured to determine the percent change in length at moult (growth increment, GI). The daily moult rate (MR) can be estimated through dividing the total number of krill at the start of the incubation into the number moulting per day. Intermoult period (IMP) is the inverse of MR.

Dividing IMP into the GI gives the percent change in length per day. Relating this back to the pre-moult length of the original krill (L_{pre}, mm) allows the daily growth rate (DGR, mm d^{-1}) to be determined using the equation of Ross et al. (2000):

$$DGR = \frac{L_{pre} \cdot GI}{IMP \cdot 100} \qquad (3.2)$$

The technique is based on three major premises: (1) that the number moulting each day is relatively constant and equal to the inverse of the moult duration; (2) that the GI is not affected by the incubation conditions for several days postcapture (Nicol et al., 1992); (3) that IMP is not affected by incubation; (4) that increase in telson/uropod length between the moulter and its moult reflects growth in total body length over 1 IMP.

So far, the method has mainly been applied to E. superba and E. pacifica and Thysanoessa species (Atkinson et al., 2006; Pinchuk and Hopcroft, 2007; Tarling et al., 2006). The technique was applied by Tarling (unpublished) on a population of M. norvegica in Gullmarsfjord during autumn 2003. DGR of juvenile animals (0.19 mm d^{-1}, range 0.07–0.33) were substantially higher than adults, with male krill (0.011 mm d^{-1}, range −0.114 to 0.194) growing faster than females (0.003 mm d^{-1}, range 0.065–0.076). This is in line with expectations from the length–frequency derived growth curves (Thomasson, 2003) although the IGR values are considerably higher in the case of the juveniles. Interestingly, some adult specimens were seen to shrink in length during the experiments, a phenomenon that has previously been reported in *E. superba* (Ikeda and Dixon, 1982) and *E. pacifica* (Marinovic and Mangel, 1999). In the present instance, Pond (unpublished) found there to be a link between those specimens that were growing well and fatty acid markers for diatoms and benthically derived algae. There appeared to be a diversity of feeding strategies of krill in the fjord, which was reflected in their body condition and growth rate. Overall, the technique is particularly useful in resolving inter-individual differences in growth rates and can also illuminate some of the potential causes behind these differences.

4.4. Growth in weight and other components

In the above sections, growth is considered in terms of change in body length. However, estimates of increases in weight and elemental composition are more important with regard to the parameterisation of the energy budget of *M. norvegica* and their biogeochemical role in the ecosystem. Furthermore, the pattern of increase in weight may differ to increase in length, particularly in adults that may continue to put on weight without lengthening. Notable considerations of growth in weight and other components have been carried out by Båmstedt (1976), Boysen and Buchholz (1984) and Falk-Petersen (1985).

Båmstedt (1976) related carapace length to dry weight, ash weight, lipid, protein and carbohydrate over 13 consecutive months in Korsfjorden, Western Norway. Overall, he found that individuals of the same length were heavier in summer and autumn than in winter and spring (Table 3.3). Growth in weight was highest during summer and autumn, with the youngest generation showing the greatest relative weight gain. The lipid content showed distinct seasonal variations, yielding highest values in winter and spring. Juveniles had higher proportions of ash and carbohydrate and lower chitin and lipid proportions than did the adults. There were no marked differences in the levels of various components between males and females.

In the Kattegat, Boysen and Buchholz (1984) found that rates of growth in males and females were approximately the same. Most variations in growth were related to prevailing food supply, which affected both sexes equally. However, there were discrepancies in certain months, such as July, where there was a high increase in both weight and length in males but no increase in females. Weight continued to increase in October in both sexes, but there was no corresponding growth in length (Table 3.3).

Falk-Petersen (1985) measured wet weight, dry weight, protein, and lipid almost monthly over 14 months in Balsfjorden, North Norway (Table 3.3). In individuals belonging to the I- and II-group, wet weight

Table 3.3 Conversion from carapace length (mm) or total body length (mm) or dry weight (mg) or wet weight (mg) for each month of the year at three different locations

	Bâmstedt (1976), Korsfjorden, W Norway	Boysen and Buchholz (1984), Kattegat, Denmark	Falk-Petersen (1985), Balsfjorden, N Norway	
Conversion	Length of carapace (C, mm) to Dry Weight (DW, mg)★	Total body length (L, mm) to Dry weight (DW, mg)	Length of carapace (C, mm) to Dry Weight (DW, mg)[a]	Length of carapace (C, mm) to Wet Weight (WW, mg)[a]
Functional form	$DW = a \times C^b$	$DW = a \times L^b$	$\log DW = \log a + b(\log)C$	$\log WW = \log a + b(\log)C$
January	$a = 0.094, b = 2.980$	$a = 1.44 \times 10^{-3}, b = 3.029$ (male)	$a = 0.0966, b = 3.0218$ (1977)	$a = 0.7797, b = 2.6931$ (1977)
February	$a = 0.097, b = 3.004$	$a = 3.59 \times 10^{-4}, b = 3.397$ (female)	$a = 0.1126, b = 2.8991$ (1977)	$a = 0.8257, b = 2.7391$ (1977)
March	$a = 0.119, b = 2.849$	$a = 1.73 \times 10^{-4}, b = 3.599$ (male) $a = 4.84 \times 10^{-4}, b = 3.261$ (female)	$a = 0.0612, b = 1.1702$ (1977)	$a = 0.5731, b = 2.8157$ (1977)
April	$a = 0.119, b = 2.920$		$a = 0.0476, b = 3.3342$ (1977)	$a = 0.3703, b = 3.0907$ (1977)
May	$a = 0.092, b = 2.954$	$a = 2.34 \times 10^{-4}, b = 3.533$ (male) $a = 1.91 \times 10^{-4}, b = 3.629$ (female)	$a = 0.0668, b = 3.0245$ (1976)	$a = 0.1512, b = 3.3107$ (1976)
June	$a = 0.101, b = 2.968$	$a = 1.74 \times 10^{-3}, b = 2.919$ (male) $a = 1.82 \times 10^{-3}, b = 2.916$ (female)	$a = 0.1181, b = 2.9554$ (1977)	$a = 0.7372, b = 2.7782$ (1977)
July	$a = 0.197, b = 2.715$ (1973) $a = 0.141, b = 2.889$ (1974)	$a = 1.66 \times 10^{-3}, b = 2.954$ (male) $a = 1.42 \times 10^{-3}, b = 3.016$ (female)	$a = 0.1089, b = 2.8325$ (1976) $a = 0.0694, b = 3.1685$ (1977)	$a = 0.4796, b = 2.8215$ (1976) $a = 0.6477, b = 2.8263$ (1977)
August	$a = 0.361, b = 2.480$	$a = 1.45 \times 10^{-3}, b = 3.028$ (male) $a = 2.19 \times 10^{-3}, b = 2.848$ (female)	$a = 0.1401, b = 2.8101$ (1976) $a = 0.0658, b = 3.1898$ (1977)	$a = 0.6982, b = 2.7141$ (1976) $a = 0.4213, b = 2.9906$ (1977)
September		$a = 4.90 \times 10^{-4}, b = 3.279$ (male) $a = 1.11 \times 10^{-3}, b = 3.041$ (female)	$a = 0.2929, b = 2.5748$ (1976)	$a = 0.6063, b = 2.8229$ (1976)

(continued)

Table 3.3 (*continued*)

	Båmstedt (1976), Korsfjorden, W Norway	Boysen and Buchholz (1984), Kattegat, Denmark	Falk-Petersen (1985), Balsfjorden, N Norway
October	$a = 0.243$, $b = 2.633$	$a = 4.01 \times 10^{-5}$, $b = 4.044$ (male) $a = 5.32 \times 10^{-4}$, $b = 3.308$ (female)	$a = 0.1299$, $b = 2.9309$ (1976) $a = 0.9307$, $b = 2.6236$ (1976)
November	$a = 0.074$, $b = 3.216$	$a = 1.45 \times 10^{-4}$, $b = 3.729$ (male) $a = 8.78 \times 10^{-5}$, $b = 3.876$ (female)	$a = 0.1693$, $b = 2.8013$ (1976) $a = 0.7449$, $b = 2.7114$ (1976)
December	$a = 0.137$, $b = 2.886$		

[a] The following can be applied to convert from carapace length to total body length: Mauchline (1960)—total body length (L), carapace length (C): immature animals—males L = 4.44 C − 0.74; females L = 4.18 C + 0.74; mature animals—males L = 3.72 C + 4.09; females L = 3.61 C + 3.60; Matthews (1973)—all stages L = 3.18 C + 2.41.

was relatively stable from January to March (~ 600 mg) until it increased to a marked maximum in April (800 mg). Dry weight decreased from January until February (175–155 mg) before increasing to 180 mg in May. Protein decreased continuously from January to April (60–40 mg) before a marked increase to 70 mg in May. Lipid peaked in October (~ 80 mg) and declined over the course of the subsequent spring and summer. The marked increase in many components during late spring/early summer was believed to reflect (1) an increase in the food quality due to the onset of feeding of the smaller zooplankton on phytoplankton and (2) an increase in food availability due to the spawning of the spring-spawning animals.

5. MORTALITY AND SEX RATIO

5.1. Larval mortality

There has been very little effort towards examining the mortality of larval stages of *M. norvegica*. Rates are difficult to determine without knowledge of the initial size of the spawned population, which will be a function of the abundance of spawning females and their rate of fecundity. Estimates of the abundance of respective larval stages are then required. An example of such an approach is detailed in Tarling *et al.* (2007) for *E. superba*, where female abundance was determined from net-catch surveys and their fecundity from examinations of the ovary to determine the number of mature oocytes and an estimation of the number of spawning cycles within a single reproductive season. Egg and early-stage larval mortality rates were extrapolated from Hirst and Kiørboe (2002) for copepod eggs and larvae, scaled to the appropriate temperature (4 °C). Furcilia mortality was derived from population analyses carried out by Brinton (1976). This gave daily exponential rates of mortality (β) of 0.04 per day for eggs and early stages and 0.034 per day for furcilia. The approach was used to estimate whether the level of spawning by the local Antarctic krill population at South Georgia was sufficient to sustain the population without additional immigration.

Tarling and Cuzin-Roudy (2003) took an alternative approach in examining the relationship between the strength of a cohort and the factors that may have influenced the survivorship of the larval stages. The study focussed on the Clyde, where it was apparent that, in 1999, there were three pulses of recruitment over the course of the productive season. Larval cohort strength showed a distinct correspondence to the level of chlorophyll *a* during the period of early furcilia stage development. The early furcilia stages of each of the three successful cohorts in 1999 developed at the start of or within phytoplankton blooms, whereas other cohorts that were unsuccessful experienced low or decreasing levels of Chl *a* during the furcilia stages. Feeding conditions during the calyptopis stages appear to have less influence on the success of a cohort, since phytoplankton levels varied greatly between successful and unsuccessful cohorts. Parallels can be drawn with the study of Rumsey and Franks (1999), who found that variability in the duration of the furcilia stage 1 and 2 in *E. pacifica* was responsible for the high mortality in larval development and hence overall recruitment success. The reasons for this could be that the pleopods and gills develop during this stage together with higher oxygen consumption and growth rates. These higher energy requirements may reduce the carrying capacity of the environment with respect to the density of larvae that can be maintained when resources are limited.

5.2. Post-larval mortality

Labat and Cuzin-Roudy (1996) estimated mortality rate in Mediterranean *M. norvegica* through fitting an exponential mortality function to time-series data. Their calculation was based on the relative size of the I-group and II-group population at a time (January–March) when there was no recruitment to the post-larval population.

The fitted exponential function was of the following form:

$$N_{t+1,\text{age}+1} = N_{t,\text{age}}e^{-\beta \cdot t} \qquad (3.3)$$

where N is population abundance, t is time in days, age is number of days since spawning and β is the mortality coefficient.

They assumed (1) that the mortality coefficient β was constant for the different age groups, allowing the use of a single exponential mortality model; (2) the fluctuations in successive recruitments were small and of random character; and (3) the studied samples represented the average population structure. With I-group comprising 71.4% of the post-larval population and II-group, 28.6%, the fitted mortality coefficient was 0.00342 per day or 1.248 per year.

A previous attempt at estimating mortality rate in *M. norvegica* was carried out by Matthews (1973) on a population in Korsfjorden, Norway. The approach was to derive an empirical expression where the time-scale was adjustable to account for a plateau in the rate of mortality during the winter months. Labat and Cuzin-Roudy (1996) re-examined these parameters and estimated that, expressed as an exponential function Eq. (3.3), the rates would have been equivalent to between 0.004 and 0.01 per day (1.46–3.65 per year) similar to those observed in the Mediterranean.

These mortality rates are mid-way between those calculated for other well-studied euphausiid species. In *E. pacifica* in Californian waters, annual mortality values of 21.84 per year in the larval stage, 4.8 per year in the juvenile stage and 6.12 per year for the post-larval stage were estimated by

Siegel (2000a) from the original data of Brinton (1976). Siegel (2000a) also calculated *E. pacifica* lifetime mortality values of 3.0 per year in Californian waters, 8.7 per year off the coast of Oregon and between 0.6 and 1.9 per year in British Columbia using the original data of Jarre-Teichmann (1996). Most reported annual mortality values for *E. superba* have ranged between 0.5 and 1 per year (Siegel, 2000a). Pakhomov (1995) considered the variation in mortality over the lifespan of *E. superba*, which appeared to have a U-shaped trajectory, declining from a rate of 1.09 per year in the first year of life to a minimum of 0.57 per year in the third year and increasing to 2.41 per year in year 6.

Figure 3.4 represents a survivorship curve in which exponential rates of mortality are 1.46 per year (0.004 per day) for post-larvae, 12.41 per year (0.034 per day) for furcilia, and 14.6 per year (0.04 per day) for eggs and early- stage larvae, and the development time is an estimate from the Clyde population in April. Almost 57% of the initial number of eggs spawned is lost by the end of the calyptopes stages, declining to 90% by the time of adolescence. Once reaching adolescence, a post-larval individual can expect to live an average of 6 months. Only 7% of adolescents would expect to survive into the II-group and achieve a second year of reproduction. Of these, around 20% could go on and reproduce for a third year, although this number represents just 4% of the total post-larval population.

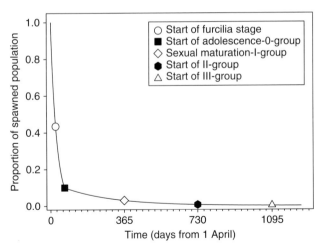

Figure 3.4 A survivorship curve representing the trajectory of decrease in cohort numbers from the point when it was spawned. Mortality rates were 0.04 d^{-1} for eggs and early-stage larvae, 0.034 d^{-1} for furcilia and 0.004 d^{-1} for post-larvae (see text for further explanation). The figure represents the situation for a cohort in the Clyde, spawned in April.

5.3. Sex ratio

A particularly noticeable aspect of adult survivorship is the fact that the rate of mortality can often differ between sexes, which affects the 'tertiary' sex ratio (i.e. sex ratio in mature individuals). There is no evidence to suggest that the primary sex ratio (i.e. at the point of zygote formation) is anything other than 1:1 in euphausiids (Timofeev, 2002). However, reports of deviations in the tertiary sex ratio are relatively common. For instance, Hollingshead and Corey (1974) report a decrease in the relative number of males in June. Mauchline (1960) reports a relative decrease in the number of males in January. In Gullmarsfjord, the proportion of males peaked at almost 70% in late summer but dropped to around 40% by mid-winter (Thomasson, 2003).

Various reasons for differential rates and timings of mortality between the sexes have been put forward. Both Hollingshead and Corey (1974) and Mauchline (1960) proposed that males generally die after spermatophore transfer. However, the process of mating itself has not been reported to have any influence on mortality rates, particularly one that disadvantages one sex more than the other. Tarling (2003) alternatively considered whether differences in behaviour and feeding may influence the relative survivorship of males and females. That year-long study of the *M. norvegica* population in the Clyde found sex ratio to be initially 1:1 when juveniles matured into adults after their first winter, but became increasingly biased towards males as the season progressed such that, by October, the male to female ratio was close to 3:1. Females were seen to migrate closer to the surface than males, and so the relative risk in terms of vulnerability to visual predation was determined for males and females using a deterministic individual-based model where predation risk was a function of the level of light. The model also calculated the relative rate of energy consumption based on measured vertical gradients in food availability. It was found that females experienced relatively more risk over the course of the summer, over which time there was a good correspondence between predicted and observed sex ratio. However, during the winter, the model predicted a continued bias in favour of males whereas observations showed that females became numerically dominant. Females gained more food resource in the process of taking more risk, so it was proposed that males were less provisioned to outlast the winter period and therefore experienced a greater extent of overwintering mortality due to unsustainable levels of weight loss.

6. Recruitment Models

Recruitment has hardly been estimated in euphausiid species, with the exception of *E. superba* (de la Mare, 1994a,b; Siegel, 2000b). However, some attempts have been made to parameterise the extent and timing of late-stage larval occurrence in *M. norvegica* populations.

Lindley (1982) examined the seasonal occurrence of *M. norvegica* furciliae over a large part of the North Atlantic from Continuous Plankton Recorder records. The study area was split into rectangular sectors, spanning approximately 4° latitude and 10° longitude. He found that furciliae were usually recorded for between 2 and 4 successive months in each of the sectors. In many sectors, this was in the period April–June. Over the American continental shelf, furciliae were not found until July. The data were used to calculate the mean time of occurrence in each sector. Also, for each sector, the following parameters were determined: (1) timing of the spring bloom of phytoplankton; (2) latitude; (3) mean surface temperature from March to June; (4) the difference in mean surface temperature in March to that in the period May–July. The data were analysed using a correlation matrix model approach (Spearman rank). The timing of larval occurrence showed a significant positive correlation with parameter 3 and a significant negative correlation with parameter 4. It was postulated that the link with temperature acts through significantly delaying the onset of breeding, particularly in areas where surface temperatures are close to 5 °C, which is believed to be the lower temperature limit for successful breeding in *M. norvegica* (Einarsson, 1945; Falk-Petersen and Hopkins, 1981; Jones, 1968; Kielhorn, 1952).

Tarling and Cuzin-Roudy (2003) modelled a situation found in the Clyde where a number of larval cohorts appeared to recruit into the population over a single productive season. Their model first considered the moult/spawn cycle of the females and, based on the fact that the females appeared to be synchronised, determined that there would be a pulse of eggs into the water column every 20–26 days. They then followed each potential cohort in terms of its environmental experience and compared the experience of those that were seen to succeed compared to those that failed. They determined that successful recruitment was only achieved when chlorophyll *a* levels were adequate during both the period of egg maturation in the ovary and the subsequent development of larvae, especially the furcilia stages. They singled out the furciliae stages as being particularly vulnerable. The fact that mortality may be particularly high in larval stages that occur a period of time after initial spawning means that the abundance and condition of the eggs and the consequent success of the cohort may show little relationship. Such a lack of correspondence has been noted in other species, such as *Thysanoessa spinifera* (Tanasichuk, 1998) and *E. pacifica* (Brinton, 1976).

7. FUTURE WORK

Despite many decades of work on *M. norvegica*, our understanding of which stocks are self-sustaining and which rely on immigration to support the adult population remains speculative. This is particularly so in the northern reaches of the species distribution, where spawning is erratic and the origin of stocks is unclear. Much of the difficulty in progressing this area of knowledge is in determining the result of interactions with the advective environment. Although the vertical migration of larvae may be more limited than the adults, it may still span over 100 m and, in so doing, enter different water masses and potentially different current regimes. The limited swimming capabilities of the larvae make them particularly vulnerable to advective export from a region of successful spawning and their behaviour will in turn impact the direction and extent to which they are displaced. A similar issue is apparent for other species of euphausiid, notably E. superba, which inhabits one of the world's strongest current regimes, the Antarctic Circumpolar Current. Work in this region has begun to consider how the behaviour and life-cycle of this species and its interactions with the prevailing currents influence distributional patterns. Thorpe et al. (2007), for instance, performed a Lagrangian modelling study at the circumpolar scale where output from the Ocean Circulation and Climate Advanced Modelling (OCCAM) project was combined with satellite data of ocean conditions and seeded with particles programmed to adopt a theoretical vertical migration behaviour of a typical krill. This allowed them to examine which regions in the Antarctic Peninsula were potential recruitment areas to the South Georgia krill population that lies downstream.

Performing such an analysis on Northern krill is a tractable and potentially illuminating proposition. Sourisseau et al. (2006) have already made some attempt, at least at a local scale. However, such models must also incorporate important aspects of the life-cycle and the response of individuals to prevailing conditions in order to improve their predictive skill. For instance, although an advective route between stocks may be identified, it may be unviable because of low food availability in the intervening region. Further work must therefore be dedicated to the response of krill to the prevailing environment, particularly starvation and thermal tolerances in larvae. The parameterisation of developmental rates derived as a function of both food and temperature is required by such models. Wider application of the IGR method across a range of environments and developmental stages would also be beneficial in this regard. Finally, greater insight into what factors permit and prevent successful spawning will allow accurate constraints to be placed around where the theoretical distributional limits of this species actually reside.

ACKNOWLEDGEMENTS

I would like to thank Ryan Saunders and Jean-Phillipe Labat for the provision of raw data to plot the size structure of populations within the Greenland/Irminger region and the Mediterranean region, respectively. Angus Atkinson gave invaluable advice on the contents of this chapter. Jack Matthews and Fred Buchholz provided the initial inspiration to undertake this work. GT carried out this work as part of the remit of the Ecosystems programme at the British Antarctic Survey.

REFERENCES

Astthorsson, O. S. (1990). Ecology of the euphausiids *Thysanoessa raschi, T. inermis* and *Meganyctiphanes norvegica* in Isafjord-Deep, northwest Iceland. *Mar. Biol.* **107,** 147–157.

Astthorsson, O. S., and Gislason, A. (1997). Biology of euphausiids in the subarctic waters north of Iceland. *Mar. Biol.* **129,** 319–330.

Atkinson, A., Shreeve, R. S., Hirst, A. G., Rothery, P., Tarling, G. A., Pond, D. W., Korb, R., Murphy, E. J., and Watkins, J. L. (2006). Natural growth rates in Antarctic krill (*Euphausia superba*): II. Predictive models based on food, temperature, body length, sex, and maturity stage. *Limnol. Oceanogr.* **51,** 973–987.

Båmstedt, U. (1976). Studies on the deep-water pelagic community of Korsfjorden, wetsern Norway. Changes in the size and biochemical composition of *Meganyctiphanes norvegica* (Euphausiacea) in relation to its life cycle. *Sarsia* **61,** 15–30.

Bargmann, H. E. (1945). The development and life history of adolescent and adult krill, *Euphausia superba. Discovery Rep.* **23,** 103–176.

Berkes, F. (1976). Ecology of euphausiids in the Gulf of St. Lawrence. *J. Fish. Res. Board Can.* **33,** 1894–1905.

Bollens, S. M., Frost, B. W., and Lin, T. S. (1992). Recruitment, growth and diel vertical migration of *Euphausia pacifica* in a temperate fjord. *Mar. Biol.* **114,** 219–228.

Boysen, E., and Buchholz, F. (1984). *Meganyctiphanes norvegica* in the Kattegat. *Mar. Biol.* **79,** 195–207.

Brinton, E. (1976). Population biology of *Euphausia pacifica* off Southern California. *Fish. Bull.* **74,** 733–763.

Cuzin-Roudy, J. (1993). Reproductive strategies of the Mediterranean krill, *Meganyctiphanes norvegica* and the Antarctic krill *Euphausia superba* (Crustacea: Euphausiacea). *Invertebr. Reprod. Dev.* **23,** 105–114.

Cuzin-Roudy, J. (2000). Seasonal reproduction, multiple spawning, and fecundity in northern krill, *Meganyctiphanes norvegica*, and Antarctic krill, *Euphausia superba. Can. J. Fish. Aquat. Sci.* **57,** 6–15.

Cuzin-Roudy, J., and Buchholz, F. (1999). Ovarian development and spawning in relation to the moult cycle in Northern krill *Meganyctiphanes norvegica* (Crustacea: Euphausiacea), along a climatic gradient. *Mar. Biol.* **133,** 267–281.

Dalpadado, P. (2006). Distribution and reproduction strategies of krill (Euphausiacea) on the Norwegian shelf. *Polar Biol.* **29,** 849–859.

Dalpadado, P., and Skjoldal, H. R. (1991). Distribution and life history of krill from the Barents Sea. *Polar Res.* **10,** 443–460.

Dalpadado, P., and Skjoldal, H. R. (1996). Abundance, maturity and growth of the krill species *Thysanoessa inermis* and *T. longicaudata* in the Barents Sea. *Mar. Ecol. Prog. Ser.* **144,** 175–183.

De la Mare, W. K. (1994a). Modelling krill recruitment. *CCAMLR Sci.* **1,** 49–54.

De la Mare, W. K. (1994b). Estimating krill recruitment and its variability. *CCAMLR Sci.* **1,** 55–69.

Einarsson, H. (1945). Euphausiacea I. Northern Atlantic Species. *Dana Report* **27,** 1–184.

Falk-Petersen, S. (1985). Growth of the Euphausiids *Thysanoessa inermis, Thysanoessa raschii* and *Meganyctiphanes norvegica* in a Subarctic Fjord, North Norway. *Can. J. Fish. Aquat. Sci.* **42,** 14–22.

Falk-Petersen, S., and Hopkins, C. C. E. (1981). Ecological investigations on the zooplankton community of Balsfjorden, northern Norway: population dynamics of the euphausiids *Thysanoessa inermis, T. raschii. J. Plankt. Res.* **3**, 177–192.

Gislason, A., and Astthorsson, O. S. (1995). Seasonal cycle of zooplankton southwest of Iceland. *J. Plankt. Res.* **17**, 1959–1976.

Harrington, S. A., and Thomas, P. G. (1987). Observations on spawning by *Euphausia crystallorophias* from waters adjacent to Enderby Land (East Antartica) and speculations on the early ontogenetic ecology of neritic euphausiids. *Polar Biol.* **7**, 93–95.

Heegard, P. (1948). Larval stages of Meganyctiphanes (Euphausiacea) and some general phylogenetic remarks. Meddelelser fra Kommissionen for Danmarks fiskeri—og. *Havundersogelser* **5**, 25.

Hirst, A. G., and Kiørboe, T. (2002). Mortality of marine planktonic copepods: global rates and patterns. *Mar. Ecol. Prog. Ser.* **230**, 195–209.

Hirst, A. G., Roff, J. C., and Lampitt, R. S. (2003). A synthesis of growth rates in marine epipelagic invertebrate zooplankton. *Adv. Mar. Biol.* **44**, 1–142.

Hollingshead, K. W., and Corey, S. (1974). Aspects of the life history of *Meganyctiphanes norvegica* (M.Sars), Crustacea (Euphausiacea), in Passamaquoddy Bay. *Can. J. Zool.* **52**, 495–505.

Hopkins, C. C. E. (1981). Ecological investigations on the zooplankton community of Balsfjorden, Northern Norway: changes in zooplankton abundance and biomass in relation to phytoplankton and hydrography, March 1976–Februrary 1977. *Kieler Meeresforsch. (Sonderh.)* **5**, 124–139.

Ikeda, T., and Dixon, P. (1982). Body shrinkage as a possible over-wintering mechanism of the Antarctic krill, *Euphausia superba* Dana. *J. Exp. Mar. Biol. Ecol.* **62**, 143–151.

Jarre-Teichmann, A. (1996). Initial estimates on krill. *Fish. Cent. Res. Rep.* **4**.

Jones, L. T. (1968). Occurrence of the larvae of *Meganyctiphanes norvegica* (Crustacea, Euphausiacea) off West Greenland. *J. Fish. Res. Board Can.* **25**, 1071–1073.

Kawaguchi, S., Candy, S. G., King, R., Naganobu, M., and Nicol, S. (2006). Modelling growth of Antarctic krill. I. Growth trends with sex, length, season, and region. *Mar. Ecol. Prog. Ser.* **306**, 1–15.

Kielhorn, W. V. (1952). The biology of the surface zone zooplankton of a boreo-Arctic Atlantic Ocean area. *J. Fish. Res. Board Can.* **9**, 223–264.

Kulka, D. W., Corey, S., and Iles, T. D. (1982). Community structure and biomass of euphausiids in the Bay of Fundy. *Can. J. Fish. Aquat. Sci.* **39**, 326–334.

Labat, J. P., and Cuzin-Roudy, J. (1996). Population dynamics of the krill *Meganyctiphanes norvegica* (Sars) (Crustacea: Euphausiacea) in the Ligurian Sea (N-W Mediterranean Sea). Size structure, growth and mortality modelling. *J. Plankton. Res.* **18**, 2295–2312.

Lacroix, G. (1961). Les migrations verticale journalières des Euphausides à l'entrée de la Baie de Chaleurs. *Contr. Dep. Pêch.* **Queb No. 83**, 257–316.

Lebour, M. V. (1924). The Euphausiidae in the neighbourhood of Plymouth and their importance as herring food. *J. Mar. Biol. Assoc. UK* **13**, 402–431.

Lindley, J. A. (1982). Population dynamics and production of euphausiids III. *Meganyctiphanes norvegica* and *Nyctiphanes couchi* in the north Atlantic ocean and the North Sea. *Mar. Biol.* **66**, 37–46.

Lucas, C. E., Marshall, N. B., and Rees, C. B. (1957). Continuous plankton records: The Faroe-Shetland Channel 1939. *Hull. Bull. Mar. Ecol.* **2**, 71–94.

Macdonald, R. (1927). Irregular development in the larval history of *Meganyctiphanes norvegica. J. Mar. Biol. Assoc. UK* **14**, 785–794.

Mackintosh, N. A. (1972). Life-cycle of Antarctic krill in relation to ice and water conditions. *Discovery Rep.* **36**, 1–94.

Makarov, R. R. (1971). Some peculiarities of life-cycles of euphausiids. *Zoologicheskii Zhurnal, Moscow* **50**, 193–198.

Makarov, R. R. (1974). Dominance of larval forms of euphausiid (Crustacea: Eucarida) ontogenesis. *Mar. Biol.* **26**, 45–56.

Marinovic, B., and Mangel, M. (1999). Krill can shrink as an ecological adaptation to temporarily unfavourable environments. *Ecol. Lett.* **2**, 338–343.

Marschall, H. P. (1983). Sinking speed, density and size of euphausiid eggs. *Meeresforschung* **30**, 1–9.

Matthews, J. B. L. (1973). Ecological studies on the deep-water pelagic community of Korsfjorden, western Norway. *Sarsia* **54**, 75–90.

Mauchline, J. (1959). The development of the Euphausiacea (Crustacea) especially that of *Meganyctiphanes norvegica* (Sars). *Proc. Zool. Soc. Lond.* **132**, 627–639.

Mauchline, J. (1960). The biology of the euphausiid crustacean, *Meganyctiphanes norvegica*. Sars. *Proc. R. Soc. Edin. B* **67**, 141–179.

Mauchline, J. (1965). The larval development of the euphausiid *Thysanoessa raschii* (Sars). *Crustaceana* **9**, 31–40.

Mauchline, J. (1968). The development of the eggs in the ovaries of euphausiids and estimation of fecundity. *Crustaceana* **14**, 155–163.

Mauchline, J. (1977). Growth and moulting of crustacea, especially euphausiids. *In* "Oceanic Sound Scattering and Prediction" (N. R. Andersen and B. J. Zahuranec, eds), pp. 401–422. Plenum Press, New York.

Mauchline, J. (1985). Growth and production of Euphausiacea (Crustacea) in the Rockall Trough. *Mar. Biol.* **90**, 19–26.

Mauchline, J., and Fisher, L. R. (1969). The biology of euphausiids. *Adv. Mar. Biol.* **7**, 1–454.

Melle, W., Ellertsen, B., and Skjoldal, H. R. (2004). Zooplankton: the link to higher trophic levels. *In* "The Norwegian Sea ecosystem" (H. R. Skjoldal, ed.), pp. 137–202. Tapir Academic, Trondheim, Norway.

Nicol, S., Stolp, M., Cochran, T., Geijsel, P., and Marshall, J. (1992). Growth and shrinkage of Antarctic krill *Euphausia superba* from the Indian Ocean sector of the Southern Ocean during summer. *Mar. Ecol. Prog. Ser.* **89**, 175–181.

Pakhomov, E. A. (1995). Demographic studies of Antarctic Krill *Euphausia superba* in the Cooperation and Cosmonaut Seas (Indian Sector of the Southern Ocean). *Mar. Ecol. Prog. Ser.* **119**, 45–61.

Pinchuk, A. I., and Hopcroft, R. R. (2007). Seasonal variations in the growth rates of euphausiids (*Thysanoessa inermis*, *T. spinifera*, and *Euphausia pacifica*) from the northern Gulf of Alaska. *Mar. Biol.* **151**, 257–269.

Quetin, L. B., and Ross, R. M. (1991). Behavioural and physiological characteristics of Antarctic krill *Euphausia superba* Dana. *Am. Zool.* **31**, 49–63.

Ross, R. M. (1981). Laboratory culture and development of *Euphausia pacifica*. *Limnol. Oceanogr.* **26**, 235–246.

Ross, R. M., and Quetin, L. B. (2000). Reproduction in Euphausiacea. *In* "Krill: Biology, Ecology and Fisheries" (I. Everson, ed.), Vol 6, pp. 150–181. Blackwell Science Ltd, Oxford.

Ross, R. M., Quetin, L. B., Baker, K. S., Vernet, M., and Smith, R. C. (2000). Growth limitation in young *Euphausia superba* under field conditions. *Limnol. Oceanogr.* **45**, 31–43.

Rumsey, S. M., and Franks, P. J. S. (1999). Influence of variability in larval development on recruitment success in the euphausiid *Euphausia pacifica*: elasticity and sensitivity analyses. *Mar. Biol.* **133**, 283–291.

Ruud, J. T. (1932). On the biology of southern Euphausiidae. *Hvalrad Skr.* **2**, 1–105.

Ruud, J. T. (1936). Euphausiacea. *Rep. Dan. Oceanogr. Exped. Mediterr.* **2**, 1–86.

Saunders, R. A., Ingvarsdottir, A., Rasmussen, J., Hay, S. J., and Brierley, A. S. (2007). Regional variation in distribution pattern, population structure and growth rates of

Meganyctiphanes norvegica and *Thysanoessa longicaudata* in the Irminger Sea, North Atlantic. *Prog. Oceanogr.* **72**, 313–342.

Siegel, V. (2000a). Krill (Euphausiacea) life history and aspects of population dynamics. *Can. J. Fish. Aquat. Sci.* **57**, 130–150.

Siegel, V. (2000b). Krill (Euphausiacea) demography and variability in abundance and distribution. *Can. J. Fish. Aquat. Sci.* **57**, 151–167.

Simard, Y., Marcotte, D., and Naraghi, K. (2003). Three-dimensional acoustic mapping and simulation of krill distribution in the Saguenay—St. Lawrence Marine Park whale feeding ground. *Aquatic Liv. Res.* **16**, 137–144.

Skjoldal, H. R., Dalpadado, P., and Dommasnes, A. (2004). Food webs and trophic interactions. *In* "The Norwegian Sea ecosystem" (H. R. Skjoldal, ed.), pp. 447–506. Tapir Academic, Trondheim, Norway.

Sourisseau, M., Simard, Y., and Saucier, F. J. (2006). Krill aggregation in the St. Lawrence system, and supply of krill to the whale feeding grounds in the estuary from the gulf. *Mar. Ecol. Prog. Ser.* **314**, 257–270.

Suh, H.-L., Soh, H. Y., and Hong, S. Y. (1993). Larval development of the euphausiid *Euphausia pacifica* in the Yellow Sea. *Mar. Biol.* **115**, 625–633.

Tanasichuk, R. W. (1998). Interannual variations in the population biology and productivity of *Thysanoessa spinifera* in Barkley Sound, Canada, with special reference to the 1992 and 1993 warm ocean years. *Mar. Ecol. Prog. Ser.* **173**, 181–195.

Tarling, G. A. (2003). Sex dependent diel vertical migration in Northern krill and its consequences for population dynamics. *Mar. Ecol. Prog. Ser.* **260**, 173–188.

Tarling, G. A., and Cuzin-Roudy, J. (2003). Synchronization in the moulting and spawning activity of northern krill (*Meganyctiphanes norvegica*) and its effect on recruitment. *Limnol. Oceanogr.* **48**, 2020–2033.

Tarling, G. A., Shreeve, R. S., Hirst, A. G., Atkinson, A., Pond, D. W., Murphy, E. J., and Watkins, J. L. (2006). Natural growth rates in Antarctic krill (*Euphausia superba*): I. Improving methodology and predicting intermolt period. *Limnol. Oceanogr.* **51**, 959–972.

Tarling, G. A., Cuzin-Roudy, J., Thorpe, S., Shreeve, R. S., and Ward, P. (2007). The recruitment of Antarctic krill (*Euphausia superba*) in the South Georgia region: adult fecundity and the fate of early developmental stages. *Mar. Ecol. Prog. Ser.* **331**, 161–179.

Thomasson, M. A. (2003). Vertical distribution, behaviour and aspects of the population biology of euphausiids, especially *Meganyctiphanes norvegica* (M. Sars), in the Gullmarsfjord, western Sweden, PhD Thesis, Goteborg University.

Thorpe, S. E., Murphy, E. J., and Watkins, J. L. (2007). Circumpolar connections between Antarctic krill (*Euphausia superba* Dana) populations: Investigating the roles of ocean and sea ice transport. *Deep Sea Res. I* **54**, 792–810.

Timofeev, S. F. (2002). Sex ratios in the population of *Thysanoessa raschii* (M. Sars, 1864) (Euphausiacea) in the Barents Sea (With some notes on thesex ratios in the order Euphausiacea). *Crustaceana* **75**, 937–956.

Wiborg, K. F. (1971). Investigations in euphausiids in some fjords on the west coast of Norway in 1966–1977. *Fiskeridir. Skr. Ser. Havunders* **16**, 10–35.

Zelickman, E. A. (1968). Some features of the evolution of the family Euphausiidae (Crustacea, Euphausiacea) in neritic and oceanic areas. *Zoologicheskii Zhurnal, Moscow* **47**, 1314–1327.

Zhou, M., Zhu, Y. W., and Tande, K. S. (2005). Circulation and behavior of euphausiids in two Norwegian sub-Arctic fjords. *Mar. Ecol. Prog. Ser.* **300**, 159–178.

CHAPTER FOUR

Physiology and Metabolism of Northern Krill (*Meganyctiphanes norvegica* Sars)

John I. Spicer* *and* Reinhard Saborowski[†]

Contents

* Marine Biology and Ecology Research Centre, School of Marine Sciences and Engineering, University of Plymouth, Plymouth, United Kingdom
[†] Alfred Wegener Institute for Polar and Marine Research, Bremerhaven, Germany

Advances in Marine Biology, Volume 57
ISSN 0065-2881, DOI: 10.1016/S0065-2881(10)57004-4

Abstract

Advances in our understanding of the physiology and metabolism of Northern krill, *Meganyctiphanes norvegica* have been sporadic but significant. Despite problems with keeping *M. norvegica* in good condition in the laboratory, those who have tried, and succeeded, have contributed to a better knowledge of krill biology and challenged our understanding of some basic biological processes. Most recent work has been concentrated in the fields of digestive physiology, lipid biochemistry, respiration and anaerobiosis, metabolic properties, and pollutants. *M. norvegica* is capable of digesting an opportunistic, omnivorous diet, showing some digestive enzyme polymorphism and high levels of enzyme activity, the latter varying with season. It also seems capable of digesting cellulose and hemicelluloses, for example, laminarin. The biochemical composition of krill is relatively well known with some recent extensive work focusing on the previously little studied lipid and fatty acid composition, particularly with reference to reproduction, overwintering energy storage and as a nutrition marker. A high aerobic metabolism (but poor anaerobic capacity) is characteristic of *M. norvegica*, and how this is affected by temperature, low O_2, and season has attracted some attention, particularly in the context of diel vertical migration (DVM) across pronounced pycnoclines. Despite determining high metabolic turnover rates and a high physiological plasticity for this species, we know little of the regulative potential of metabolites, particularly their modulative effect on enzyme activity. Certainly a modest ability to maintain aerobic metabolism when encountering hypoxia, and little or no ability to osmoregulate in hyposaline conditions, does not prevent DVM in adults of this species. The ability to maintain aerobic metabolism develops early in ontogeny at about furcilia III (i.e. concurrent with first DVM behaviour). The respiratory pigment of *M. norvegica*, haemocyanin, has a low O_2 affinity and high temperature sensitivity (although temperature has the opposite effect on O_2 binding than found for nearly every other haemocyanin). Also surprising is the apparent use of haemocyanin as an energy source/store. While recent work has focused on physiological effects, the ecophysiological effects of transuric elements and trace metals, the effects of pollution generally are widely understudied.

1. INTRODUCTION

Advances in our understanding of the physiology and metabolism of *M. norvegica* since the reviews of Mauchline [Mauchline, J., Fisher, L.R., 1969. The biology of euphausiids. *Adv. Mar. Biol.* 7, 1–454; Mauchline, J., 1980. Part II: The biology of euphausiids. *Adv. Mar. Biol.* 18, 373–623] have been both sporadic and patchy, but there have certainly been advances. We review the literature published since the last review of krill biology, highlighting such advances but also noting areas which, despite their obvious importance, still require attention.

2. THE PHYSIOLOGY OF DIGESTION

The digestive systems of euphausiids are broadly similar to those of other eucarid crustaceans (Mauchline and Fisher, 1969). Since no histological and cytological studies are available for the digestive organs of *Meganyctiphanes norvegica*, a brief description based on the extensive work done on decapods (e.g. Dall and Moriarty, 1983; Loizzi, 1971) and Antarctic krill, *Euphausia superba* (e.g. Ikeda *et al.*, 1984; Ullrich *et al.*, 1991), is given as background to what follows on the digestive system and its function.

2.1. The digestive organs

The digestive system consists of an ectodermal foregut with the oesophagus and the stomach, an endodermal midgut which forms the digestive midgut gland, and the again ectodermal hindgut (Fig. 4.1). While the stomach has a primarily mechanical function in the maceration of the ingested food (see Chapter 5), the midgut gland (or hepatopancreas) is the principal organ of digestive enzyme production as well as nutrient resorption. The gland is a bunch of numerous blind ending tubules each of which are constructed from a monolayer epithelium of specialised cells. This cell layer develops from so-called embryonic cells (E-cells) at the distal tip of the tubules. The E-cells develop into at least two cell types: the R-cells which perform

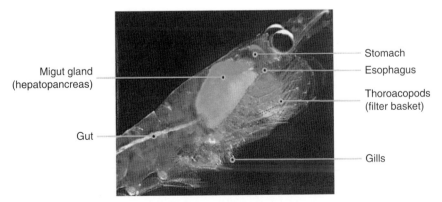

Figure 4.1 The cephalothorax of *Meganyctiphanes norvegica* showing the location of the digestive organs. Photograph provided by Uwe Kils.

nutrient resorption or the F-cells (fibrillar cells) which facilitate enzyme synthesis and secretion. The origin of a third cell type in decapods, which is characterised by the presence of a huge vacuole (blister cells or B-cells), is still under debate.

The stomach and the midgut gland form a functional unit. Enzymes synthesised in the midgut gland are released into the lumen of the tubules. The proximal ends of the tubules merge to form larger channels which, ultimately, end up as a pair of funnels into the posterior part of the stomach. The enzymes accumulate in the stomach. Food, which is ground in the stomach by the gastric mill (Ullrich *et al.*, 1991), is simultaneously mixed with the digestive enzymes initializing the extracellular digestion of the food. The chyme is then pressed through a fine filter system which allows the liquid fraction to pass into the midgut gland for nutrient resorption but retains larger solid particles such as diatom frustles, copepod mouthparts, or cuticle fragments for defecation.

2.2. The digestive enzymes

Investigations on enzyme polymorphism in *M. norvegica* included some digestive enzymes (Fevolden, 1982). Studies on the properties of digestive enzymes were primarily focused on chitinolytic and proteolytic enzymes. Spindler and Buchholz (1988) identified, in the extracts of whole animals, different forms of endo- and exochitinases with broad pH optima and high stabilities. Separate analyses of the digestive organs and the cuticle revealed the presence of specific chitinases which are involved either in moulting or in digestion (Peters *et al.*, 1998, 1999). The stomach and the midgut gland of *M. norvegica* are also rich in proteolytic enzymes (Buchholz, 1989). According to inhibition assays, the majority of endopeptidase activity results from serine-endopeptidases, including trypsin-like enzymes (Dittrich, 1992b; Kreibich *et al.*, 2010). A set of various digestive hydrolases was detected by Donachie *et al.* (1995) in the stomach and in the midgut gland (Table 4.1).

The pH of the gastric fluid of *M. norvegica* ranges between 6.5 and 6.9. Maximum *in vitro* activity of trypsin-like enzymes occurs across a broad pH-range from pH 7 to pH 10. Eighty to ninety percent of activity in trypsin-like enzymes is inhibited by soybean inhibitor (SBI) and N-tosyl-L-lysine chloro-methyl ketone (TLCK). Salts containing trace metal ions ($HgCl_2$, $AgNO_3$, and $CuSO_4$) reduce the tryptic activity of *M. norvegica* by 40–50% (Dittrich, 1992b).

2.3. Endosymbionts

Bacteria have been detected in the stomach and in the midgut gland of *M. norvegica* from the Kattegat (Donachie *et al.*, 1995). Laboratory experiments showed that these bacteria seem to contribute to a lesser extent to the production of a range of enzymes, including chitinolytic and proteolytic

Table 4.1 Digestive enzymes detected in the stomach and in the midgut gland of *Meganyctiphanes norvegica*

Enzyme	Substrate	Reference
Proteolytic enzymes		
Total proteinase (alk.)	Azocasein	Spindler and Buchholz (1988), Buchholz (1989), Donachie et al. (1995), Kreibich et al. (2010), Kreibich (unpublished), Saborowski (unpublished)
Trypsin	BAPNA, TAME	Bämstedt (1988), Dittrich (1992a,b), Kreibich et al. (2010), Kreibich (unpublished), Saborowski (unpublished)
	N-Benzoyl-DL-arginine-2-naphthylamide	Donachie et al. (1995)
Chymotrypsin	SAAPNAA	Saborowski (unpublished)
	N-Glutaryl-phenylalanine-2-naphthylamide	Donachie et al. (1995)
Alanine aminopeptidase	Ala-*p*-nitrophenol	Kreibich (unpublished), Saborowski (unpublished)
Leucine arylamidase	L-Leucyl-2-naphthylamide	Fevolden (1982), Donachie et al. (1995)
Valine arylamidase	L-Yalyl-2-naphthylamide	Donachie et al. (1995)
Cysteine arylamiadse	L-Cystyl-2-naphtylamide	Donachie et al. (1995)
Esterases and lipases		
Esterase (C2)	1-Naphthyl acetate	Fevolden (1982)
Esterase (C4)	2-Naphthyl butyrate	Donachie et al. (1995)
Esterase lipase (C8)	2-Naphty caprylate	Donachie et al. (1995)
Lipase (C14)	2-Naphthyl myristate	Donachie et al. (1995)
Triacylglycerolacylhydrolase	1,2-Diglycide	Kreibich (unpublished)

(*continued*)

Table 4.1 *(continued)*

Enzyme	Substrate	Reference
Phosphatases		
Alkaline phosphatase	1-Naphthyl phosphate (pH 8.6)	Fevolden (1982)
	2-Naphthyl phosphate (pH 8.5)	Donachie et al. (1995)
Acid phosphatase	1-Naphthyl phosphate (pH 5.0)	Fevolden (1982)
	2-Naphthylphosphate (pH 5.4)	Donachie et al. (1995)
Naphthol-AS-BI-phosphohdrolase	6-Naphthol-AS-BI-phosphate	
Glycoside hydrolases		
α-Glucosidase	2-Naphthyl-α-D-glucopyranoside	Donachie et al. (1995)
β-Glucosidase	6-Br-Naphthyl-β-D-glucopyranoside	Donachie et al. (1995)
α-Galactosidase	6-Br-naphthyl-α-D-galactopyranoside	Donachie et al. (1995)
β-Galactosidase	2-Naphthyl-β-D-glucopyranoside	Donachie et al. (1995)
β-Glucuronidase	Naphthol-AS-BI-βD-glucuronide	Donachie et al. (1995)
α-Mannosidase	6-Br-naphthyl-βD-mannopyranoside	Donachie et al. (1995)
α-Fucosidase	2-Naphthyl-α-L-fucopyranoside	Donachie et al. (1995)
Amylase	Amylose, starch, CM-starch-RBB	Båmstedt (1988), Donachie et al. (1995), Saborowski (unpublished)
Cellulase	CM-cellulose-RBB	Saborowski (unpublished)
Laminarinase	CM-curdlan-RBB	Saborowski (unpublished)
Chitinase	CM-chitin-RBB, chitin (cryst.)	Spindler and Buchholz (1988), Buchholz (1989), Donachie et al. (1995)

enzymes. However, none of the enzymes studied in the digestive organs of krill were exclusively of bacterial origin. Moreover, rapid gut transit times of the krill and frequent moulting of the foregut, including the stomach, prevents the formation of a persistent population of symbiotic bacteria.

2.4. Ecological implications

M. norvegica is well suited to digesting food of both plant and animal origin. Moreover, it shows a certain degree of enzyme polymorphism (Fevolden, 1982) and comparatively high levels of enzyme activity (Kreibich *et al.*, 2010). Seasonally elevated levels of activity of trypsin and amylase appear during the spring (Båmstedt, 1988), probably reflecting a higher metabolic energy demand at this time of year. Northern krill is also capable of digesting cellulose and hemicelluloses such as laminarin.

3. Biochemical Composition

The biochemical composition of various euphausiid species was intensively studied in the 1970s. The results were thoroughly summarised by Mauchline and Fisher (1969) and Mauchline (1980). The data comprise the biochemical gross composition (water, ash, lipid, carbohydrate, protein, and chitin), the carbon and nitrogen amounts, the elemental composition including trace metals and pollutants, the amino acid composition, as well as the ATP and the RNA concentrations. Only a few additional studies on these parameters in *M. norvegica* have been published since. In contrast, extensive work has been carried out since on the relatively unstudied lipid and fatty acid composition. This will be reviewed in a separate section below.

3.1. Protein, lipid, ash

Falk-Petersen (1981) examined in detail some major biochemical parameters in a population of *M. norvegica* from the Norwegian Balsfjorden (70° N). Increase in weight and changes in protein and lipid contents were closely related to the seasonal cycles of primary production. A decrease in lipid content in I-group *M. norvegica* during winter was likely related to the use of energy for overwintering and the growth of gonads. The relative amount of proteins ranged from 23% to 36% DW and the amount of lipids from 30% to 47%. The ash content varied around 15% of the body dry weight. Buchholz and Prado-Fiedler (1987) studied the seasonal changes of biochemical parameters of a krill population in the Danish Kattegat. These data were also presented as percentage of the dry weight. If an average water content of 80% is assumed (Mauchline and Fisher, 1969), then the recalculated values related to wet weight would amount to ash (2.2–5.8% WW), lipid (1.6–9.6% WW), carbohydrates (0.2–0.4% WW), protein (9.2–13.8% WW), and chitin (0.6–1.1% WW). These data for *M. norvegica* from the Kattegat are in the same range as those data from krill from various locations previously summarised by Mauchline and Fisher (1969).

3.2. Nucleic acids

The RNA concentration, or the RNA/DNA-ratio, is a widely used index to determine the physiological condition of organisms. It can serve as a proxy for the performance in terms of, for example, growth or reproduction (Chícharo and Chícharo, 2008). RNA values from *M. norvegica* sampled in the Norwegian Korsfjorden ranged from 2.5 to 20.8 μg mg^{-1} DW in February and 3.8 to 21.3 μg mg^{-1} DW in September (Båmstedt and Skjoldal, 1980). The authors established an allometric relationship between the RNA concentration and the dry weight of krill and presented a relationship between growth rate and RNA concentration to estimate the average growth rate. A more detailed seasonal cycle of the RNA content was determined by Båmstedt (1983) on a krill population from the Kosterfjorden, Swedish west coast. The values were highest in spring and early summer (16–18 μg mg^{-1} DW), decreased to almost 4 μg mg^{-1} DW in July and increased again in August to 11.5 μg mg^{-1} DW. The variation was high and the seasonal changes were suggested to reflect the variation in food supply and gonad growth.

3.3. Lipids

In the past three decades, intensive research has been carried out on the lipid biochemistry of krill and other pelagic crustaceans. The functions of lipids and fatty acids in the physiology and the ecology of *M. norvegica* were studied either in relation to reproduction, as an energy store for overwintering, or as markers for nutrition. The amount of total lipids in whole body extracts of *M. norvegica* can range from as low as 7–8% DW (Kreibich *et al.*, 2010; Saether *et al.*, 1986), or 3.5% DW in krill from oligotrophic waters (Mayzaud *et al.*, 1999), to up to more than 40% DW (Sargent and Falk-Petersen, 1981). The lipid content and its composition depends strongly on season, location, sex, maturation, and, particularly, on the nutritive state.

3.3.1. Major lipid storage organs

The cephalothorax is the site of major lipid accumulation in *M. norvegica*. The thorax of krill from the Trondheimfjord contained 15.8% lipids on a dry weight basis and the abdomen just 9.1% (Saether *et al.*, 1986). Albessard and Mayzaud (2003) reported that the lipid content in the cephalothorax from Ligurian, Kattegat, and Clyde Sea krill was 2–4 times greater than in the abdomen.

The cephalothorax contains, besides other organs, the stomach, the digestive or midgut gland, the ovaries, and the so-called fat body consisting of connective tissue (Cuzin-Roudy, 1993). The lipid concentration in the stomach can amount to 17% DW and that of the midgut gland 65% DW (Albessard *et al.*, 2001) which confirms the relevance of the midgut gland as the major lipid storage organ, for example, in Antarctic krill (*E. superba*) starvation for 19 days entailed a significant decrease in total lipids from 21% to 9% DW (Virtue *et al.*, 1993). The gonads and the fat body contain 24% DW and 20% DW, respectively.

The abdomen consists mainly of muscular tissue. It contains 8% lipids on a dry weight basis. The lipid composition reflects the predominance of phospholipids from biomembranes and shows low levels of neutral lipids like TAG (Albessard and Mayzaud, 2003). The relative amount of the body lipids in resting males from the Ligurian Sea is summarised in Table 4.2.

The lipid content in the separate organs of *M. norvegica* varied significantly with season and the reproductive cycle; details are given by Albessard *et al.* (2001) and Albessard and Mayzaud (2003).

3.3.2. Lipid classes

A basic classification of lipids can be made through separating neutral from polar lipids. The neutral lipids are generally separated further into the major classes of triacylglycerols (TAG), wax esters (WE), and sterols (ST). The polar lipid (PL) fraction comprises the phospholipids mainly represented by phosphatidylcholine (PC). Depending on the instrumentation and analytical performance, these lipid classes can be further separated and identified (Mayzaud *et al.*, 1999).

TAG: The major lipids in *M. norvegica* are triacylglycerols (TAG). Sargent and Falk-Petersen (1981) found that more than the half of total lipids in *M. norvegica* caught during winter (November/December) in Balsfjorden near Tromsø occurred as TAG and Saether *et al.* (1986) found more than 70% TAG in the total lipids. Their contents varied strongly and

Table 4.2 Lipid content of different organs in resting males from the Ligurian Sea (after Albessard *et al.*, 2001)

	Stomach	Midgut gland	Fat body	Gonads	Abdomen
% DW	14.5 ± 2.4	65.0 ± 11.2	19.7 ± 5.3	24.4 ± 10.3	8.1 ± 1.5
% of total body lipid	9.4 ± 1.7	35.1 ± 6.1	18.6 ± 4.8	9.7 ± 4.8	27.1 ± 6.1

correlated with the total lipid content (Kreibich *et al.*, 2010; Saether *et al.*, 1986) which shows that TAG serve as the major fraction of storage lipids in *M. norvegica*. Other lipid classes were present in amounts of less than 5% of the body dry weight. Their amount did not change with the total lipid content but remained constant (Fig. 4.2). This indicates that these lipids are more involved in defined physiological functions, for example, as membrane compounds.

WE: Wax esters (WE) are important and frequently occurring storage lipids in many pelagic crustaceans including euphausiids (Falk-Petersen *et al.*, 1981; Saether *et al.*, 1986). It has also been suggested that they play an important role as a 'long-term' lipid store in deep sea and high-latitude zooplankton (Lee *et al.*, 2006). However, only small amounts of WE have been detected in *M. norvegica* (Fig. 4.2). Morris (1972) determined 5% in relation to the total lipid amount in krill sampled northeast of the Azores. Krill from the Greenland Sea contained 6% WE which accounts for 1.6% DM (Kreibich *et al.*, 2010). No WE were found in krill from the Norwegian Balsfjorden (Falk-Petersen *et al.*, 1981), or the Ligurian Sea, the Scottish Clyde Sea, and the Danish Kattegat (Albessard and Mayzaud, 2003; Mayzaud *et al.*, 1999). Thus, *M. norvegica* seems not to follow a typical high-latitude lipid storage strategy where WE are preferentially accumulated. Sargent and Falk-Petersen (1981) suggested that traces of WE may be derived from copepods consumed by the krill. Mayzaud *et al.*

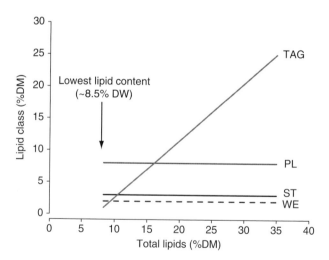

Figure 4.2 Content of the major lipid classes in relation to the total lipid content of *Meganyctiphanes norvegica*. The theoretically lowest lipid content was calculated from data given by Saether *et al.* (1986) and Kreibich *et al.* (2010) for krill from higher latitudes. It does not apply for krill from the Ligurian Sea which contained as low as 3.5% DW of lipids (Mayzaud *et al.*, 1999). PL = polar lipid fraction, ST = sterols, TAG = triacylglycerols, WE = wax esters.

(1999) found no WE in krill from the Ligurian Sea and concluded that those krill prey predominantly upon organisms that are poor in WE.

PL: The largest share of the polar lipid fraction derives from biomembranes. These polar lipids predominantly consist of phospholipids of which PC is a major component. PC is thought to play an important role as storage lipid in high-latitude euphausiids (Lee *et al.*, 2006). In krill from northern Norwegian fjords, the PL-fraction accounted for 15–20% of total lipids (Saether *et al.*, 1986) or more than 30% of total lipids, respectively (Falk-Petersen *et al.*, 1981) which is approximately the same amount as in Ligurian krill (Mayzaud *et al.*, 1999). PL in krill from the Greenland Sea made up 29% of total lipids (Kreibich *et al.*, 2010). This accounts for about 8% on a dry weight basis or 2–3% on a wet weight basis (Albessard and Mayzaud, 2003). The relative amount of PC remains constant at different total lipid amounts because, besides the formation of biomembranes, PC is involved in egg production (Albessard *et al.*, 2001; Cuzin-Roudy *et al.*, 1999).

ST: Ballantine and Roberts (1980) studied the sterol (ST) content of some pelagic marine crustaceans and found in *M. norvegica* from the Atlantic (25°N, 17°W) 3.4% ST in total lipid extracts. The major ST compound was cholesterol accounting for 98% of the ST fraction. Cholesterol was present in krill samples from various other locations usually not exceeding 1–2% of the animal's dry weight (Albessard and Mayzaud, 2003; Kreibich *et al.*, 2010; Mayzaud *et al.*, 1999).

FFA: Large amounts of free fatty acids (FFA) are probably the result of rapid lipolysis post mortem. Sargent and Falk-Petersen (1981) reported an FFA-amount of almost 20% and Falk-Petersen *et al.* (1981) reported a seasonal maximum of even about 45%. The high FFA amounts most likely result from inappropriate handling and drying of the krill samples. Saether *et al.* (1986) took special precautions to avoid *post-mortem* lipolysis. The FFA contents in his samples were low, in the range 0.6% of the dry weight of the krill. Similar values were reported by Albessard and Mayzaud (2003) and Kreibich *et al.* (2010). These latter values are probably the closest to the level of natural occurrence in krill tissues.

3.3.3. Fatty acid composition

Depending on the scientific aim of the study, the fatty acid compositions were investigated in the total lipid fraction or in separated lipid classes of either whole animals or of different body sections such as the cephalothorax or the abdomen.

The fatty acid composition of the total lipid fraction of complete individuals of *M. norvegica* was analysed by Mayzaud *et al.* (1999) and Kreibich *et al.* (2010). A selection of the most abundant fatty acids, accounting together for 80% or more of total fatty acids, is given in Table 4.3. The bulk of saturated fatty acids (SFA) were represented by myristic acid (14:0)

Table 4.3 Approximate amount of major fatty acids in *M. norvegica*

Fatty acid	% of total FA
SFA	
14:0	1.4–7.6
16:0	9.4–22.6
Σ SFA	16.2–40.3
MUFA	
$16:1(n-7)$	<1.0–13.2
$18:1(n-9)$	8.2–17.4
$18:1(n-7)$	2.2–9.2
$20:1(n-9)$	<1.0–21.9
$22:1(n-11)$	<1.0–26.6
Σ MUFA	15.1–48.3
PUFA	
$20:5(n-3)$	2.6–24.8
$22:6(n-3)$	4.0–37.5
Σ PUFA	20.1–61.5

and, particularly, by palmitic acid (16:0). The proportion of single mono-unsaturated fatty acids (MUFA) varied strongly between studies indicating a strong dependence on altering intrinsic factors or nutrition. The major polyunsaturated fatty acids (PUFA) were the ⍵-3 PUFA eicosapentaenoic acid [EPA, $20:5(n-3)$] and the docosahexaenoic acid [DHA, $22:3$ $(n-6)$]. Pronounced differences in the amount of single fatty acids were noted between lipid fractions, sampling sites, and seasons, for example, in krill from the Ligurian Sea, the Clyde Sea, and the Kattegat, the MUFA $20:1(n-9)$ was almost absent in the polar lipid fraction but amounted to 5% in the TAG-fraction (Virtue *et al.*, 2000). In the whole lipid extracts of krill from the Greenland Sea, $20:1(n-9)$ accounted for 18% of fatty acids (Kreibich *et al.*, 2010).

3.4. Factors of lipid and fatty acid variation

The lipid content and the lipid composition may be influenced by several factors including food supply, sex, maturity, spawning, season, and locality (Saether *et al.*, 1986).

3.4.1. Reproduction

Falk-Petersen (1981) concluded that variation in lipid content was related to gonad maturation in winter. Båmstedt (1976) reached the same conclusion but also suggested that the loss of lipids cannot be explained solely by gonad maturation but also by lipid catabolism. No clear relation between the lipid content and gonad maturation could be established by Buchholz and Prado-Fiedler (1987). Moreover, those authors stated, for animals from the Kattegat that lipid accumulation is not in phase with gonad maturation. By August, 60% of the females have laid their eggs and by October, virtually all the females have spawned. During that time the lipid content was still increasing.

In the Ligurian Sea, the Kattegat, and the Clyde Sea, males and ready-to-spawn females did not show significant differences in the lipid content of the cephalothorax. However, ready-to-spawn females displayed higher lipid levels than post-spawn females. The loss of lipids from the cephalothorax after spawning can amount to 55% (Albessard and Mayzaud, 2003). Conse-quently, there must be a significant increase of lipids between vitellogenic and ready-to-spawn females through the synthesis or allocation of lipids into the maturing eggs.

The dynamics and the amplitude of lipid store variation during the reproductive season depend on the capacity of lipid accumulation either due to food availability or due to the duration of the spawning season, for example, the changes in the lipid amount were distinct in krill from the Ligurian Sea which live in an oligotrophic environment and do not accu-mulate high lipid reserves. Krill from the Scottish Clyde Sea generally had higher lipid stores than Ligurian krill. Though the absolute loss of lipids due to spawning was similar in both populations, the relative change was lower in Clyde Sea krill (33%). Krill from the Kattegat did not seem to lose significant amounts of lipids compared to krill from both the other loca-tions. Kattegat krill may perform more successive spawning cycles with a more continuous lipid uptake and less distinct change of lipid store (Albessard and Mayzaud, 2003).

3.4.2. Season, latitude, and trophic conditions

The seasonal changes in trophic conditions are major factors which deter-mine the dynamics of lipid storage in krill populations. Different lipid levels in krill from different latitudes reflect the variation in primary production with regard to the accumulation of overwintering lipid stores. Thus, they are related to the seasonal pattern of primary and secondary production rather than directly influenced by climate.

The total lipid content of *M. norvegica* from central and northern Norwegian fjords followed a distinct seasonal cycle. It was greatest in autumn and early winter and least in spring (Saether *et al.*, 1986). A similar seasonal cycle of lipid contents was found in krill in the Danish Kattegat (Buchholz and Prado-Fiedler, 1987). The values ranged from as low as 7.8% DW in July, up to 47.8% DW in late November. The seasonal lipid variations reflect the deposition and utilisation of winter reserves.

In the Ligurian Sea, krill showed apparently a reverse pattern with maximum values in early summer (20% DW) and lowest values ($\sim$3–4% DW) in early winter (Mayzaud *et al.*, 1999). The trophic situation in the Ligurian Sea consists of a short phytoplankton bloom in late spring (April–May) and a subsequent period of zooplankton production. In summer and fall, the biomass is low in the Ligurian. Accordingly, krill accumulate lipids during the short bloom but have to utilise them again when facing the oligotrophic summer conditions (Mayzaud *et al.*, 1999). The spring-peak of lipid accumulation is also closely linked with reproduction.

A comparative study on the influence of environmental and physiological factors on the distribution of lipids was carried out on three populations of *M. norvegica* from the Kattegat, the Clyde Sea, and the Ligurian Sea (Albessard and Mayzaud, 2003; Mayzaud *et al.*, 1999, 2000). Krill at sexual rest and early in the reproductive period showed lowest lipid values in the Ligurian Sea. The lipid content increased in the Clyde Sea and was greatest in Kattegat krill. The tendency of increasing lipid amounts towards the higher latitudes was previously emphasised by Mayzaud *et al.* (1999).

3.5. Fatty acids as trophic markers

The fatty acid composition of *M. norvegica*, particularly the TAG-fraction (Virtue *et al.*, 2000), reflects to a certain extent the composition of the prey and, thus, may serve as a biochemical indicator of trophic interactions. This has been applied to Northern krill in a number of studies (Dalsgaard *et al.*, 2003; Falk-Petersen, 1981; Falk-Petersen *et al.*, 2000; Mayzaud *et al.*, 1999; Saether *et al.*, 1986; Stübing *et al.*, 2003, Virtue *et al.*, 2000).

There is no single fatty acid (FA) which is unique to any species. However, a rough distinction between primarily herbivorous or carnivorous feeding habits can readily be made. Major fatty acids in diatoms include $16:1(n-7)$, C16PUFA, and $20:5(n-3)$. Dinoflagellates are characterised by $18:4(n-3)$, $18:5(n-3)$, and $22:6(n-3)$. FA $18:4(n-3)$ is present in other flagellates (Virtue *et al.*, 2000). In contrast, the FAs $20:1(n-9)$ and $22:1(n-11)$ appear only in traces in phytoplankton. Accordingly, a low content of $16:1(n-7)$ and high contents of $20:1(n-9)$ and $22:1(n-11)$ as found in krill from Ullsfjorden may reflect a carnivorous or omnivorous diet with copepods as the major food component (Saether *et al.* 1986). Similarly, high levels of $20:1(n-9)$ and $22:1(n-11)$ are indicative of heavy predation on copepods by, for example, krill from the Kattegat (Virtue *et al.*, 2000), as confirmed by stomach analysis (Lass *et al.*, 2001). Virtue *et al.* (2000) distinguished the predominant diet of krill from the Ligurian Sea, the Kattegat, and the Clyde Sea by means of their fatty acid composition. The higher latitude krill contained greater diatom and copepod signals. The Ligurian krill fed more opportunistically, probably as an adaptation to the oligotrophic conditions in Ligurian waters.

Fatty alcohols derived from WE have been found in the stomachs of krill (Lass *et al.*, 2001) but only trace amounts of WE occur in the rest of the body. Apparently, *M. norvegica* catabolizes WE and fatty alcohols from prey both quickly and efficiently. Only in krill from the Azores and from the Greenland Sea have WE been found to occur in more than trace amounts, comprising up to 6% of total lipids (Morris, 1972; Kreibich *et al.*, 2010). Both studies were carried out on total lipid extracts of whole animals so we cannot determine whether the WE were from food items or stored in the tissues.

Promising approaches to study the lipid uptake of *M. norvegica* from the food were carried out on the stomach content and faeces of krill. Virtue *et al.* (2000) analysed the FA composition of the faeces and compared it with the fatty acid composition of zooplankton catches. In the faeces, the amount of PUFAs was reduced but the amount of saturated fatty acids was higher than in the food. Apparently, dietary PUFA were selectively metabolised by krill. Lass *et al.* (2001) analysed the FA composition of the stomach contents and found high amounts of still undigested fatty alcohols from copepods and phytol from microalgae. Unfortunately, this approach is tedious due to the small amount of sample available for analysis.

4. Respiratory Gas Exchange

4.1. Effects of temperature and season on O_2 uptake

Krill may encounter both temporal (over the course of a few hours during DVM) and spatial (different locations and seasonal changes over the course of months) differences in temperature, so it is not surprising that the effect of temperature on rates of O_2 uptake (as a measure of metabolism) has been subject to much attention (Mauchline, 1980; Mauchline and Fisher, 1969). Saborowski *et al.* (2002) investigated the effect of temperature on individuals from three geographically separate populations of *M. norvegica* which experienced markedly different patterns of spatio-temporal temperature variation (Fig. 4.3) and different trophic conditions (rich to poor; Clyde > Kattegat, Ligurian Sea). Using a specially engineered device which allows the measurement of respiration in swimming krill (Saborowski and Buchholz, 1998), Saborowski *et al.* (2002) found that there was a pronounced effect of acute temperature change on O_2 uptake (over the temperature range 4–16 °C) for all individuals, of roughly the same magnitude (Fig. 4.3); Q_{10} values ≤ 2, indicating little thermal adaptation with metabolism doubling for each 10 °C increase in temperature much like the effect of temperature on chemical reactions. Such a pronounced acute temperature effect on metabolism had been noted previously for *M. norvegica* from the Alkor Deep, Kattegat (Saborowski *et al.*, 2000) and from the Gullmarsfjord (Hirche, 1984; also see Strömberg and Spicer,

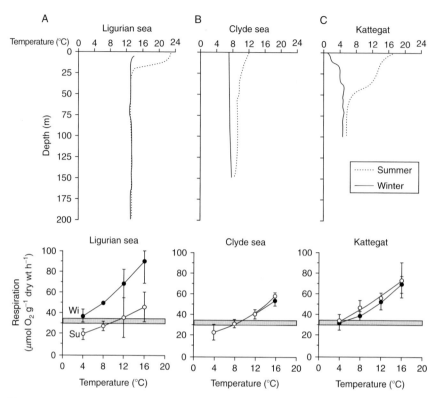

Figure 4.3 Temperature profiles of sampling locations (summer and winter) and rates of respiration (O$_2$ uptake: means $\pm$ SD) at different environmental temperatures of individuals of *M. norvegica* from three geographically separate populations. The shaded line represents the rate of O$_2$ uptake at each of the respective environmental temperatures (From Saborowski *et al.*, 2002).

2000). What is especially interesting in Saborowski *et al.* (2002) is that individuals from each of the three different populations displayed near identical rates of O$_2$ uptake (30–35 μmol g^{-1} DW h^{-1}) when tested at the environmental temperatures they were each experiencing (Fig. 4.3.). This indicates that the metabolism of *M. norvegica* can acclimatise to local temperature conditions over a time course greater than the period of acute temperature change used in the experiments described above, more likely in the order of days to weeks. However, there was an important exception to this phenomenon. Rates of O$_2$ uptake of Ligurian *M. norvegica* in the winter were twice as great as those in summer despite similar thermal regimes prevailing which Saborowski *et al.* (2002) interpreted as being in response to increased food ability. Certainly Salomon *et al.* (2000) showed that the starving of *M. norvegica* resulted in a reduction in O$_2$ uptake. The use of

DVM as behavioural thermoregulation of metabolism has been raised by a number of authors (e.g. Saborowski *et al.*, 2000, 2002; Strömberg and Spicer, 2000) and warrants further study. Interestingly, Mayzaud *et al.* (2005) found there was no significant difference in CO_2 excretion between *M. norvegica* kept at 12.5 °C and those kept at 17.8 °C, although there was an increase in respiratory quotient ($CO_2:O_2$) from 1.29 to 1.62 at the higher temperature, this increase due to rates of O_2 uptake seeming to be more sensitive to temperature change than rates of CO_2 elimination, indicating that both the physiological and ecological effects of temperature on metabolism still requires further elucidation.

In passing, comparison of rates of O_2 uptake between studies is notoriously difficult because of methodological and ecological differences and the thermal history of the animal. Not surprisingly there is a range of O_2 uptake values reported for *M. norvegica* (Table 4.4.). Certainly in few of the studies has activity, a feature that will dramatically affect rates, been quantified, though it may have been standardised. Methodological differences such as the use of closed respirometry compared with flow-through respirometry may well contribute to the large differences. Saborowski *et al.* (2002) found lower rates of O_2 uptake in closed compared with open systems. What can be said is that the metabolism of the pelagic and actively swimming *M. norvegica* is generally relatively high compared to other crustaceans.

4.2. Effects of hypoxia on O_2 uptake and anaerobic metabolism

van den Thillart *et al.* (1999) were the first to investigate the effect of exposure to acutely declining O_2 tensions, or PO_2, (hypoxia) on oxygen uptake (as a measure of metabolism) of *M. norvegica*, in individuals from the Gullmarsfjord, Sweden. Despite an earlier report which seemed to suggest that *M. norvegica* would not traverse a pycnocline into an area of low O_2 (Bergström and Strömberg, 1997), excursions into severely hypoxic bottom water ($PO_2 = 3–10$ kPa at a depth of 65–85 m, roughly 15–50% O_2 saturation; the waters are hypoxic because of a delay, or sometimes cessation, in annual water renewal as a result of altered currents) in the Gullmarsfjord meant that krill during the day resided at such depths (Spicer *et al.*, 1999). van den Thillart found that *M. norvegica* was able to regulate its O_2 uptake down to approx. 30% saturation (critical O_2 tension, or $P_c = \sim6–7$ kPa, $T = 10$ °C). So in common with a number of other krill species *M. norvegica* displays some ability to maintain O_2 uptake in the face of declining O_2 tensions but the ability does not seem to be any more developed than that found in krill species which do not encounter hypoxia on a regular basis, if at all. Furthermore, *M. norvegica* is characterised by one of the poorest anaerobic capacities of any crustacean, surviving not more than 1 h in anoxia, and accumulating large concentrations of L-lactic acid quickly (Spicer *et al.*, 1999). So reliance on a shift from aerobic

Table 4.4 *Meganyctiphanes norvegica* (some literature data of respiration rates; from Saborowski *et al.*, 2002)

Region	Season	Temperature (°C)	Respiration rate (μmol O_2 g^{-1} dry wt h^{-1})
Ligurian Sea	Nov	5–20	23.7–43.9
Ligurian Sea	Winter	13	56.7
	Spring	13	58.0
Gulf of St. Lawrence	Feb–Aug	2–10	60.3–92.9[a]
Kosterfjorden (Sweden)	Dec–Sep	5–6	21.9–40.6
Kattegat (Alkor Deep)	Jun–Sep	5–10	50.1–73.4
Gullmarsfjorden (Sweden)	Sep	6.5	46.9
Gullmarsfjorden	Summer	10	38.7
Gullmarsfjorden	Sep	7–15	16.0–30.3[b]

[a] Rates calculated for 10-mg animals.
[b] Rates were recalculated from wet weight to dry weight assuming dry weight equals 25% of wet weight.

to anaerobic metabolism when exposed to low O_2 can barely be seen as an adaptation to migrating into hypoxic layers during DVM; indeed if krill are caged in deep hypoxic water for a greater period than they would normally reside there, they have extremely high levels of L-lactate if they survive, but survival itself is poor (Spicer *et al.*, 1999). Given this poor anaerobic capacity, it would be interesting to know the functional significance (if any) of the low levels of polymorphism of the enzyme LDH (lactate dehydrogenase— responsible for converting pyruvate to lactate) found in *M. norvegica* (Mulkiewicz *et al.*, 2001).

Strömberg and Spicer (2000) confirmed this modest regulatory ability in *M. norvegica* from the same population but also found that exposure temperature dramatically affected regulatory ability. The P_c decreased from 8–11 kPa at 15 °C to 4–6 kPa at 7 °C. Their experiments were carried out during September when 15 °C was the temperature of the upper water layers and 7 °C was the temperature below the pycnocline between 40 and 50 m deep. Thus, they suggested that a reduction in temperature dramatically improves regulatory ability (linked to a reduction in overall metabolism) and could allow excursions into cold hypoxic water. If the hypoxic deep water below the pycnocline had been characterised by the same (high) temperature as found at the surface, then krill would have been unable to regulate their metabolism, and the poor anaerobic capacity of these animals (Spicer *et al.*, 1999) would have been insufficient to permit residence in those waters for more than a few hours.

The modest regulatory ability of *M. norvegica* when challenged by acutely declining PO_2s was not present in the earliest life stages investigated (calyptopis III/early furcilia I) but began to be detectable in furcilia III (Spicer and Strömberg, 2003). Furcilia III displayed a P_c of 15.4 kPa ($\sim$75% saturation), which improved until furcilia V, $P_c = 12.6$ kPa ($\sim$63% saturation). Clearly this regulatory ability, measured at 10 °C, was not as good as the adult with a $P_c = 6$–7 kPa, from which we may surmise that regulatory ability continues to develop through the remainder of the furcilia stages, and possibly beyond. Hypoxia-related hyperventilation was achieved by an increase in pleopod (but not thoracic limb) beating frequency which appeared at or just before furcilia V. It is reasonable to suppose that the development of regulatory ability in furcilia V is intimately linked with this hyperventilation which fails at oxygen tensions lower than 11 kPa. However, the earlier appearance of regulation in furcilia III cannot be attributed to pleopod beating (even though this stage has a full complement of functional setose pleopods) as beat rate declines with declining PO_2s. The timing of the onset of regulation could not be modified, as it can in other crustaceans, by pre-exposing larvae to hypoxia, and indeed such pre-exposure resulted in significant mortality leading Spicer and Strömberg (2003) to conclude that the development of respiratory regulation in *M. norvegica* was not open to environmental influence as is the case in other crustaceans. They further speculated that, as pleopod ontogeny is intimately associated with the ontogeny of DVM behaviour and the ontogeny of respiratory regulation, this co-occurrence is fortuitous in the Gullmarsfjord as krill do not descend into hypoxic deep water until they have developed the physiology that will allow them to regulate their metabolism there.

4.3. Respiratory pigments

To our knowledge, *M. norvegica*, and krill generally, utilises just the one extracellular respiratory pigment, the copper-based haemocyanin (Hc), that is found in some crustaceans and some molluscs (Brix *et al.*, 1989; Spicer and Strömberg, 2002). Brix *et al.* (1989) was the first to characterise the O_2 binding properties of *M. norvegica* haemocyanin (Fig. 4.4). Using haemolymph pooled from 766 specimens collected from the North Sea, he investigated O_2 binding ability of the haemolymph at two different temperatures (5 and 10 °C). He found that at 5 °C, *M. norvegica* haemolymph had a very low affinity for oxygen (P_{50} or half saturation value = 50.1 mmHg [6.66 kPa], at pH = 7.9). This was very similar to P_{50} values of 6.12–6.31 kPa, pH = 7.80, $T = 7$ °C, recorded by Spicer and Strömberg (2002) for dialysed haemolymph from individuals collected from the Gullmarsfjord, Sweden. Brix *et al.* (1989) also found that the cooperativity (the 'sigmoidness' of the O_2 binding curve) was high, but not exceptional ($n_{50} = 2.5$–3.0) although the haemolymph did exhibit a marked Bohr effect

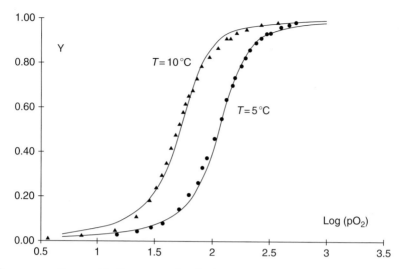

Figure 4.4 Oxygen binding curves for the haemolymph of *M. norvegica* at 5 (closed symbols) and 10 (open symbols) °C, pH = 7.7 (Brix *et al.*, 1989).

($\Delta \log PO_2 / \Delta \log$ pH $= -1.99$), amongst one of the highest recorded for crustaceans. What was quite exceptional, however, was that when O_2 binding curves were constructed at a higher temperature ($T = 10$ °C), the affinity for O_2 increased markedly ($P_{50} = 18.2$ mmHg [2.42 kPa], pH $= 7.8$). In other words, there was a strong effect of temperature, but in the opposite direction from that recorded for every other crustacean species examined. The reaction is normally exothermic, but in this case it was endothermic as evidenced by the positive value for the heat of oxygenation ($\Delta H = 133.76$ kJ mol^{-1}, pH $= 7.9$). Although Brix *et al.* (1989) speculated that this increase in Hc-O_2 affinity with increasing temperature may be related to passing through thermoclines during DVM, it is difficult to be definitive about the adaptive nature (if any) of this unique feature.

Concentrations of Hc ([Hc]) in individuals of *M. norvegica* collected from the Gullmarsfjord, Sweden, were extremely variable (0.39–0.89 mmol l^{-1}), but at the upper end are some of the highest recorded in aquatic crustaceans (Spicer and Strömberg, 2002). Spicer and Strömberg (2002) found that [Hc] varied during DVM. This is one of the most exciting findings of their study, that [Hc] could and did change over a timescale of hours, rather than days, as generally believed. The [Hc] decreased with increasing depth, when measured in individuals trawled or caged at different depths. Laboratory experiments showed that this pattern could not easily be explained by differences in O_2, temperature, or salinity affecting Hc concentration. However, starvation had a dramatic effect on Hc concentration over <10 h, and this was exacerbated by an increase in temperature. Spicer and Strömberg (2002) suggested that when *M. norvegica* migrates into deep water during the day, for whatever reason, they cannot secure enough energy to meet routine metabolic demands and so they resort to breaking down Hc and using it as an energy source. This notion, that there could be a trade-off between the respiratory function of Hc and its importance in nutrition when krill migrate into deeper, nutritionally poorer water during DVM, was further supported by Dawdry (2004) who investigated feeding and [Hc] concentrations during an actual DVM, and in the laboratory. Her work also highlighted that both sex and moult stage influences [Hc], with females having a significantly greater [Hc] than males.

5. METABOLIC PROPERTIES

Due to its wide geographical distribution in the North Atlantic, and so covering different climatic zones from sub-tropical to sub-polar, *M. norvegica* is a valuable tool for comparative physiological investigations. Northern krill are found in high productive as well as in oligotrophic waters and live at temperatures from 2 to 16 °C. Thereby, krill may be exposed to persistently low temperatures at higher latitudes, constantly moderate temperatures in the Mediterranean, or seasonally variable temperatures, for example, in the Kattegat. Northern krill may even experience almost the whole range of its thermal spectrum within several hours when migrating vertically through different water strata, as happens in the Kattegat (Matthews *et al.*, 1999; Saborowski *et al.*, 2000). Accordingly, a European Union research programme (PEP) in the second half of the 1990s used this species as a model for examining adaptive metabolic responses to biotic and physical factors (Buchholz, 2003; Buchholz and Saborowski, 2000; Buchholz *et al.*, 1998), of which the results are discussed alongside others below.

5.1. Moult and digestion

Moulting: Spindler and Buchholz (1988) examined the potential adaptive properties of biocatalysts in the chitinolytic enzymes of *M. norvegica* and *E. superba*. In both species, the authors found similar temperature optima around 40–50 °C. Although enzyme activity was still high at 0 °C. In both euphausiids, activation energies were reduced at lower temperatures. The authors concluded that the enzymes showed a functional adaptation to a low temperature range. Nevertheless, the fact that the activity profiles were the same in both species, despite the fact that *M. norvegica* occurs in waters that are between 4 and 12 °C warmer than those inhabited by *E. superba*, indicates that the enzymes can operate in a relatively wide range of environments.

Digestive proteases: High apparent temperature optima appeared also in proteolytic digestive enzymes of *M. norvegica* (Dittrich, 1990). Again no shift of the optimum towards lower temperatures was evident. Michaelis–Menten constants (K_M) of trypsin-like enzymes hydrolyzing the artificial substrate BAPA were low at 0 °C and continuously increased towards 20 °C. Since low K_M-values indicate higher affinity of the enzyme towards the substrate, *M. norvegica* seems to partly compensate the rate limiting effects of low temperatures for this reaction (Dittrich, 1992a). In contrast, the activation energy of the trypsin reaction was surprisingly high and similar to tropical species (Dittrich, 1992b).

NAGase isoforms: Buchholz and Vetter (1993) isolated the different isoforms of the chitinolytic enzyme *N*-acetyl-β-D-glucosaminidase (NAGase) and studied the kinetic properties of each isoform separately. The enzymes displayed different temperature maxima and a reduced K_M-value in the range of the ambient environmental temperature. This characteristic can be interpreted as an adjustment of the species to the temperature regime (Buchholz and Vetter, 1993). The authors assumed from their study that hydrolases such as NAGase may not be regulated to a great extent but instead there is a fine tuning, with respect to temperature, of the complex molecular pathways that regulate metabolism.

5.2. Key metabolic enzymes

CS and PK: The enzymes citrate synthase (CS) and pyruvate kinase (PK) are key enzymes in metabolic pathways. The mitochondrial citrate synthase (CS), also referred to as a condensing enzyme, initiates and regulates the citric acid cycle by catalysing the condensation of acetyl-Coenzyme A (acetyl-CoA) and oxaloacetate (OA) to citric acid. The K_M-value for acetyl-CoA was up to seven times greater than that of OA. Accordingly, the catalytic rate of CS is mainly controlled by the supply of acetyl-CoA.

The cytosolic pyruvate kinase (PK) is one of the major regulatory enzymes of the glycolytic pathway; it catalyses the conversion of phosphoenolpyruvate (PEP) to pyruvate and the phosphorylation of ADP to ATP.

No distinct adaptive properties of Northern krill CS with respect to steno- or eurythermy were found in terms of specific activities, activation energy, pH/activity profiles or Michaelis–Menten constants (Vetter, 1995a). However, short-term acclimation experiments with krill from the Swedish Gullmarsfjord indicated reduced K_M-values as well as elevated specific activities at ambient maintenance temperatures (Vetter, 1995a). The author suggested that *M. norvegica* is capable of increasing CS synthesis or producing alternative CS-isoforms with a higher specific activity to compensate enzyme activities when the animals are exposed to temperature changes.

A more detailed study on the CS and PK in *M. norvegica* from three different locations was unable to confirm adjustment of specific activities to temperature. But that study revealed that both enzymes were not equally distributed within the body and the organs of krill and that each enzyme showed a negative (CS) or positive (PK) allometric relationship (Saborowski and Buchholz, 2002). Considering this, variations in activity is most probably a result of metabolic scaling and thus only depends on the size of the animals. When the CS/PK-ratio was plotted against the sample weight, the data points of each population appeared on the same regression line. However, one exception was evident in summer krill from the Ligurian Sea which showed reduced CS/PK-ratios and stood apart from other populations. The reduced CS/PK-ratio might be due to a reduction of muscle tissue or mitochondria to cope with food-limiting conditions during the summer months in the Ligurian Sea (Saborowski and Buchholz, 2002).

CS-inhibition: Citrate synthase is controlled by cellular energy levels. Increasing ATP concentrations inhibit the CS-activity. Vetter (1995b) reported a peculiarity of the CS of *M. norvegica*: low ATP concentrations caused an increase of the maximum reaction velocity (v_{max}). The physiological consequence is that high energy demands accelerate the channelling of acetyl-CoA into the citric acid cycle. This mechanism can be understood as an adaptation to the energy demanding pelagic life style of krill.

PK-isoforms: In order to study physiological mechanisms of seasonal temperature adaptation Vetter and Buchholz (1997) isolated two isoforms of pyruvate kinase (PK) of *M. norvegica* from the Danish Kattegat. One isoform (PKI) dominated in the abdominal muscle while the other isoform, PKII, was mainly present in the cephalothorax. The K_M-values of PKI for phosphoenolpyruvate (PEP) was in summer twice as high as in winter which indicates a higher enzyme–substrate affinity to compensate for low turnover rates during the cold season. Only PKI showed features of temperature adaptation by decreasing activation energy (E_a) and K_M-values. Apparently, the requirement for temperature adaptation is more important in the energetically intensive swimming muscles of the abdomen (Vetter and Buchholz, 1997).

Dietary influence on a key metabolic enzyme: The influence of nutrition on the two PK-isoforms (PKI and PKII) of *M. norvegica* was studied by Salomon *et al.* (2000) in specimens from the Ligurian Sea. Similar to Vetter and Buchholz (1997), the authors found two PK-isoforms. A common feature of Ligurian and Kattegat krill was the inhibition of PKI and PKII by ATP and the activation of PKII by fructose-1,6-bisphosphate (FBP), a key metabolite in the glycolytic pathway. Moreover, in the presence of FBP, the sigmoidal kinetics of PKII was shifted to hyperbolic kinetics (Vetter and Buchholz, 1997). Differences were apparent in the share of both isoforms in whole body extracts: PKI amounted to 80% in Ligurian krill, but only 44% in Kattegat krill. It is not clear yet whether this difference is intrinsic or results from differences in the chromatographic separation procedure. Neither in feeding experiments nor in field samples could the authors identify properties of PK which indicate enzyme modifications in relation to food supply.

Tissue specificity of PK-isoforms: Almost all organs contained PKI and PKII. However, PKI prevailed in the locomotive organs, that is, the abdominal muscles, the pleopods, the thoracopods, and in the thoracic muscles. PKII dominated in the eyes, the midgut glands and in the ovaries (Salomon and Saborowski, 2006). Since PKII is activated by FBP, the authors determined the adenylate energy charge and the concentrations of FBP. ATP levels did not change significantly during six days of starvation but the concentrations of FBP decreased by 30%. Moreover, FBP-values were lower in cold-acclimated than in warm-acclimated krill. The tissue-specific distribution of the two different PK-isoenzymes seems to improve the krill's physiological flexibility to successfully cope with low temperatures or limited food supply. As a consequence of food deprivation, or decreased temperature, the glycolytic energy turnover may be reduced in some organs such as the gonads and the midgut gland. Simultaneously, the locomotive organs maintain high glycolytic turnover rates due to the presence of the active PK-isoform.

6. Osmotic/Ionic Regulation and Excretion

6.1. Osmotic and ionic regulation

Osmotic, and to a lesser extent ionic, regulation by *M. norvegica* was first investigated by Forward and Fyhn (1983). They found that krill, like many other oceanic animals, were osmoconformers, at least over the salinity range 40–24 PSU ($T = 3$–7 °C). Equilibration to test salinities occurred within a few hours: while haemolymph sodium was iso-ionic within the range of experimental salinities, chloride was consistently hypo-ionic (by 50–70 mmol l^{-1}) pointing to some degree of regulation of chloride but not sodium. Amino acid regulation, probably as a means of regulating cell volume, was detected in the abdominal flexor muscle for proline (over the range 35–28 PSU) and glycine (over the range 32–24 PSU). Exposure to salinities below 24 PSU resulted in irreversible damage and death.

6.2. Fluoride accumulation and regulation

We have some knowledge of fluoride accumulation (and to some extent regulation) in the tissues of *M. norvegica*. Fluorides derive principally from the weathering of fluoride minerals and volcanic activity although a small proportion is contributed by the use of aerosols. The fluoride content of Northern krill, in agreement with what is known of other euphausiids, is quite high compared to other oceanic invertebrates with whole body burdens of 2360 μg F$^-$ g^{-1} dry weight (DW) (Soevik and Braekkan, 1979) and 2153 μg F$^-$ g^{-1} DW (Adelung *et al.*, 1987). Adelung *et al.* (1987) showed that most of the fluoride was incorporated into the exoskeleton (3343 μg F$^-$ g^{-1} DW), with relatively small amounts found in the tissues (e.g. muscle 5.7 μg F$^-$ g^{-1} DW). Tissue fluoride appears to be regulated at these low concentrations, concentrations not unlike those of vertebrates. Having found the highest concentrations of fluoride in the mouthparts of *E. superba* (12,876 μg F$^-$ g^{-1} DW), Sands *et al.* (1998) suggested that as the mouthparts were the hardest part of the exoskeleton, that inorganic fluoride was important in hardening krill exoskeleton. The fact that the period of maximum fluoride uptake is post-moult adds weight to this idea. Exactly how fluoride is bound into the exoskeleton of *M. norvegica* remains to be elucidated.

6.3. Excretion

Excretion occurs as the result of catabolism in animals. Rogers (1978) found that excretion rates in individuals from a Mediterranean population of *M. norvegica* were low (0.07–0.11 μg-at NH$_4$–N mg^{-1} day^{-1}, and 0.009–0.010 μg-at PO$_4$–P mg^{-1} day^{-1}) in summer, autumn, and early winter but rose sharply through the next 2 months to peak in spring (0.25 μg-at NH$_4$–N mg^{-1} day^{-1} and 0.026 μg-at PO$_4$–P mg^{-1} day^{-1}, $T = 13$ °C). A study of 15 zooplankton species from Kosterfjorden, west Sweden by Båmstedt (1985) included a breakdown of the different nitrogen and phosphorus excretion products for *M. norvegica*, at three different times of the year. Although not strictly comparable with Rogers (1978), the general patterns are interesting. While individuals collected in the autumn (averaged over September and October) had an ammonia excretion of 1.85 nmol mg protein^{-1} h^{-1}, individuals collected in the spring (March) showed a doubling of this rate (3.66 nmol mg protein^{-1} h^{-1}; both rates were determined in darkness, $T = 4$–6 °C). However, spring rates of urea excretion were about one-third of autumn rates (1.33 compared with 5.35 nmol mg protein^{-1} h^{-1}). This pattern requires further elucidation. Both Båmstedt (1985) and Rogers (1978) record a greater value for the nitrogen:phosphorus excretion ratio in autumn (September) compared with spring (March) despite using different methods.

Saborowski *et al.* (2002) measured rates of ammonia excretion (and O:N ratios) at a number of different environmental temperatures for *M. norvegica*

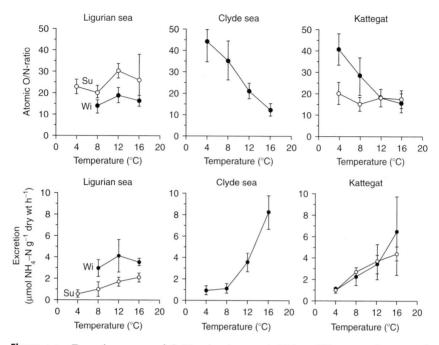

Figure 4.5 Excretion rates and O:N ratios (means ± SD) at different environmental temperatures of individuals of *M. norvegica* from three geographically separate populations (Saborowski *et al.*, 2002).

from three geographically separate populations (Fig. 4.5). There was tremendous variation in excretion rates between populations, in their response to temperature, and in seasonal modification. While the patterns obtained were quite different for different locations (Fig. 4.5), some generalisations can be made. For instance, in *M. norvegica* from the Kattegat and Clyde Sea, O:N ratios decreased with increasing temperature (winter) indicating a shift from the use of lipids (which occurs at winter environmental temperatures) to the use of protein in metabolism. O:N was equal to 35–40 in both the Kattegat and the Clyde, which Saborowski *et al.* (2002) suggest indicates equally favourable trophic conditions and similar feeding strategies of *M. norvegica* in the two locations.

7. Pollution and Trace Metals

7.1. Transuric elements

Exposure to, and accumulation of, transuric elements has commanded most attention with respect to pollution in *M. norvegica*. Fisher *et al.* (1983) demonstrated that Americium-241 (Am-241), a key radionuclide from waste disposal, was accumulated from contaminated sea water by passive adsorption onto the exoskeleton of *M. norvegica*. The exoskeleton showed a 100-times increase in Am-241 concentration over the course of about a week. However, 96% of the total Am-241 body burden was associated with the exoskeleton, and that burden was shed during moult. There was some assimilation by internal tissues (3% after 4 days feeding on radiolabelled— 2.1 kBq Am-241—diatoms *Thalassiosira pseudonana*). Thorium-234 (Th-234) is also taken up (exposure conditions = 0.03 Bq ml^{-1}) and concentrated by *M. norvegica*, reaching a steady state ($\sim$180 concentration factor) within 3 or 4 days (Baena *et al.*, 2008). As with Am-241 much of Th-234 is found in association with the exoskeleton, but not quite to the same extent. Only about 53% of the Th-234 is lost through the cast exoskeleton.

The transfer of Polonium-210 (Po-210) and Lead-210 along a food chain was examined by Stewart *et al.* (2005) who fed *M. norvegica* (NW Mediterranean) with contaminated *Artemia* adults, themselves fed on radiolabelled diatoms *T. pseudonana* and *Isochrysis galbana*. Po-210 is interesting because it is both a useful geochemical tracer and a source of high-energy alpha-emitter in marine organisms and humans: Po-210 is a naturally occurring radionuclide formed by beta decay of Pb-210. Unlike Am-241 and Th-234, it is thought that Po-210 accumulates mainly in the internal tissues *M. norvegica*, namely the hepatopancreas (midgut gland) and alimentary tract (Heyraud *et al.*, 1976).

The Po-210:Pb-210 ratio within animals increased 5–12 times with each trophic level indicating preferential bioaccumulation of Po-210 (44% assimilated by *M. norvegica*) over Pb-210 (3.5% assimilated by *M. norvegica*). The poor assimilation efficiency of Pb-210 means that the Po-210:Pb-210 ratio was 1–2 orders of magnitude smaller in krill faecal pellets than in the krill producing them. Stewart *et al.* (2005) suggest that in surface waters Po-210 has the potential to accumulate and concentrate in krill, and in the food chains of which they are a part, and be biologically recycled, whereas Pb-210 does not, making *M. norvegica* an important source of Po-210 to those predators consumed as seafood by humans. This study also appears to be the first to investigate the trophic transfer of these key radionuclides to any carnivorous planktonic organism.

7.2. 'Biomarkers'

The particle spectrum and feeding behaviour of *M. norvegica* were affected by the addition of Venezuelan crude oil ($\sim$6 μl l^{-1}) to sea water (Hebert and Poulet, 1980). Oil exposure had the same effect as starvation in reducing growth and survival.

Mixed function oxidase (MFO) has been proposed as a biomarker to assess the health status of krill. Fossia *et al.* (2002) measured MFO activity in Mediterranean populations by assaying benzo[a]pyrene monooxygenase

(BPMO) activities. However, there was no difference between populations with a mean value for BPMO of 1.83 (range 0.47–3.20) A.F.U. mg protein^{-1} h^{-1}. The same study also considered poly-aromatic hydrocarbons (PAH), which are byproducts of burning oil or coal. Mean PAH was 2624 (range 963.8–5038 ng g^{-1} DW) and the mean concentration of carcinogenic PAHs was 88.4 (range 61.84–141.7) ng g^{-1} DW with highest concentrations detected adjacent to the Ligurian coast. Amongst other measures, mean DDT (dichlorodiphenyltrichloroethane) levels were 72.83 (range 41.02–163.2) ng g^{-1} DW; mean PCBs (polychlorinated biphenyls), 145.4 (range 84.60–210.2) ng g^{-1} DW and mean HCB (hexachlorobenzene) = 6.08 (range 3.5–11.56) ng g^{-1} DW.

7.3. Trace metals

Rainbow (1989, 1993) provides baseline data for trace metals (essential as well as non-essential) from geographically separate populations of *M. norvegica* (Table 4.5). The first thing to notice is that there appears to be substantial spatial variability in trace metal concentrations, but also, where investigated, equally marked variability within a given population. Rainbow (1989) demonstrated a clear size dependency on whole body concentrations of copper and cadmium, but not zinc, iron, or manganese, in two geographically separate populations. This casts some doubt on mean values of copper and zinc presented in the table. However, it should be noted that Ridout *et al.* (1989) and Zauke and Schmalenbach (2006) also tested for size dependency in many of the same elements but were unable to detect any relationship. Some caution is therefore necessary when considering mean values, particularly when all size classes of krill have not been considered.

Nevertheless, there are clear population differences in cadmium and copper, particularly when we compare the North Atlantic (oceanic) and Firth of Clyde (coastal) populations (Rainbow 1989). The Atlantic population consistently had a greater concentration of both of those metals, across a range of body sizes. Furthermore, as the concentration of cadmium and copper was greater in the oceanic compared with the coastal individuals, we can discount the difference being attributable to enriched dissolved metal concentrations in coastal water, or other anthropogenic effects.

What is interesting is that the nature of the size dependency of these two metals is quite different. While copper concentration increased with increasing body mass (*b* is positive), cadmium concentration decreased with increasing body mass (*b* is negative). The negative relationship as found for cadmium might indicate that surface adsorbed metal contributes a significant proportion of total body metal burden, which is enhanced in small individuals with surface area to body mass ratios decreasing with increasing size. However, one might expect such a negative relationship for metals such as iron and manganese which show tendencies in sea water

Table 4.5 Trace metal concentrations ($\mu g\ g^{-1}$ DW) of whole *M. norvegica* collected from different locations

Location	Cd	Cu	Pb	Zn	Fe	Mn	V	Cr	Ni	Co	As	Reference
Barents Sea	0.2	47	—	73	—	—	—	—	—	—	—	Zauke and Schmalenbach (2006)
Greenland Sea	0.4	35	<0.3	42	—	—	—	—	—	—	—	Ritterhoff and Zauke (1997)
N.E. Atlantic	0.54–6.06[1]	8.8–67.2[2]	—	43★	38.9★	5★	—	—	—	—	—	Rainbow (1989)
N.E. Atlantic	0.39★	71.9★	—	96.5★	25.6★	2.9★	0.17★	0.27★	0.80★	0.16★	59.3★	Ridout et al. (1989)
North sea/Atlantic	0.5	26	1.0	45	—	—	—	—	—	—	—	Zauke et al. (1996)
Firth of Clyde, Scotland	0.14–1.83[3]	30.8–72.6[4]	—	102★	31.8★	1.14★	—	—	—	—	—	Rainbow (1989)
Corsica, Mediterranean	0.5	25.4	4.03	59	—	—	—	—	—	—	—	Romeo and Nicolas (1986)

All values are mean values, except from those marked with a superscript (1–4) where the range of concentrations measured is presented. This is because in those cases a relationship was detected between body weight and trace metal concentration, described by the equation $\log C = {}_{\log}a + b_{\log}d.w.$, where C = trace metal concentration, d.w., dry weight; and a and b are constants, the values for which are as follows.

[1] $a = -1.646$, $b = -1.126$.
[2] $a = 2.591$, $b = 0.639$.
[3] $a = -0.989$, $b = -0.779$.
[4] $a = 2.031$, $b = 0.367$.

Values marked with ★ denotes where a relationship between dry mass and metal concentration was tested for, but none was found.

to adsorb onto resuspended particles, and no such significant relationships were detected where it was tested for in *M. norvegica*. What is clear is that the cadmium concentrations are amongst some of the lowest for planktonic crustaceans and the copper values are amongst some of the highest, the latter fact perhaps related to the comparatively high concentrations of the copper-based respiratory pigment, haemocyanin, found in the haemolymph of *M. norvegica* (Brix *et al.* 1989; Spicer and Strömberg, 2002; see Section 4.3).

The concentrations of zinc, iron, and manganese were not atypical for planktonic crustaceans in general. As there was no size dependency detected for these elements, a comparison of mean values seems valid and shows that there is considerably spatial variation in zinc (means range from 42 to 109 μg g^{-1}) and manganese (means range from 1.14 to 5 μg g^{-1}), but not so much spatial variation in iron (means range from 25.6 to 38.9 μg g^{-1}). Experiments on the uptake, accumulation, and excretion of trace metals by *M. norvegica* exposed to natural and enriched concentrations of trace metals remain to be performed, as does an investigation on the physiology and pathology of exposure to enriched concentrations.

8. PERSPECTIVES

With respect to our understanding of the physiology of digestion, further biochemical and cytological studies are needed to investigate the assimilation process and, in particular, the membrane transfer of nutrients within the midgut gland. Furthermore, additional information is also required about the properties of key enzymes such as those involved in lipid digestion and lipid conversion. The cellulase and laminarinase activities in Northern krill need to be investigated as a matter of urgency, and it should be determined whether they are of endogenous origin or from microbial symbionts. Unfortunately, much of this work (and a number of the studies proposed below) requires long-term maintenance of krill in captivity and Northern krill are not easy to keep either in good condition, or at all, in the laboratory. Nevertheless, more experiments are required to study the processes of lipid digestion, conversion, and synthesis in *M. norvegica*.

Our understanding of key elements of the respiratory system of krill generally, and in some cases of *M. norvegica* specifically, have advanced markedly over the past 30 years, although mainly due to a handful of papers. Krill haemocyanin has a low O_2 affinity and a high sensitivity to temperature, although in the case of *M. norvegica*, the temperature effect is the opposite to that found for the respiratory pigments of nearly every other organism, including the closely related Antarctic krill. The positive heat of oxygenation discovered for the haemocyanin of *M. norvegica*, and the fact that the haemocyanin appears to be used as much as an energy source/store, as an O_2 transporter, deserve further, more detailed study. We are beginning to understand the effects of changing environmental factors on the physiology of *M. norvegica*, but to date no study has attempted to investigate synergistic or additive effects as a result of varying two or more environmental parameters. Furthermore as we realise that larval krill are not merely small adults in terms of their physiological capacity and tolerance, the need to consider development as central to our understanding of the ecophysiology of *M. norvegica* becomes clear.

When raising the notion of synergistic interactions, it is worth noting again that there are few studies, and none of them very detailed, on the effects of pollutants on the ecophysiology of *M. norvegica*. Certainly, we only know a little about the physiological effects and responses to transuric elements and trace metals. There is also one of the potentially most important (perceived) challenges to marine life in the twenty-first century, ocean acidification, and in particular its interaction with increasing temperature (Widdicombe and Spicer, 2008).

We have an appreciation of the high metabolic turnover rates and a high physiological plasticity of *M. norvegica*. However, very few investigators have probed the regulative potential of metabolites. Future studies should, therefore, focus more on the modulative effect of metabolites on enzyme activity levels.

M. norvegica has a physiology which is as interesting and surprising as it is beautiful to behold. Despite the problems of keeping this species in good condition in the laboratory for prolonged periods of time, those who have tried, and succeeded, have increased our knowledge of krill biology. Even more so those who have paid some attention to *Meganyctiphanes* biology have in doing so challenged our understanding of some basic biological processes, such as how respiratory pigments work and what they can be used for. There is no reason why this should not continue particularly given the key ecological role of krill in the world's oceans, and the importance of understanding their physiological responses to the pressing environmental problems of those oceans, particularly increasing CO_2 and temperature.

REFERENCES

Adelung, D., Buchholz, F., Culik, B., and Keck, A. (1987). Fluoride in tissues of krill *Euphausia superba* Dana and *Meganyctiphanes norvegica* M. Sars in relation to the moult cycle. *Polar Biol.* **7**, 43–50.

Albessard, E., and Mayzaud, P. (2003). Influence of tropho-climatic environment and reproduction on lipid composition of the euphausiid *Meganyctiphanes norvegica* in the Ligurian Sea, the Clyde Sea and the Kattegat. *Mar. Ecol. Prog. Ser.* **253**, 217–232.

Albessard, E., Mayzaud, P., and Cuzin-Roudy, J. (2001). Variation of lipid classes among organs of the Northern krill, *Meganyctiphanes norvegica*, with respect to reproduction. *Comp. Biochem. Physiol. A* **129**, 373–390.

Baena, A. M. R., Fowler, S. W., and Warnau, M. (2008). Could krill schools significantly bias 234Th-based carbon flux models? *Limnol. Oceanogr.* **53**, 1186–1191.

Ballantine, J. A., and Roberts, J. C. (1980). Marine sterols. XII. The sterols of some pelagic marine crustaceans. *J. Exp. Mar. Biol. Ecol.* **47**, 25–33.

Båmstedt, U. (1976). Studies on the deep-water pelagic community of Korsfjorden, western Norway. Changes in the size and biochemical composition of *Meganyctiphanes norvegica* (Euphausiacea) in relation to its life cycle. *Sarsia* **61**, 15–30.

Båmstedt, U. (1983). RNA concentration in zooplankton: seasonal variation in boreal species. *Mar. Ecol. Prog. Ser.* **11**, 291–297.

Båmstedt, U. (1985). Seasonal excretion rates of macrozooplankton from the Swedish west coast. *Limnol. Oceanogr.* **30**, 607–617.

Båmstedt, U. (1988). Interspecific, seasonal and diel variations in zooplankton trypsin and amylase activities in Kosterfjorden, western Sweden. *Mar. Ecol. Prog. Ser.* **44**, 15–24.

Båmstedt, U., and Skjoldal, H. R. (1980). RNA concentration of zooplankton: relationship with size and growth. *Limnol. Oceanogr.* **25**, 304–316.

Bergström, B., and Strömberg, J.-O. (1997). Behavioural differences in relation to pycno-clines during vertical migration of the euphausiids *Meganyctiphanes norvegica* (M. Sars) and *Thysanoessa raschii* (M. Sars). *J. Plankton Res.* **19**, 255–261.

Brix, O., Borgund, S., Barnung, T., Colosimo, A., and Giardina, B. (1989). Endothermic oxygenation of hemocyanin in the krill *Meganyctiphanes norvegica*. *FEBS* **247**, 177–180.

Buchholz, F. (1989). Moult cycle and seasonal activities of chitinolytic enzymes in the integument and the digestive tract of the Antarctic krill, *Euphausia superba*. *Polar Biol.* **9**, 311–317.

Buchholz, F. (2003). Experiments on the physiology of Southern and Northern krill, *Euphausia superba* and *Meganyctiphanes norvegica*, with emphasis on moult and growth - A review. *Mar. Fresh. Behav. Physiol.* **36**, 229–247.

Buchholz, F., and Prado-Fiedler, R. (1987). Studies on the seasonal biochemistry of the Northern krill *Meganyctiphanes norvegica* in the Kattegat. *Helgol. Meeresunters.* **41**, 443–452.

Buchholz, F., and Saborowski, R. (2000). Metabolic and enzymatic adaptations in northern krill, *Meganyctiphanes norvegica*, and Antarctic krill, *Euphausia superba*. *Can. J. Fish. Aquat. Sci.* **57**, 115–129.

Buchholz, F., and Vetter, R.-A. H. (1993). Enzyme kinetics in cold water: Characteristics of N-acetyl-β-D-glucosaminidase in the Antarctic krill, *Euphausia superba*, compared with other crustacean species. *J. Comp. Physiol. B* **163**, 28–37.

Buchholz, F., David, P., Matthews, J., Mayzaud, P., and Patarnello, T. (1998). Impact of a climatic gradient on the Physiological Ecology of a Pelagic crustacean (PEP). *In* "Proc. 3rd Europ. Mar. Sci. Technol. Conf." Vol I: Marine Systems, pp. 39–48.

Chícharo, M. A., and Chícharo, L. (2008). RNA:DNA ratio and other nucleic acid derived indices in marine ecology. *Int. J. Mol. Sci.* **9**, 1453–1471.

Cuzin-Roudy, J. (1993). Reproductive strategies of the Mediterranean krill, *Meganyctiphanes norvegica* and the Antarctic krill, *Euphausia superba* (Crustacea: Euphausiacea). *Invertebr. Reprod. Dev.* **23**, 105–114.

Cuzin-Roudy, J., Albessard, E., Virtue, P., and Mayzaud, P. (1999). The scheduling of spawning with the moult cycle in Northern krill (Crustacea: Euphausiacea): A strategy for allocating lipids to reproduction. *Invertebr. Reprod. Dev.* **36**, 163–170.

Dall, W., and Moriarty, D. J. W. (1983). Functional aspects of nutrition and digestion. *In* "The biology of Crustacea" (L. E. Mantel, ed.), Internal anatomy and physiological regulation **5**, pp. 215–261. New York, Academic Press.

Dalsgaard, J., St. John, M., Kattner, G., Müller-Navarra, D., and Hagen, W. (2003). Fatty acid trophic markers in the pelagic marine environment. *Adv. Mar. Biol.* **46,** 225–340.

Dawdry, N. E. (2004). Diel vertical migration and feeding by krill *Meganyctiphanes norvegica.* Ph.D. thesis, University of Plymouth.

Dittrich, B. (1990). Temperature dependence of the activities of trypsin-like proteases in decapods crustaceans from different habitats. *Naturwissenschaften* **77,** 491–492.

Dittrich, B. (1992a). Life under extreme conditions: Aspects of evolutionary adaptation to temperature in crustacean proteases. *Polar Biol.* **12,** 269–274.

Dittrich, B. (1992b). Thermal acclimation and kinetics of a trypsin-like protease in eucarid crustaceans. *Comp. Biochem. Physiol.* B **162,** 38–46.

Donachie, S. P., Saborowski, R., Peters, G., and Buchholz, F. (1995). Bacterial digestive enzyme activity in the stomach and hepatopancreas of *Meganyctiphanes norvegica* (M. Sars, 1875). *J. Exp. Mar. Biol. Ecol.* **188,** 151–165.

Falk-Petersen, S. (1981). Ecological investigations on the zooplankton community of Balsfjorden, Northern Norway: Seasonal changes in body weight and the main biochemical composition of *Thysanoessa imeris* (Krøyer), *T. raschii* (M. Sars), and *Meganyctiphanes norvegica* (M. Sars) in relation to environmental factors. *J. Exp. Mar. Biol. Ecol.* **49,** 103–120.

Falk-Petersen, S., Gatten, R. R., Sargent, J. R. S., and Hopkins, C. C. E. (1981). Ecological investigations on the zooplankton community in Balsfjorden, Northern Norway: Seasonal changes in the lipid class composition of *Meganyctiphanes norvegica, Thysanoessa raschii* and *Thysanoessa inermis. J. Exp. Mar. Biol. Ecol.* **54,** 209–224.

Falk-Petersen, S., Hagen, W., Kattner, G., Clarke, A., and Sargent, J. (2000). Lipids, trophic relationships, and biodiversity in Arctic and Antarctic krill. *Can. J. Fish. Aquat. Sci.* **57,** (Suppl 3), 178–191.

Fevolden, S. E. (1982). Feeding habits and enzyme polymorphism in *Thysanoessa raschi* and *Meganyctiphanes norvegica* (Crustacea; Euphausiacea). *Sarsia* **67,** 1–10.

Fisher, N. S., Bjerregaard, P., and Fowler, S. W. (1983). Interactions of marine plankton with transuranic elements. 3. Biokinetics of americium in euphausiids. *Mar. Biol. (Berl.)* **75,** 261–268.

Forward, R. B. Jr, and Fyhn, H. J. (1983). Osmotic regulation of the krill *Meganyctiphanes norvegica. Comp. Biochem. Physiol.* **74,** 301–305.

Fossia, M. C., Borsanib, J. F., Di Mentob, R., Marsilic, L., Casinic, S., Neric, G., Moric, G., Ancorac, S., Leonzioc, C., Minutolia, R., and Notarbartolo di Sciarab, G. (2002). Multitrial biomarker approach in *Meganyctiphanes norvegica*: A potential early indicator of health status of the Mediterranean "whale sanctuary". *Mar. Environ. Res.* **54,** 761–767.

Hebert, R., and Poulet, S. A. (1980). Effect of modification of particle size of emulsions of venezuelan crude oil on feeding, survival and growth of marine zooplankton. *Mar. Environ. Res.* **4,** 121–134.

Heyraud, M., Fowler, S. W., Beasley, T. M., and Cherry, R. D. (1976). Polonium-210 in euphausiids: a detailed study. *Mar. Biol.* **34,** 127–136.

Hirche, H.-J. (1984). Temperature and metabolism of plankton—I. Respiration of antarctic zooplankton at different temperatures with a comparison of Antarctic and Nordic krill. *Comp. Biochem. Physiol.* A **77,** 361–368.

Ikeda, T., Nash, G. V., and Thomas, P. G. (1984). An observation of discarded stomach with exoskeleton moult from Antarctic krill *Euphausia superba* Dana. *Polar Biol.* **3,** 241–244.

Kreibich, T., Hagen, W., and Saborowski, R. (2010). Food utilization of two pelagic crustaceans from Greenland Sea waters: *Meganyctiphane norvegica* (Euphausiacea) and *Hymenodora glacialis* (Decapoda, Caridea). *Mar. Ecol. Prog. Ser.* **413,** 105–115.

Lass, S., Tarling, G. A., Virtue, P., Matthews, J. B. L., Mayzaud, P., and Buchholz, F. (2001). On the food of Northern krill *Meganyctiphanes norvegica* in relation to its vertical distribution. *Mar. Ecol. Prog. Ser.* **214,** 177–200.

Lee, R. F., Hagen, W., and Kattner, G. (2006). Lipid storage in marine zooplankton. *Mar. Ecol. Prog. Ser.* **307**, 273–306.

Loizzi, R. F. (1971). Interpretation of crayfish hepatopancreatic function based on fine structural analysis of epithelial cell lines and muscle network. *Z. Zellforsch.* **113**, 420–440.

Matthews, J. B. L., Buchholz, F., Saborowski, R., Tarling, G. A., Dallot, S., and Labat, J. P. (1999). On the physical oceanography of the Kattegat and the Clyde Sea area, 1996–98, as background to ecophysiological studies on the planktonic crustacean, *Meganyctiphanes norvegica* (Euphausiacea). *Helgol. Mar. Res.* **53**, 70–84.

Mauchline, J. (1980). Part II: The biology of euphausiids. *Adv. Mar. Biol.* **18**, 373–623.

Mauchline, J., and Fisher, L. R. (1969). The biology of euphausiids. *Adv. Mar. Biol.* **7**, 1–454.

Mayzaud, P., Virtue, P., and Albessard, E. (1999). Seasonal variations in the lipid and fatty acid composition of the euphausiid *Meganyctiphanes norvegica* from the Ligurian Sea. *Mar. Ecol. Prog. Ser.* **186**, 199–210.

Mayzaud, P., Albessard, E., Virtue, P., and Boutoute, M. (2000). Environmental constraints on the lipid composition and metabolism of euphausiids: The case of *Euphausia superba* and *Meganyctiphanes norvegica*. *Can. J. Fish. Aquat. Sci.* **57**, (Suppl 3), 91–103.

Mayzaud, P., Boutoute, M., Gasparini, S., and Moussea, L. (2005). Respiration in marine zooplankton—the other side of the coin: CO_2 production. *Limnol. Oceanogr.* **50**, 291–298.

Morris, R. J. (1972). The occurrence of wax esters in crustaceans from the north-east Atlantic Ocean. *Mar. Biol.* **16**, 102–107.

Mulkiewicz, M. S., Ziętara, M. S., Strömberg, J.-O., and Skorkowski, E. F. (2001). Lactate dehydrogenase from the northern krill *Meganyctiphanes norvegica*: Comparison with LDH from the Antarctic krill *Euphausia superba*. *Comp. Biochem. Physiol. B* **128**, 233–245.

Peters, G., Saborowski, R., Mentlein, R., and Buchholz, F. (1998). Isoforms of an *N*-acetyl-b-D-glucosaminidase from the Antarctic krill. *Euphausia superba*: Purification and antibody production. *Comp. Biochem. Physiol. B* **120**, 743–751.

Peters, G., Saborowski, R., Buchholz, F., and Mentlein, R. (1999). Two distinct forms of the chitin-degrading enzyme *N*-acetyl-*β*-D-glucosaminidase in the Antarctic krill: specialists in digestion and moult. *Mar. Biol.* (*Berl.*) **134**, 697–703.

Rainbow, P. S. (1989). Copper, cadmium and zinc concentrations in oceanic amphipod and euphausiid crustaceans, as a source of heavy metals to pelagic seabirds. *Mar. Biol.* (*Berl.*) **103**, 513–518.

Rainbow, P. S. (1993). The significance of trace metal concentration in marine invertebrates. *In* "Ecotoxicology of metals in invertebrates" (R. Dallinger and P. S. Rainbow, eds), pp. 4–23. Lewis Publishers, Boca Raton USA.

Ridout, P. S., Rainbow, P. S., Roe, H. S. J., and Jones, H. R. (1989). Concentrations of V, Cr, Mn, Fe, Ni, Co, Cu, Zn, As and Cd in mesopelagic crustaceans from the North East Atlantic Ocean. *Mar. Biol.* (*Berl.*) **100**, 465–471.

Ritterhoff, J., and Zauke, G. P. (1997). Trace metals in field samples of zooplankton from the Fram Strait and the Greenland Sea. *Sci. Total Environ.* **199**, 255–270.

Roger, C. (1978). Azote et phosphore chez un Crustacé macroplanctonique, *Meganyctiphanes norvegica* (M. Sars) (Euphausiacea): Excrétion minérale et constitution. *J. Exp. Mar. Biol. Ecol.* **33**, 57–83.

Romeo, M., and Nicolas, E. (1986). Cadmium, copper, lead and zinc in three species of planktonic crustaceans from the east coast of Corsica. *Mar. Chem.* **18**, 359–367.

Saborowski, R., and Buchholz, F. (1998). Internal current generation in respiration chambers. *Helgoland. Meeresunter.* **52**, 103–109.

Saborowski, R., and Buchholz, F. (2002). Metabolic properties of Northern krill, *Meganyctiphanes norvegica*, from different climatic zones. II. Enzyme characteristics and activities. *Mar. Biol.* (*Berl.*) **140**, 557–565.

Saborowski, R., Salomon, M., and Buchholz, R. F. (2000). The physiological response of Northern krill (*Meganyctiphanes norvegica*) to temperature gradients in the Kattegat. *Hydrobiologia* **426**, 157–160.

Saborowski, R., Brohl, S., Tarling, G. A., and Buchholz, F. (2002). Metabolic properties of Northern krill, *Meganyctiphanes norvegica*, from different climatic zones. I. Respiration and excretion. *Mar. Biol. (Berl.)* **140**, 547–556.

Saether, O., Eilingsen, T. E., and Mohr, V. (1986). Lipids of north Atlantic krill. *J. Lipid Res.* **27**, 274–285.

Salomon, M., and Saborowski, R. (2006). Tissue-specific distribution of pyruvate kinase isoforms improve the physiological plasticity of Northern krill, *Meganyctiphanes norvegica*. *J. Exp. Mar. Biol. Ecol.* **331**, 82–90.

Salomon, M., Mayzaud, P., and Buchholz, F. (2000). Studies on metabolic properties in the Northern Krill, *Meganyctiphanes norvegica* (Crustacea, Euphausiacea): Influence of nutrition and season on pyruvate kinase. *Comp. Biochem. Physiol. A* **127**, 505–514.

Sands, M., Nicol, S., and McMinn, A. (1998). Fluoride in Antarctic marine crustaceans. *Mar. Biol. (Berl.)* **132**, 591–598.

Sargent, J. R., and Falk-Petersen, S. (1981). Ecological investigations on the zooplankton community in Balsfjorden, northern Norway: Lipids and fatty acids in *Meganyctiphanes norvegica*, *Thysanoessa raschi* and *T. inermis* during mid winter. *Mar. Biol. (Berl.)* **62**, 131–137.

Soevik, T., and Braekkan, O. R. (1979). Fluoride in Antarctic krill (*Euphausia superba*) and Atlantic krill (*Meganyctiphanes norvegica*). *J. Fish. Res. Bd. Can.* **36**, 1414–1416.

Spicer, J. I., and Strömberg, J.-O. (2002). Diel vertical migration and the haemocyanin of Norway krill *Meganyctiphanes norvegica*. *Mar. Ecol. Prog. Ser.* **238**, 153–162.

Spicer, J. I., and Strömberg, J.-O. (2003). Developmental changes in the responses of O_2 uptake and ventilation to acutely declining O_2 tensions in larval krill *Meganyctiphanes norvegica*. *J. Exp. Mar. Biol. Ecol.* **295**, 207–218.

Spicer, J. I., Thomasson, M. A., and Strömberg, J.-O. (1999). Possessing a poor anaerobic capacity does not prevent the diel vertical migration of Nordic krill *Meganyctiphanes norvegica* into hypoxic waters. *Mar. Ecol. Prog. Ser.* **185**, 181–187.

Spindler, K.-D., and Buchholz, F. (1988). Partial characterization of chitin degrading enzymes from two euphausiids, *Euphausia superba* and *Meganyctiphanes norvegica*. *Polar Biol.* **9**, 115–122.

Stewart, G. M., Fowler, S. W., Teyssié, J.-L., Cotret, O., Cochran, J. K., and Fisher, N. S. (2005). Contrasting transfer of polonium-210 and lead-210 across three trophic levels in marine plankton. *Mar. Ecol. Prog. Ser.* **290**, 27–33.

Strömberg, J.-O., and Spicer, J. I. (2000). Cold comfort for krill? Respiratory consequences of diel vertical migration of *Meganactyphanes norvegica* into deep hypoxic waters. *Ophelia* **53**, 213–217.

Stübing, D., Hagen, W., and Schmidt, K. (2003). On the use of lipid biomarkers in marine food web analysis: An experimental case study on the Antarctic krill, *Euphausia superb*. *Limnol. Oceanogr.* **48**, 1685–1700.

Ullrich, B., Storch, V., and Marschall, H. P. (1991). Microscopic anatomy, functional morphology, and ultrastructure of the stomach of *Euphausia superba* Dana (Crustacea, Euphausiacea). *Polar Biol.* **11**, 203–211.

van den Thillart, G., George, R. Y., and Strömberg, J.-O. (1999). Hypoxia sensitivity and respiration of the euphausiid crustacean *Meganyctiphanes norvegica* from Gullmarn Fjord, Sweden. *Sarsia* **84**, 105–109.

Vetter, R.-A. H. (1995a). Ecophysiological studies on citrate-synthase: (I) enzyme regulation of selected crustaceans with regard to temperature adaptation. *J. Comp. Physiol. B* **165**, 46–55.

Vetter, R.-A. H. (1995b). Ecophysiological studies on citrate-synthase: (II) enzyme regulation of selected crustaceans with regard to life-style and the climatic zone. *J. Comp. Physiol. B* **165,** 56–61.

Vetter, R.-A. H., and Buchholz, F. (1997). Catalytic properties of two pyruvate kinase isoforms in Nordic krill, *Meganyctiphanes norvegica*, with respect to seasonal temperature adaptation. *Comp. Biochem. Physiol. A* **116,** 1–10.

Virtue, P., Nicol, S., and Nichols, P.D. (1993) .Changes in the digestive gland of Euphausia superba during short-term starvation: lipid class, fatty acid and sterol content and composition. *Mar. Biol.* **117,** 441–448.

Virtue, P., Mayzaud, P., Albessard, E., and Nichols, P. (2000). The use of fatty acids as dietary indicators in krill, *Meganyctiphanes norvegica*, from North Eastern Atlantic, Kattegat, and Mediterranean during summer and winter. *Can. J. Fish. Aquat. Sci.* **57,** (Suppl. 3), 104–114.

Widdicombe, S., and Spicer, J. I. (2008). Predicting the impact of ocean acidification on benthic biodiversity: What can physiology tell us? *J. Exp. Mar. Biol. Ecol.* **366,** 187–197.

Zauke, G.-P., and Schmalenbach, I. (2006). Heavy metals in zooplankton and decapod crustaceans from the Barents Sea. *Sci. Total Environ.* **359,** 283–294.

Zauke, G.-P., Krause, M., and Weber, A. (1996). Trace metals in mesozooplankton of the North Sea: Concentrations in different taxa and preliminary results on bioaccumulation in copepod collectives (*Calanus finmarchicus/C. helgolandicus*). *Int. Rev. Gesamten Hydrobiol.* **81,** 141–160.

CHAPTER FIVE

FOOD AND FEEDING IN NORTHERN KRILL (*MEGANYCTIPHANES NORVEGICA* SARS)

Katrin Schmidt

Contents

Abstract

Early feeding studies on *Meganyctiphanes norvegica* described the morphology of the feeding appendages and the actual process of food uptake and digestion. Insights into diurnal, seasonal and ontogenetic pattern in feeding activity and diet were derived from field studies on the Clyde Sea population. Since then, technical advances have confirmed some of the early assumptions and rejected others. Submersible, remotely operated vehicles and echosounders, for instance, proved that *M. norvegica* stay often close to the seabed and feed on particles in the epibenthic layer and sediment-water interface. Scanning electron microscopy showed that mandibles of the so-called carnivorous *M. norvegica* have an elaborated grinding region, which allows efficient feeding

British Antarctic Survey, Natural Environment Research Council, High Cross, Cambridge, United Kingdom

Advances in Marine Biology, Volume 57
ISSN 0065-2881, DOI: 10.1016/S0065-2881(10)57005-6

on diatoms. Three-dimensional silhouette video imaging revealed mechanoreception, not vision, as the main sensory modality involved in proximity prey detection by *M. norvegica*. Fatty acid analysis and stomach content microscopy have now been conducted on *M. norvegica* across a range of environments including the Gulf of Maine, Greenland Sea, Barents Sea, Scandinavian fjords, the Kattegat and Mediterranean Sea. Regional and seasonal differences in the trophic environment are reflected in their daily ration and in the relative importance of copepods versus phytoplankton in their diet. Overall, phytoplankton is an important food source for *M. norvegica* during the spring bloom and part of the summer, but copepods are dominant in autumn and winter. Depending on their vertical co-occurrence, *M. norvegica* can feed on a range of copepods from early stages of *Oithona* spp. up to adult *Calanus* spp. There are clear ontogenetic differences in diet, with adults feeding more on copepods and benthic food items than early post-larvae. Future studies should link diet to simultaneously measured growth and reproduction and emphasise comparison across the spectrum of environments inhabited by this versatile species.

1. INTRODUCTION

Most of an animal's activity is in some way related to feeding. First, food delivers the required energy and material to run all physiological processes (e.g. movements, adaptations, moulting, growth, reproduction). Second, many behavioural patterns reflect evolutionary adaptations either to ensure sufficient food is acquired (e.g. schooling, vertical migration, seasonal migration) or to allow some length of survival without food (e.g. build up of lipid reserves, dormancy). Predation avoidance and reproduction can modify this behaviour, but on a population level, feeding will always be a primary process. Moreover, feeding and being preyed on is the way of energy-and material transfer from lower to higher trophic levels within communities, so its quantification will be a key factor when food web relationships are studied. Further, feeding leads to remineralisation and/ or export of carbon and essential nutrients (nitrogen, phosphate, iron, silica), which has an effect on primary production and bentho-pelagic coupling.

The diet of *Meganyctiphanes norvegica* has been studied for more than 100 years (Holt and Tattersall, 1905; Paulsen, 1909). Anything from diatoms, dinoflagellates, coccolithophores, chrysomonads, filamentous algae, pine pollen, fern sporangia, lithogenic sediment, tintinnids, radiolarians, foraminifera, sponge spicules, insect egg cases to fragments of terrestrial plants, cnidaria, chaetognaths, copepods and polychaetes has been found in the stomach of *M. norvegica*, even ommatidia from other euphausiids (Fisher and Goldie, 1959; Mauchline, 1980; Lass *et al.*, 2001).

Some authors suggest that the species feeds mainly on copepods or other animal prey (i.e. it is 'carnivorous'; Artiges *et al.*, 1978; Falk-Petersen *et al.*, 2000; Kaartvedt *et al.*, 2002; Petursdottir *et al.*, 2008; Kreibich *et al.*, 2010; Sargent and Falk-Petersen, 1981), while others suggested a balance of autotrophic and heterotrophic sources (i.e. 'omnivorous'; Båmstedt and Karlson, 1998; Berkes, 1973; Dalpadado *et al.*, 2008; Fevolden, 1982; Lass *et al.*, 2001; Mauchline and Fisher, 1969; Mayzaud *et al.*, 1999; Onsrud and Kaartvedt, 1998; Sameoto, 1980; Suh, 1996; Virtue *et al.*, 2000) or even a mostly phytophagous diet (i.e. 'herbivorous', De Jong-Moreau *et al.*, 2001). This reflects the very versatile feeding biology of *M. norvegica*, but also regional and seasonal specifics, differences in methodological approaches and in the interpretation of data.

Predation on copepods is certainly the aspect of the feeding ecology of *M. norvegica* that has received most attention in recent years (Abrahamsen *et al.*, 2010; Båmstedt and Karlson, 1998; Kaartvedt *et al.*, 2002; Lass *et al.*, 2001; Torgersen, 2001; Petursdottir *et al.*, 2008). Other topics such as ontogenetic differences in feeding behaviour or potential competition between euphausiid species have not been re-investigated since Berkes (1973), Fisher and Goldie (1959) and Mauchline (1960). To represent comprehensively the current knowledge on feeding of *M. norvegica*, this review focuses on recent publications but includes older work if essential. It leads from the morphology of the feeding appendages, via the sensory and mechanics of food uptake, to a range of *in situ* observations and some quantitative estimates of feeding under laboratory conditions. Three aims are followed: (1) to give background information on which to base future studies, (2) to show quantitative and qualitative aspects important for food web models and (3) to draw attention to open questions.

2. FEEDING BASKET, MANDIBLES AND GASTRIC MILL

Investigation of foraging requires accurate knowledge of the morphology and function of the feeding appendages. Structures directly involved in feeding exhibit morphological adaptations to the diet of a species. Such insights are usually gained when comparing the morphology of feeding appendages across a range of species and considering additional results from stomach content analysis (Berkes, 1973; De Jong-Moreau *et al.*, 2001; Mauchline, 1980; Nemoto, 1977). In euphausiids, feeding involves the integrated action of externally located thoracopods and mouthparts and the internally placed armature of the stomach wall (Suh and Nemoto, 1988). The thoracopods gather the food, the mouthparts handle it and the mandibles pierce, cut and grind the items before they are swallowed. The internal armature of the stomach breaks the food up into even smaller pieces until they are fine enough to enter the digestive gland for final digestion and absorption (Suh, 1996).

In *M. norvegica*, seven pairs of thoracic legs collectively form a feeding basket (Artiges *et al.*, 1978; Mauchline, 1989; Fig. 5.1), which is used to capture copepods or filter suspended material. The fine structure of this basket is a three-dimensional filter of primary, secondary and tertiary setae (Suh and Nemoto, 1987). Primary setae are present on the coxa, basis, ischium and merus of the legs (Artiges *et al.*, 1978). Along the primary setae, there are two rows of secondary setae inserted at an angle of 90° and again a single row of tertiary setae along the secondary setae (Suh and Nemoto, 1987). The filter area and the intersetal distance determine the filtering efficiency of the feeding basket, which can differ between species and size classes (Boyd *et al.*, 1984; Suh and Choi, 1998; Suh and Nemoto, 1987). McClatchie (1985) compared the feeding baskets of *M. norvegica* and *Euphausia superba*, and found that the increase in filter area with body length was less pronounced in *M. norvegica*. Young *M. norvegica* had a similar filter area than predicted for *E. superba* of the same size, but that of adults was smaller by almost 50% (McClatchie, 1985). If the two species apply their feeding baskets in the same way and at the same rate, 22–39 mm sized *E. superba* would have three times higher clearance rates than *M. norvegica* of the same body length (McClatchie, 1985). In addition, the minimum spacing between tertiary setae is about 25 μm for *M. norvegica*

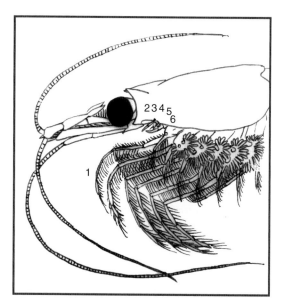

Figure 5.1 Feeding appendages of *M. norvegica*. Lateral aspect showing the relative position of the feeding basket and mouthparts. 1, thoracic legs; 2, labrum; 3, mandible with mandibular palp; 4, labium; 5, maxillule; 6, maxilla. Drawing after Einarsson (1945).

(Berkes, 1973; Artiges *et al.*, 1978), but only 1–2 μm for *E. superba* (McClatchie and Boyd, 1983; Suh and Nemoto, 1987), indicating that *E. superba* is more efficient in retaining small particles. However, in comparison to other euphausiids from the Ligurian Sea or North Atlantic (e.g. *Euphausia krohnii*, *Thysanoessa longicaudata*), *M. norvegica* has a well-developed, heavily setose feeding basket (Berkes, 1973; Artiges *et al.*, 1978; Mauchline, 1989).

Once the food is enclosed in the feeding basket, various mouthparts (mandibular palps, maxillules and maxillae) are involved in passing it on to the oral cavity built by the upper lip (labrum), mandibles and lower lip (labia). The mandibles are hard and have strong cusps, the *incisor process*, in the ventral region and grinding surfaces, the *molar process*, in the dorsal region. It is assumed that large grinding areas of the mandible are associated with a phytophagous tendency, whereas a pronounced cutting region indicates carnivorous feeding (Nemoto, 1977; Mauchline, 1980). Mandibles of *M. norvegica* have a longer cutting than grinding region, which suggests a more carnivorous habit (Mauchline, 1980). The sharpness of the cutting region and the keen edge also indicate carnivorous possibilities in *M. norvegica* (Artiges *et al.*, 1978). However, De Jong-Moreau *et al.* (2001) gave a different interpretation of the mandible morphology. First, they state that when looking at the mandible area instead of mandible length, the molar and incisor processes are equally developed in *M. norvegica*. Second, they suggest that the specialisation seen in the molar process is a better predictor of a species' diet than the relative size of the molar and incisor processes (De Jong-Moreau *et al.*, 2001). In the case of *M. norvegica*, the molar process is an elaborated grater of parallel lamellae of strong spines in the middle and seta-like spines at the periphery (De Jong-Moreau *et al.*, 2001; Fig. 5.2). From the seven species studied, *M. norvegica* and the mysid *Boreomysis inermis* showed the most advanced molar process, which lead the authors to the conclusion that these species are mostly phytophagous (De Jong-Moreau *et al.*, 2001).

After crushing and grinding by the mandibles, the stomach is an additional place for maceration of food particles (Suh, 1996). The gastric mill is the main grinding structure in the stomach. *M. norvegica* has a well-developed gastric mill with a pair of lateral teeth and a pair of cluster spines (Suh and Nemoto, 1988; Fig. 5.3). A complex system of muscles enables movements of the stomach wall, which compress the food between the armoured areas. In species that feed mainly on soft-bodied food, lateral teeth are usually absent and the internal armature of the stomach is less pronounced (Mauchline, 1980; Suh and Nemoto, 1988). However, it is suggested that the morphology of the gastric mill is more related to the phylogeny of the euphausiids than to its feeding pattern (Suh and Nemoto, 1988).

In conclusion, the setose feeding basket, the elaborated molar processes and the well-developed gastric mill indicate that *M. norvegica* is able to feed efficiently on diatoms. The large intersetal distance points to an omnivorous component in its diet, more so than, for example, in *E. superba*. However,

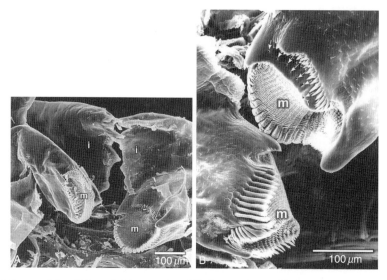

Figure 5.2 Mandibles of *M. norvegica*. (A) Inner view of left and right mandible showing relative size of incisor process (i) and molar process (m). (B) Molar process of left and right mandible. The convex ridge of the left molar fits into the broad concavity of the right molar. The grater is composed of parallel lamellae of strong spines surrounded by a row of pores (arrows). Alignments bring the left and right grinding surface into close contact, enabling efficient crushing of hard-shelled food items such as diatoms, silicoflagellates and foraminifera. Photos: B. Casanova, published in De Jong-Moreau *et al.* (2001).

M. norvegica is certainly not a predatory (carnivorous) species, as mouthparts of exclusive predators are clearly different from those of filter feeders and omnivorous species (Mauchline, 1998). The surprising conclusion of De Jong-Moreau *et al.*, (2001) that *M. norvegica* is mainly a phytophagous species, might have a simple explanation. Considering that it requires an enormous amount of force to break silicified diatoms frustles (Hamm *et al.*, 2003), feeding on diatoms was probably a stronger challenge for the morphology of mandibles than feeding on soft copepods. Therefore, the mandibles might show strong signs of herbivory in an omnivorous species, while adaptation to feeding on copepods might be more reflected in the species' sensory abilities.

3. Foraging Tactics

3.1. Filter feeding

Euphausiids are able to sense odours of phytoplankton along a diffusion gradient (Hamner and Hamner, 2000; Price, 1989). In a positive response, the scent trail will be tracked, filtration rates increase and the krill try to

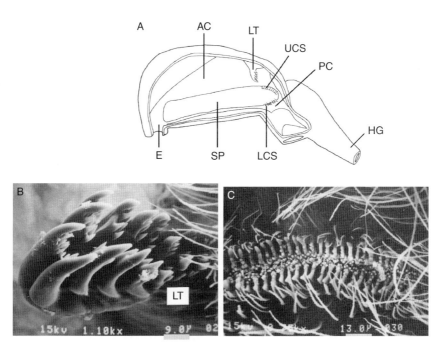

Figure 5.3 Gastric mill of *M. norvegica*. (A) Lateral dissected view of the external and internal structures of the stomach (foregut). AC, anterior chamber; E, oesophagus; HG, hindgut; LCS, lower process of cluster spines; LT, lateral teeth; PC, posterior chamber; SP, side plate; UCS, upper process of cluster spines. (B) Left lateral teeth. (C) Upper process of cluster spines. Drawing and photos: H. L. Suh, published in Suh (1988).

remain within the area of highest phytoplankton concentration (Hamner and Hamner, 2000; Price, 1989). Negative responses of chemoreceptors cause rejection or avoidance of the particles in the water. Most studies on the mechanisms of filter feeding have been carried out on *E. superba*, but it is assumed that processes are identical in all euphausiids that have thoracic appendages similar to *E. superba* and filter feed on phytoplankton (Hamner, 1988). During filter feeding, the paired thoracic legs move downward and outward in a metachronal rhythm to enclose a body of water from the front (Hamner, 1988; McClatchie and Boyd, 1983). The legs are then closed and pressed to the ventral side of the body, and the water is forced laterally through the setae. This compression pumping increases water velocity and the Reynolds number of the euphausiid setae, while other suggested methods of filter feeding such as stationary filtration or fishing with the feeding basket are ineffective in a viscous environment (McClatchie and Boyd, 1983). As the water is strained through the setae, phytoplankton remains inside the filter basket. The filter setae are then scraped and cleaned by a second set of comb setae and the particles passed forward to the mouth

(Hamner, 1988). The metachronal rhythm of the thoracic legs during expansion and contraction of the basket enables the passage of food from posterior to anterior legs (Mauchline, 1989).

3.2. Predatory feeding

Four processes are involved in successful predatory feeding: detection of the prey, attack, capture and ingestion. The sensory basis for detection and localising of prey by *M. norvegica* include vision, mechanoreception and chemoreception. Based on diet studies, Kaartvedt *et al.* (2002) concluded that feeding on dormant, motionless *Calanus* spp. in 120 m water depth during the day is based on vision, while mechanoreception is more important when hunting shallow living, more active copepods at night. Torgersen (2001) studied the relative importance of vision versus mechanoreception in *M. norvegica* by feeding them a selection of copepods in total darkness and at typical daytime light intensity. Predation rates were three times higher in the illuminated containers than in the dark ones, leading to the conclusion that *M. norvegica* is primarily a visual predator on copepods (Torgersen, 2001). However, as the krill were free-swimming in the experiments, it cannot be excluded that the light simply triggered a higher swimming activity and therefore higher prey encounter rate of *M. norvegica*.

In an alternative experiment, *M. norvegica* were tethered and their responses to *Calanus* spp. were observed in 3D silhouette video imaging (Abrahamsen *et al.*, 2010). The authors confirmed that vision plays an important role in predatory feeding of *M. norvegica* (Abrahamsen *et al.*, 2010). First, in light, krill detected copepods from a further distance than in the dark (mean detection distances 26.5 and 19.5 mm, respectively). Second, the greatest increase in detection distance with light occurred in front and slightly above the predator, suggesting that *M. norvegica* benefit from down-welling light due to the maximal contrast of potential prey (Abrahamsen *et al.*, 2010). Third, in light, *M. norvegica* feeding rates on copepods were higher (Torgersen, 2001) even though copepods showed an earlier escape reaction (Abrahamsen *et al.*, 2010), which underlines the effectiveness of the visual activity (Abrahamsen *et al.*, 2010). Overall, however, these authors conclude that mechanoreception, not vision, is the main sensory modality involved in proximity prey detection by *M. norvegica* (Abrahamsen *et al.*, 2010; Browman, 2005). This is because in 80% of the cases, prey were detected when below the krill's body axis and, given the placement and orientation of the compound eyes, presumably outside its visual field (Abrahamsen *et al.*, 2010). The morphology of the *M. norvegica* eye is such that its spatial resolution is inadequate to detect small objects at close range, but they might be able to pick up clusters of prey at a distance and movements via changes in light intensity (Browman, 2005; Nilsson, 1996). Most likely, the role of mechanoreception increases when distances and/or light level decrease (Abrahamsen *et al.*, 2010). Alternatively, prey might be detected via their bioluminescence. Fisher and Goldie (1959) found high numbers of euphausiid eyes in stomachs of *M. norvegica* in late autumn and winter, when other food was rare. They assumed that predators were attracted to the luminescent photophore on the eyestalk (Fisher and Goldie, 1959).

Attack responses are initiated well before the copepods reach the feeding appendances of *M. norvegica*. During an attack, antennae are moved towards the target, followed by a propulsion and opening of the feeding basket (Abrahamsen *et al.*, 2010). If successful, the copepod is sucked into the basket with the inward flow of water. Once the copepod is captured, the cusps of the mandibles and the spines of the maxillules can pierce the integument (Mauchline and Fisher, 1969). Ponomareva (1954) reported from various euphausiid species that only the soft internal tissue of the copepods is extracted through the ruptured prosome while the remains are discarded. Beyer (1992) made a similar observation on *M. norvegica* from the Oslofjord. He reported that small copepods, up to the size of adult *Paracalanus* spp. and *Calanus* spp. copepodid stage II, were swallowed whole without grinding. Larger copepods, from about the size of *Calanus* spp. stage III, were held against the mouth, and only soft parts of the prey were ingested (Beyer, 1992). This would suggest that the consumption of tissue from large *Calanus* spp. remains unrecognised by stomach content analysis (Beyer, 1992). To test this hypothesis, Båmstedt and Karlson, (1998) did some feeding incubations with *M. norvegica* and noticed that indeed ~20% of the copepods caught by *M. norvegica* were only partly eaten (Båmstedt and Karlson, 1998). However, different handling of small and large copepods as described by Beyer (1992) was not confirmed. About 70% of the *Calanus* spp. mandibles in stomachs of *M. norvegica* derived from copepodid stage III and older (prosome length > 0.8 mm), including ~5% adult *Calanus* spp. (prosome length ~1.8 mm) (Båmstedt and Karlson, 1998). Complete ingestion of late stage *Calanus* spp. and *Metridia* spp. has also been observed in other feeding experiments with *M. norvegica* (Torgersen, 2001). Båmstedt and Karlson (1998) concluded that prey damage will not cause a major bias in stomach content analysis as ~80% of the copepods are ingested whole and for the remaining 20%, which are only partly ingested, there is no preference for the front or back.

3.3. Feeding on the sediment surface

In shelf and upper slope areas, euphausiids can have access to the seabed. *M. norvegica* has been caught in epibenthic sledges, their stomachs filled with lithogenic sediment (Mauchline, 1989). The species has also been observed feeding on sediment in the bottom of experimental tanks (Mauchline and Fisher, 1969; their Fig. 64). The authors described the feeding behaviour as follows: *M. norvegica* approach the sediment at an angle of ∼ 45° and material is swept into suspension by the action of their pleopods. The animals then move backwards up off the sediment and filter the resuspended material. Alternatively, they approach the bottom at a more acute angle and plough up the surface layers with their antennae. Accumulated material is transferred to the filtering basket by lateral flailing of the thoracic legs (Mauchline and Fisher, 1969). This matches the description of benthic feeding in Antarctic krill, *E. superba*, observed *in situ* between 200 and 3500 m depth (Clarke and Tyler, 2008).

4. FOOD AND FORAGING IN THE ENVIRONMENT

4.1. Regional and seasonal specifics

M. norvegica has a wide distribution across the North Atlantic, from the coasts of the USA and Canada, over the shelf south of Greenland and Iceland, to the British Isles and the Norwegian coast. The Greenland Sea and the Barents Sea are the northern limits, while in the south, *M. norvegica* is found as far as the Mediterranean Sea and Canary Islands (Mauchline and Fisher, 1969). Within this geographical range, *M. norvegica* is exposed to a variety of trophic conditions. In fjords and estuaries, there is usually plenty of phytoplankton from spring to autumn, while in the open ocean conditions are less favourable (Fig. 5.4). Between the nutrient-rich St. Lawrence Estuary and the nutrient-depleted Ligurian Sea, there is an almost 30-fold difference in chlorophyll *a* concentration during the summer. The onset of the spring bloom varies from February/March in the Kattegat to May/June in the most northern habitats. Some areas have only one phytoplankton peak during spring (e.g. Barents Sea and Ligurian Sea), while in other areas there is a second distinct bloom in autumn (e.g. Korsfjorden and Norwegian Shelf) or a continuum of elevated food concentrations from spring to autumn (e.g. St. Lawrence Estuary, Oslofjorden, Clyde Sea). The zooplankton abundance and seasonality is usually closely linked to the phytoplankton availability (Boysen and Buchholz, 1984; Mayzaud *et al.*, 1999), except in high latitudes where copepod species can survive long periods of food shortage due to their lipid stores.

The various populations of *M. norvegica* reflect the regional differences in their trophic environment in a number of parameters, for example, body morphometry, amount of lipid, metabolic properties, the onset and length of the spawning season and the number of juvenile cohorts (Cuzin-Roudy *et al.*, 2004; Mayzaud *et al.*, 1999; Saborowski *et al.*, 2002; Siegel, 2000). In the Clyde Sea, for instance, the conditions for *M. norvegica* are designated as 'extremely favourable' (Mauchline, 1960). Krill from this location had summer and winter a stomach fullness of at least 50%, in some cases even

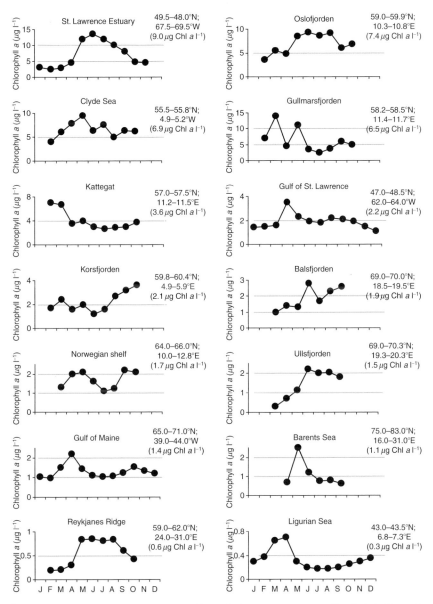

Figure 5.4 Monthly average surface chlorophyll *a* concentrations (SeaWiFS climatology, September 1997–March 2010, http://reason.gsfc.nasa.gov/giovanni/ courtesy of NASA) for various locations inhabited by *M. norvegica*. The coordinates to the right give the area for which the SeaWiFS data have been extracted. The locations are listed in order of decreasing phytoplankton abundance (values in brackets give the average chlorophyll *a* concentration from March to September). Note that ocean colour data can be compromised in near-shore habitats and values might deviate from those of traditional chlorophyll *a* measurements. However, these SeaWiFS data are presented to reflect the general pattern of phytoplankton abundance across the greatly contrasting habitats.

100%, while values were lower in the Kattegat (Lass *et al.*, 2001). As a consequence, Clyde Sea *M. norvegica* are characterised by a greater body size and high amounts of the storage lipid triacylglycerol (Saborowski *et al.*, 2002; Virtue *et al.*, 2000). However, even krill from the Kattegat showed some sign of excess food availability, as their faecal pellets contained ~17% of essential polyunsaturated fatty acids (PUFA). In krill faecal pellets from the Ligurian Sea, only 7% PUFA were present (Virtue *et al.*, 2000).

Another pattern, which has been suggested, is that carnivory in *M. norvegica* increases from southern to northern habitats (Båmstedt and Karlson, 1998; Kaartvedt *et al.*, 2002; Mayzaud *et al.*, 1999). As regional comparisons of the krill stomach content are rare (Lass *et al.*, 2001), this hypothesis was tested using the fatty acid composition of the krill tissue (Table 5.1). Carnivory indicating fatty acids are the 20:1 and 22:1 isomers, which are synthesised *de novo* by calanoid copepods (Falk-Petersen *et al.*, 2000). The relative abundance of these fatty acids shows clear seasonal differences in the Ligurian Sea, the Clyde Sea, Kattegat, Balsfjorden/Ullsfjorden and the Greenland Sea, with higher values outside of phytoplankton blooms. This means that feeding on copepods is most important during the autumn–winter season in temperate and subarctic regions, and during summer-autumn in the Mediterranean Sea. The exceptionally high proportions of 20:1 and 22:1 fatty acids in krill from the Ullsfjorden in March and April might be interpreted as intensive feeding on copepods that had returned from deep water dormancy to surface waters before the spring bloom (Saether *et al.*, 1986). However, the same study also found *M. norvegica* lipid content to have dropped by >50% during the transition from winter to spring (Saether *et al.*, 1986). The 20:1 and 22:1 isomers appeared in only low proportions in the free fatty acid pool, which suggests that other fatty acids were more affected by the ongoing mobilisation of storage lipids. Thus, high proportions of the 20:1 and 22:1 isomers might not exclusively be related to feeding, but also to their selective retention in krill tissue.

As the 20:1 and 22:1 fatty acids and fatty alcohols are not equally abundant in zooplankton communities from different habitats, this indicator is not suitable for a comparison across regions (Virtue *et al.*, 2000). Alternatively, the $18:1(n-9)/18:1(n-7)$ fatty acid ratio can be used as a carnivory index, with $18:1(n-7)$ reflecting dietary input from phytoplankton and $18:1$ $(n-9)$ animal sources (Falk-Petersen *et al.*, 2000). This ratio shows highest values for *M. norvegica* sampled during spring in Ullsfjorden, but gives an almost equally high value for krill from the Ligurian Sea in autumn. Thus, when phytoplankton is rare, heterotrophic food sources seem to play an important role in the diet of *M. norvegica* in northern as well as southern habitats. The $18:1(n-9)/18:1(n-7)$ ratio also indicates that during summer krill fed more carnivorously in the Kattegat than in the Clyde Sea, which confirms results from the stomach content analysis

Table 5.1 Regional and seasonal differences in the relative abundance (%) of trophic-indicator fatty acids in triacylglycerol of *M. norvegica*

Fatty acid	Ligurian Sea		Clyde Sea		Kattegat		Reykjanes R.	Balsfjorden/Ullsfjorden			Greenland S.
	Apr	Sep	Feb	Jun	Mar	Jul	Jun	Mar	Apr	Nov	Jun
16:1(n-7)	2.4	2.8	2.9	8.8	10.4	11.3	6.0	3.6	3.7	4.5	5.0
18:1(n-9)	11.9	12.3	12.1	12.2	11.3	14.2	11.8	8.8	10.2	14.2	10.4
18:1(n-7)	5.3	4.3	5.5	8.3	7.6	6.9	7.0	2.6	3.3	7.4	3.5
18:4(n-3)+18:5(n-3)	4.4	3.5	2.0	3.0	1.7	1.3	1.6	0.8	0.9	1.8	0.8
20:5(n-3)	7.2	8.5	17.9	11.3	8.9	10.1	9.1	2.6	3.2	6.6	9.2
22:6(n-3)	19	20.2	12.3	8.5	7.4	8.4	6.9	4.0	4.3	6.4	10.7
20:1+22:1	2.4	7.3	15.7	9.7	15.8	6.1	15.7	49.6	48.0	18.0	29.7
18:1(n-9)/18:1(n-7)	2.2	2.9	2.2	1.5	1.5	2.1	1.7	3.4	3.1	1.9	3.0
Phytoplankton bloom	B	C	A	B	AB	B	BC	A	A	B	BC
Reference	1	1	1	1	1	1	2	3	3	4	5

Note: A – pre-bloom, B – bloom, C – post-bloom.
The fatty acids indicate diatoms [16:1(n-7); 20:5(n-3)], flagellates [18:4(n-3); 18:5(n-3); 22:6(n-3)], heterotrophs in general [18:1(n-9)], phytoplankton in general [18:1(n-7)], and wax ester-rich copepods [20:1 and 22:1 isomers]. The locations are listed in order of increasing latitude.
References: 1 –Virtue *et al.* (2000); 2 – Petursdottir *et al.* (2008); 3 – Saether *et al.* (1986); 4 –Sargent and Falk-Petersen (1981); 5 – Kreibich *et al.* (2010).

(number of copepod mandibles and proportion of lipid trophic marker; Lass et al., 2001). However, in March, when the spring bloom had started, the diatom indicator $16:1(n-7)$ and its elongation product $18:1(n-7)$ were prominent in krill tissue and stomach content from the Kattegat. M. norvegica from the Reykjanes Ridge seems to have fed on both diatoms and copepods at the time of sampling, while diatom markers were scarce in krill from the Ligurian Sea. High proportions of the fatty acids $18:4(n-3)$, $18:5(n-3)$ and $22:6(n-3)$ suggest that flagellates rather than diatoms were the main autotrophic food for M. norvegica in the Ligurian Sea (Virtue et al., 2000). However, there might be interannual differences, as an earlier study recorded increased amounts of diatom markers in krill tissue in association with the spring bloom in the Ligurian Sea (Mayzaud et al., 1999).

In summary, it appears that M. norvegica can switch between herbivory and carnivory quite opportunistically depending on what food is available. Feeding plasticity and a broad food spectrum allow the species to inhabit a large range of trophic environments and to remain active throughout the year. Even though food uptake is most intensive during the months of high phytoplankton abundance, the species keeps feeding in winter (Fisher and Goldie, 1959; Kaartvedt et al., 2002; Lass et al., 2001; Mauchline and Fisher, 1969; Onsrud and Kaartvedt, 1998; Sameoto, 1980).

4.2. Vertical migration and diel periodicity in feeding

M. norvegica performs diel vertical migrations with residence at depth during daytime and ascent into upper water after sunset. This behaviour is believed to reflect the trade-off between gaining sufficient nutrition from the food-rich surface layer, whilst minimising predation risk through avoiding habitats with high light levels (Tarling et al., 2000). During a phytoplankton bloom in the St. Lawrence Estuary, the chlorophyll a content in stomachs of M. norvegica increased from 10 to 100 and occasionally even to >1000 ng per individual during the night and dropped back to the initial value over several hours after dawn descent (Sourisseau et al., 2008). Similar observations in earlier studies have led to the conclusion that M. norvegica feed mainly herbivorously when migrating to shallow depths at night and carnivorously during the day (Mauchline, 1960; Onsrud and Kaartvedt, 1998; Sameoto, 1980; Simmard et al., 1986).

However, the upper water column is not only home to the most actively growing phytoplankton, but also to protozoa and small copepod species, which dwell among the algae. Therefore, distinction between herbivorous and carnivorous feeding cannot be based on fluorescence estimates alone. Lass et al. (2001) counted and identified copepod mandibles in the stomachs of M. norvegica from the Clyde Sea and Kattegat and found the genera Temora and Pseudocalanus to be an important part of the nocturnal diet.

Even krill sampled during daytime in deep water contained mainly mandibles from copepods dominant in the upper water column and not from the deep-living *Calanus* spp. (Lass *et al.*, 2001). The authors therefore concluded that feeding activity ceases during the day and mandibles found in the stomach are retained from the night before (Lass *et al.*, 2001). According to this result, the apparent switch from night-time herbivory to daytime carnivory might just reflect a faster digestion of algae pigments versus copepod mandibles in krill stomachs after simultaneous ingestion during the night.

The theory of Lass *et al.* (2001) that significant food uptake by *M. norvegica* is restricted to surface water during the night, was examined on krill from the Gullmarsfjord/Sweden (Schmidt, unpublished data). A range of autotrophic and heterotrophic items was enumerated from stomachs of krill sampled in autumn. Similar to the results of Lass *et al.* (2001), mandibles of copepod species from the upper water column (e.g. *Oithona* spp., *Acartia* spp. and *Paracalanus* spp.) were far more frequent in the stomachs than those from deep living *Calanus* spp. and *Metridia* spp., during both midnight and midday (Fig. 5.5A). In addition, almost all food items were found in lower numbers during the day (Fig. 5.5B). However, when looking at the exact day:night ratio, items could be divided into three groups (Fig. 5.5C): Those which were significantly lower during the day by about two-thirds, those which were insignificantly lower during the day by about one-third and those which were higher during the day. The first group contains items most likely ingested in the upper water column or shallow parts of the fjord (e.g. tintinnids, pollen, lithogenic particles) and therefore sets the background value for what had been retained in the stomachs from the previous night. Against this background value, items in the second group (filamentous algae, the dinoflagellate *Ceratium* spp. and younger stages of *Calanus* spp.) were partly retained and partly eaten during the day, while items in the last group (Cnidarian nematocysts and late stages of *Calanus* spp. and *Metridia* spp.) were mainly eaten during the day. Even when krill ingest only one large *Calanus* spp. or *Metridia longa* at depth during the day, in terms of prey volume this can match $\sim 50\%$ of the nocturnal intake via small copepods (Fig. 5.5D). Similar daytime feeding on *Calanus* spp. has been observed in several studies (Buchholz *et al.*, 1995; Kaartvedt *et al.*, 2002; Lass *et al.*, 2001; Sameoto, 1980).

The most comprehensive stomach content analysis was done on *M. norvegiva* from the Oslofjord/Norway (Kaartvedt *et al.*, 2002). The authors converted both copepod mandibles and gut fluorescence into carbon units and assessed the seasonal and diurnal importance of copepod versus algal food (Fig. 5.6). In spring, most carbon intake came from feeding on sedimenting phytoplankton in deep water during the day. On this occasion, mean stomach chlorophyll *a* content exceeded 1000 ng ind^{-1} even 11 h after sunrise. In summer, food uptake was low both during night and day. In autumn, nocturnal feeding on *Temora* spp. supplied most of the carbon, while in winter *M. norvegica* gained sufficient nutrition from dormant *Calanus* spp. caught at 120 m during the day (Kaartvedt *et al.*, 2002).

In conclusion, several studies suggested that *M. norvegica* has a higher feeding activity during the night when it has migrated into the upper water column (Fig. 5.5B, Båmstedt and Karlson (1998), Buchholz *et al.* (1995), Lass *et al.* (2001)). However, the results of Onsrud and Kaartvedt (1998) and Kaartvedt *et al.* (2002) show that this pattern does not necessarily hold for all seasons and locations. If the preferred food source is available in deep water (e.g. settling phytoplankton or overwintering copepods), then the main feeding time will be during the day and therefore unrelated to the upward migration. The species opportunistic behaviour towards food sources and feeding habitat can also cause a flexible time of feeding.

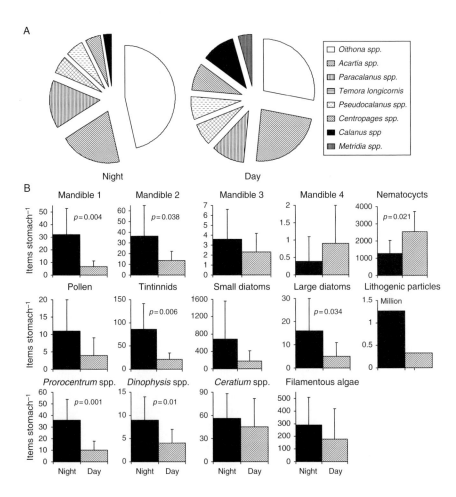

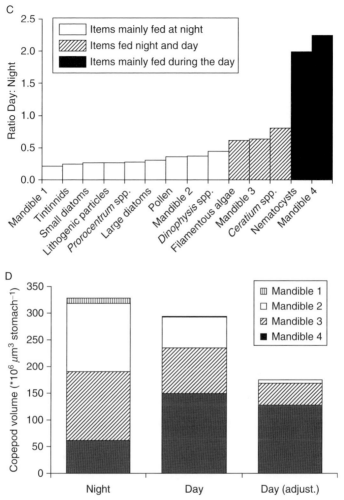

Figure 5.5 Night–day differences in the stomach content of *M. norvegica* from Gull-marsfjorden, Sweden, September 2003. (A) numerical importance of various copepod species; (B) absolute amount of food items; (C) day–night ratio of food items in the stomach; (D) volume-specific importance of different size copepods. *Note:* Ten adult krill have been analysed for each, 'Night' (00:00) and 'Day' (12:00), and significant differences are indicated (two-sample *t*-test). Lithogenic particles were analysed from a pooled sample of 10 krill. Mandible 1: blade width 10–20 μm (mainly *Oithona* spp.); Mandible 2: 25–65 μm (mainly small calanoid copepod species); Mandible 3: 70–125 μm (early copepodids of *Calanus* spp. and *Metridia longa*); Mandible 4: 130–175 μm (late copepodids/adults of *Calanus* spp. and *Metridia longa*). (C) Most of the food items typical for the upper water column had a day–night ratio of ∼0.3 (white columns), suggesting that about 30% of the food is retained until the next day. (D) The daytime stomach content on copepods has been corrected for potentially retained items, 'Day (adjust.)', by subtracting 30% of the night values ('Night') from the observed day values ('Day'). The copepod volume was estimated from the mandible width via a transformation into prosome length (Karlson & Båmstedt, 1994; Mauchline, 1998).

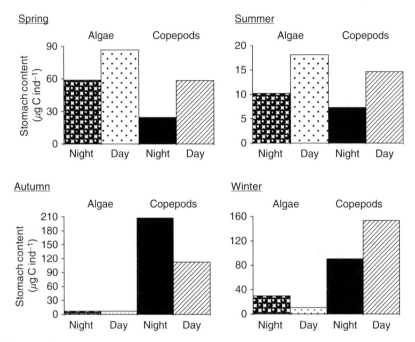

Figure 5.6   Night–day and seasonal differences in the abundance of copepods and algae in the stomach content of *M. norvegica* from Oslofjorden, Norway. After Kaartvedt *et al.* (2002, their Fig. 8, Stn S).

4.3. Quantifying the role of copepods as a food source

Based on the morphology of their mouthparts, their biochemical composition and stomach content, a number of studies have suggested that *M. norvegica* is primarily a carnivorous species (Mauchline, 1980; Falk-Petersen *et al.*, 2000; Kaartvedt *et al.*, 2002). However, proper quantitative estimates are rare. Karlson and Båmstedt (1994) presented a method for the quantitative assessment of planktivorous predation on copepods. This method is based on the following observations: (1) the size and shape of copepod mandibles is species-specific; (2) there is a linear relationship between mandible width and prosome length of the copepod; (3) a wide-ranging, interspecific relationship between copepod prosome length and dry weight can be defined. Thus, by counting and measuring copepod mandibles in euphausiid stomachs, the copepod prey mass can be estimated and related to the total food mass in the stomach (Karlson and Båmstedt, 1994). Important assumptions of this method such as complete ingestion of most copepods and no accumulation of mandibles in the stomach have been demonstrated in the laboratory (Båmstedt and Karlson, 1998). Applying this method to *M. norvegica* from the northwest coast of Norway and the

Skagerrak, Båmstedt and Karlson (1998) found that predation on copepods was substantial, but not dominant. In spring and summer, copepods accounted on average for 23% (range 3–44%) of the total stomach content (Båmstedt and Karlson, 1998). This 'mass'-based approach might underestimate the role of copepods if non-digestible items such as lithogenic particles are abundant in the stomach.

Kaartvedt *et al.* (2002) slightly modified the approach of Båmstedt and Karlson (1998) by using 'carbon' instead of 'mass' as a unit and relating copepod prey to ingested algae, not to the total stomach content. These data for *M. norvegica* from the Oslofjord (Kaartvedt *et al.*, 2002; their Fig. 8) indicate a higher degree of carnivory than the values from Båmstedt and Karlson (1998) and show clear seasonal differences. The role of copepods in the diet varied from 29 to 49% in spring and summer, was highest in autumn (> 90%) and slightly less in winter (75–93%). However, the role of copepods might have been overestimated if other non-fluorescent food items, for example, protozoa, were also common in the diet. Another modified procedure is to measure mandibles and hard-structured algae and heterotrophs under the microscope and estimate their respective volumes (Schmidt *et al.*, 2006). The copepod volume was calculated from mandible width via a transformation to prosome length (Karlson and Båmstedt, 1994; Mauchline, 1998). For *M. norvegica* sampled in the Gullmarsfjord in autumn, copepods accounted on average for 81%, algae for 11%, protozoan for 6% and other items (e.g. cnidarians nematocysts, pollen) for 2% of the total identified food volume (Schmidt, unpublished data).

Overall, the results suggest that copepods are the most common food source for *M. norvegica* in autumn and winter, but are equally or less important than phytoplankton in spring and summer (Table 5.2). There are two different explanations for this pattern, which might each be valid to some extent. First, the degree of carnivory in *M. norvegica* might be inversely related to the phytoplankton abundance, as has been reported for other euphausiids (Stuart and Pillar, 1990). The transition from winter to spring supports this theory. The nocturnal pigment content in krill stomachs rises from 100 to over 1000 ng per individual in concert with the increase of chlorophyll *a* in the water column (Onsrud and Kaartvedt, 1998). However, in autumn, *M. norvegica* fed mainly on copepods even when the phytoplankton abundance was high (Kaartvedt *et al.*, 2002).

A second possible explanation for a seasonal change in diet relates to the reproductive cycle. Cuzin-Roudy *et al.* (2004) observed that, with the onset of autumn, the reproductive activity of *M. norvegica* stopped and ovaries regressed even though phytoplankton concentrations were often higher than found in spring and summer, when spawning and recruitment were active. This suggests that there might be a seasonal shift in food preference for *M. norvegica* concomitant with physiological changes from growth and reproduction in spring and summer to accumulation of body reserves

Table 5.2 Seasonal differences in the contribution (%) of copepods and other food sources to the diet of *M. norvegica*

Season	Location	Copepods	Algae	Protozoans	Others	Reference
Spring	Norwegian Shelf	30				Båmstedt and Karlson (1998); their Fig. 10
Spring	Skagerrak	12				Båmstedt and Karlson (1998); their Fig. 10
Spring	Oslofjorden	40	60			Kaartvedt *et al.* (2002); their Fig. 8
Summer	Norwegian Shelf	15				Båmstedt and Karlson (1998); their Fig. 10
Summer	Oslofjorden	43	57			Kaartvedt *et al.* (2002); their Fig. 8
Autumn	Oslofjorden	97	3			Kaartvedt *et al.* (2002); their Fig. 8
Autumn	Gullmarsfjord	81	11	6	2	Schmidt, unpublished data (*n* = 20)
Winter	Oslofjorden	87	13			Kaartvedt *et al.* (2002); their Fig. 8

Values are given as percentage of (a) the total weight of the stomach content (Båmstedt and Karlson, 1998), (b) the copepod and algae-derived carbon equivalents in the stomach content (Kaartvedt *et al.*, 2002) or (c) the total volume of identifiable food in the stomach (Schmidt *et al.*, 2006).

during autumn. The first process demands high food quality and involves mainly the build up of polar lipids while, during the second process, triacylglycerols are accumulated (Cuzin-Roudy et al., 1999; Hirano et al., 2003). Laboratory experiments could test whether phytoplankton is a better source of polar lipids for M. norvegica and therefore more essential during spring and summer, while copepods are the ideal food for winter preparation. Support for this theory comes from the fact that triacylglycerols in M. norvegica usually contain high amounts of 20:1 and 22:1 fatty acids typical for calanoid copepods (Sargent and Falk-Petersen, 1981).

Finally, it should be noted that the overwhelming biomass of copepods in the stomach of M. norvegica seen in autumn does not exclude a substantial feeding pressure on phyto- or proto-zooplankton. In the Gullmarsfjord, for instance, 50 tintinnids and 10 thecate heterotrophic dinoflagellates were found on average in each M. norvegica, but this represented only ~6% of the total identified volume in the stomachs (Schmidt, unpublished data). The maximum numbers of 192 tintinnids and 20 Dinophysis spp. per stomach are similar to what has previously been found in the larger, more filter feeding Antarctic krill, E. superba (Schmidt et al., 2006).

4.4. Preferences among copepod species

During stomach content analyses on M. norvegica, copepod mandibles are often identified to species level (Båmstedt and Karlson, 1998; Dalpadado et al., 2008; Kaartvedt et al., 2002; Lass et al., 2001), which can provide insights into the species-specific predation pressure. Most studies found either Calanus spp. or Temora longicornis to be the major prey (Table 5.3). M. norvegica might benefit from the strong hydrodynamic signal of Calanus spp. or the particularly low escape capacity of T. longicornis (Viitasalo et al., 1998). However, it seems that most crucial is a spatial overlap between these copepod species and M. norvegica. Sameoto (1980) found high numbers of copepod remain in the stomachs of M. norvegica during the day and pointed out that encounter probability with Calanus finmarchicus is much higher at daytime, when both species concentrate at similar depth compared to the night, when copepods are spread over the water column. Likewise, M. norvegica could only feed on overwintering Calanus spp. at a shallow site, where the copepods occurred within their daytime habitat (Kaartvedt et al., 2002). At the deeper site (> 120 m), visual senses of M. norvegica seem to become insufficient and dormant copepods are safe (Kaartvedt et al., 2002). The smaller copepods T. longicornis and Pseudocalanus spp./Paracalanus spp. were the most common prey for M. norvegica in the Kattegat, Clyde Sea and Oslofjord, whenever they dominated the copepod assemblage of the upper water column (Kaartvedt et al., 2002; Lass et al., 2001).

Less consistent are the results with cyclopoid copepods. Even though abundant in the water column, this group was under-represented in the krill

Table 5.3 Most common copepod prey of *M. norvegica* based on mandible identification in the stomach content

Location	Season	Dominant copepod genus in water column	Dominant copepod genus in krill stomach	Explanation for preferred copepod genus	Ref.
Norwegian Shelf	Spring/Sum.	*Calanus* [1]	*Calanus*	*Calanus* spp.: Large prey → easy to detect	1
Skagerrak	Spring	*Calanus* [1]	*Calanus*		
Clyde Sea	Summer	*Temora* [2]	*Temora*	*Pseudocalanus* spp.: Slow swimmer → easy to catch	2
Clyde Sea	Winter	*Pseudocalanus* [2]	*Temora, Pseudocalanus*		
Kattegat	Summer	*Temora* [2]	*Pseudocalanus*		
Kattegat	Winter	*Pseudocalanus* [2]	*Temora, Pseudocalanus*		
Oslofjorden	Spring	Cyclopoid [1]	*Calanus*	*Temora longicornis*: Pigmentation and low escape capacity → easy to detect and catch	3
Oslofjorden	Summer	Cyclopoid [1]	*Calanus*		
Oslofjorden	Autumn	*Temora* [1]	*Temora*		
Oslofjorden	Winter	Cyclopoid [1]	*Calanus*, Cyclopoid		
Barents Sea	Summer	*Calanus* [1]	*Calanus*		4
Gullmarsfjorden	Autumn	Cyclopoid [2]	*Oithona*	*Oithona* spp.: Small prey → easy to catch	5

Note: [1] Integrated sample from the whole water column, [2] depth-specific samples, species only dominant in upper water column.
References: 1 – Båmstedt and Karlson, 1998; 2 – Lass et al., 2001; 3 – Kaartvedt et al., 2002; 4 – Dalpadado et al., 2008; 5 – Schmidt, unpublished data.

diet from the Oslofjord (Kaartvedt et al., 2002), but was the dominant prey in autumn 2003 in the Gullmarsfjord (Schmidt, unpublished data). In the latter study, the appearance of Oithona spp. mandibles in the stomach coincided with a high content of the dinoflagellate Prorocentrum spp. ($R^2 = 0.42$; $p < 0.01$) and tintinnids ($R^2 = 0.2$; $p < 0.05$), while no such relationships were found for mandibles from larger copepod species. This suggests that M. norvegica ingested Oithona spp. while feeding on small plankton. High uptake of Oithona spp. was also observed by the primarily filter feeding Euphausia lucens and E. hanseni and explained by the low ability of these small copepods to withstand the negative pressure caused by the feeding beats (Barange et al., 1991; Gibbons et al., 1991). Our second observation was that most of the Oithona spp. mandibles derived from early copepodid stages, as only two or three of the mandible teeth were well developed. These stages usually aggregate in the top ~ 10 m of the water column (Titelman and Fiksen, 2004). Thus, if the krill ascent terminates at about 20 m depth, as found in the study of Kaartvedt et al. (2002), then feeding on these young stages is unlikely. In conclusion, predation risk is highest for Oithona spp. if krill migrate right to the surface and perform filter feeding.

Petursdottir et al. (2008) looked at the role of C. finmarchicus as a food source for various crustacean and fish species in the pelagic ecosystem over the Reykjanes Ridge. The authors quantified the amount of Calanus-derived fatty acids, $20:1(n − 9)$ and $22:1(n − 11)$, in the lipids of five potential predators including M. norvegica. Even though stable isotope data suggested that M. norvegica had a mainly carnivorous diet, the amounts of $20:1(n − 9)$ and $22:1(n − 11)$ were lower than in the other four species (Petursdottir et al., 2008). The authors concluded that Calanus spp. is rather unimportant in the diet of M. norvegica (Petursdottir et al., 2008). However, this cross-species comparison is based on two assumptions, which might not have been fulfilled in this study. The first is that all of the studied species have the same ability to digest and incorporate high-energy lipids of cope-pods such as the 20:1 and 22:1 isomers. The second assumption is that all species are in the same state of lipid metabolism, where storage lipids become accumulated and reflect the recent diet. For both assumptions, contradicting evidence has been found in other studies (Saether et al., 1986; Stübing et al., 2003). Therefore, it would be beneficial to estimate the maximum incorporation of the 20:1 and 22:1 isomers in M. norvegica, for example, by offering a pure diet of C. finmarchicus in the laboratory. Similar experiments have been conducted with Antarctic krill, E. superba (Stübing et al., 2003). The experiment might reveal whether, under such feeding conditions, M. norvegica can attain 20:1 and 22:1 proportions as high as seen for the mesopelagic fish Benthosema glaciale ($\sim$ 40% of total fatty acids; Petursdottir et al., 2008). It could also be tested if starvation might lead to increased proportions of the 20:1 and 22:1 isomers in storage lipids due to selective catabolism of other fatty acids. In field studies such as that of Petursdottir *et al.* (2008), fatty acid data should ideally be supported by visual or genetic stomach content analysis.

4.5. Males, females and young stages

In both *M. norvegica* and *E. superba*, there is little difference in body size between males and females of the same age. Even though reproduction is costly for females, their growth seems not to be compromised (Atkinson *et al.*, 2006; Buchholz *et al.*, 2006). The simple conclusion is that females have to feed more than males to provide for both energy-rich oocytes and their somatic growth. In an energy budget for *E. superba*, Clarke and Morris (1983) estimated that the female energy intake should exceed that of males by 50% during the summer months. Likewise, Tarling (2003) suggested that a 40% higher energy intake in female *M. norvegica* would be sufficient to cover reproduction and somatic growth. Field observations to back up this theory are rare. During one sampling event in March, Fevolden (1982) found that 65% of female *M. norvegica*, but only 35% of the males had food in their stomachs. In contrast, Priddle *et al.* (1990) did not observe any differences in gut fullness and feeding capacity of mature male and female *E. superba* when analysing the gut fluorescence in over 3000 individuals from 38 swarms. Stable isotope ratios and visual stomach content analysis suggest that male and female *E. superba* are very similar in their trophic level and diet (Schmidt *et al.*, 2004a; Schmidt *et al.*, 2006).

A potential mechanism for a higher food uptake in female *M. norvegica* lies in the segregation of males and females within the water column (Buchholz *et al.*, 2006; Tarling, 2003). A shallower nocturnal ascent of females has been explained by their effort to gain maximal food intake in the rich upper layer, even when risking higher morality from visual predation (Tarling, 2003). This hypothesis was examined by comparing the stomach content of male and female *M. norvegica* from the Gullmarsfjord in September 2003 (Schmidt, unpublished data). While most food items showed similar numbers in males and females, the amount of mandibles was clearly higher in females, especially at night (97 $\pm$ 52 SD in females, 48 $\pm$ 20 SD in males). This means that on average females ingested twice as many copepods than males, but due to high individual variability and low sample size, the difference was not significant (two-sample t-test: $p = 0.109$, $n = 5$). For both genders, the majority of the mandibles were of small size (Fig. 5.7A), deriving from *Oithona* spp., *Acartia* spp., *T. longicornis* and other neritic species. However, the proportion of larger mandibles from *Calanus* spp. and *Metridia* spp. was somewhat higher in males than females (night: 5% in females, 10% in males; day: 9% in females, 15% in males). This agrees with observations from the Norwegian coast and Skagerrak, where *Calanus* copepodids were the main prey for male and female

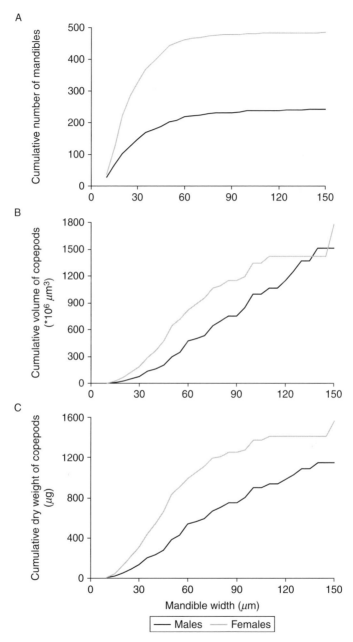

Figure 5.7 Male–female differences in nocturnal feeding on copepods, *M. norvegica* from Gullmarsfjorden/Sweden, September 2003. Five males and five females have been analysed and results are cumulatively presented for increasing mandible size. (A) Number of mandibles; (B) Volume of copepods; (C) Dry weight of copepods. Copepod volume and dry weight were calculated from the mandible width via a transformation into prosome length (Karlson & Båmstedt, 1994; Mauchline, 1998).

M. norvegica, but females contain a higher proportion of smaller copepods such as *Acartia* spp. and *T. longicornis* in their stomachs (Båmstedt and Karlson, 1998).

Calculating the volume of the ingested copepods from mandible width, it becomes obvious that even though females contained far more mandibles in their stomachs, males widely compensated by feeding on a few larger copepods (Fig. 5.7B). A 100% difference in number of mandibles reduces to a 17% difference in volume of copepods. Overall, there was no clear indication that females fed more than males. The total volume of identifiable food items in the stomachs of females and males was very similar (females: $425 \pm 263 \times 10^6 \ \mu m^3$; males: $377 \pm 182 \times 10^6 \ \mu m^3$). Females contained on average slightly more copepods and protozoans, while males contained more algae. However, neither of these differences was significant due to the high individual variability.

Another problem is the conversion from number of items into biomass. Volume, dry weight or carbon content can be used as a biomass indicator. Using the volume is most straightforward, as for many items the dimensions can directly be measured under the microscope and the volume calculated. Exceptions are copepods, where the hardened mandibles are often the only recognisable remains. Therefore, the volume or dry weight of the copepod is estimated from the mandible width using established equations (Karlson and Båmstedt, 1994; Mauchline, 1998). However, such relationships depend on the shape or lipid content of the copepods and can vary with species, region and season. As seen from Fig. 5.7C, the dry weight calculation after Karlson and Båmstedt (1994) results in a more gentle increase for large mandible sizes than the volume calculation after Mauchline (1998), leaving a final $\sim 35\%$ difference between females and males. To avoid uncertainty about which equation to use when looking for subtle differences as those between males and females, the copepod weight or volume relationship with mandible size should directly be estimated at the time of sampling. In conclusion, a significant male–female difference in energy uptake is still unproven for *M. norvegica*. Tracking such a potential difference would require the examination of many specimens and special attention to the number versus energy content of ingested copepod prey. Alternatively, the energy budgets of the two genders have to be reinvestigated to explain why males might have a similar nutritional demand than females. Records on low lipid reserves and elevated mortality of *E. superba* males point to increased energetic expenses during the breeding season (Kawaguchi *et al.*, 2007; Pond *et al.*, 1995; Virtue *et al.*, 1996).

The natural diet of larval and juvenile *M. norvegica* has received very little attention. It has only been studied in the Clyde Sea/Scotland (Fisher and Goldie, 1959; Mauchline, 1960). The authors summarised their observations as follows: Nauplius and metanauplius of *M. norvegica* probably do not feed, as the mouthparts are not functional yet. Calyptopis and early furciliae can filter-feed on phytoplankton and suspended material. Later furciliae additionally feed on copepods. Overall, larval stages seem to be mainly herbivorous, while later stages become more omnivorous and their diet basically resembles that of adults (Mauchline and Fisher, 1969). Morphological studies have shown that the feeding basket of young *M. norvegica* has a proportionally larger filtering area than that of adults, which suggests that efficiency for filter feeding drops with age in this species (McClatchie, 1985). Differences in the diet between juvenile and adult krill might also result from a vertical segregation in the water column. Older stages occupy gradually deeper daytime levels and spent less time in the surface layer during the diurnal vertical migration (Mauchline, 1960).

Three important differences were observed when comparing the stomach content of juvenile (~14 mm body length) and adult *M. norvegica* from the Gullmarsfjord (Schmidt, unpublished data). First, copepods contributed on average much less to the diet of juveniles than adults (21% *vs* 82% of the identifiable stomach content, Fig. 5.8), supporting the notion that predation in *M. norvegica* increases with body size (Båmstedt and Karlson, 1998; Fisher and Goldie, 1959; Mauchline, 1960; McClatchie, 1985). The largest copepod mandibles found in a juvenile *M. norvegica* were 80 μm, in which case the copepod accounted for ~85% of the stomach content. However, most

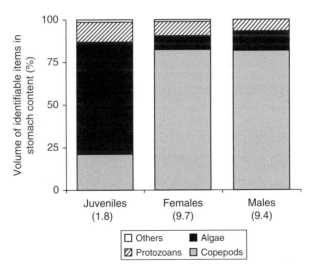

Figure 5.8 Contribution of different food sources to the total volume of identifiable items in the stomach content for juvenile (33 ± 5 SD mg wet weight, $n = 10$), adult female (441 ± 71 SD mg wet weight, $n = 5$) and male *M. norvegica* (408 ± 14 SD mg wet weight, $n = 5$) from Gullmarsfjorden/ Sweden, September 2003. Values for the weight-specific total volume ($\times 10^5$ μm^3 mg^{-1} wet weight) are given in brackets underneath the plot.

juveniles contained either up to four small mandibles (<20 μm) or none at all. Second, the dinoflagellate *Ceratium* spp. was the main autotrophic food for juveniles (98%), but only of moderate importance in adults ($\sim$34%). Instead, adults were feeding on filamentous algae. Mauchline (1960) also found that, with increased body size, *M. norvegica* feed less on dinoflagellates but more on filamentous algae. Finally, the total volume of identifiable food in the stomachs was about five times higher in adults than juveniles, even when normalising for differences in body weight (Fig. 5.8). This could be caused by a faster gut passage in juveniles after being caught. Alternatively, it might indicate that the feeding conditions were not optimal for juveniles. Copepods and filamentous algae comprised almost 90% of the stomach content in adults, but these food sources seem not to be equally accessible or preferred by juveniles.

4.6. Feeding at the seabed

Several studies have indicated that *M. norvegica* interacts with the seabed and that feeding happens at the benthos or sediment-water interface (Table 5.4). Most of these observations are from relatively shallow locations near land, but there are also a few recordings from deeper water. Some of the studies are unambiguous, for example, when krill are directly seen alive near the seabed (Hudson and Wigham, 2003; Jonsson *et al.*, 2001; Rosenberg *et al.*, 2005; Youngbluth *et al.*, 1989). In the Gulf of Maine and along the southern edge of Georges Bank, Youngbluth *et al.* (1989) encountered epibenthic aggregations of *M. norvegica* with a manned submersible. The species was seen 5–12 m above the seabed on 65 of the 72 dives. Individuals in this benthic boundary layer had usually full stomachs and appeared to be foraging when observed, possibly on the flocculent 'fluff' layer at the sediment-water interface (Youngbluth *et al.*, 1989). The authors concluded that *M. norvegica* plays a significant role in the exchange of nutrients and material between benthic and pelagic food webs due to their daily vertical migrations, enormous densities and widespread distribution in nearshore waters (Youngbluth *et al.*, 1989).

Remotely operated vehicles repeatedly recorded *M. norvegica* swimming among the epibenthos (Hudson and Wigham, 2003; Jonsson *et al.*, 2001; Rosenberg *et al.*, 2005). The species is found in the stomach content of benthic predators such as basket stars (Emson *et al.*, 1991) and has been caught in significant numbers with a hyperbenthic sledge (>300 individuals 100 m^{-3} of water filtered; Beyer, 1992). Acoustic studies in fjords and estuaries also revealed aggregations of krill in the vicinity of the seabed (Cotté and Simard, 2005; Falk-Petersen and Hopkins, 1981; Lavoie *et al.*, 2000; Liljebladh and Thomasson, 2001; Onsrud *et al.*, 2004; Simmard *et al.*, 1986). Information from the 120 kHz echograms indicated that at about noon, the krill Sound Scattering Layers came into direct contact with the

Table 5.4 *Meganyctiphanes norvegica.* Compilation of evidence for their aggregations near the seabed and feeding on benthic food sources

Method	Observations	Location	Depth (m)	Season	Ref.
Manned submersible	– Epibenthic stocks of krill, feeding on particles in epibenthic layer and sediment-water interface	Gulf of Maine Georges Bank	200–800	Summer, Autumn	1
Remotely operated vehicle	– Swimming near anthozoan, *Bolocera tuediae*	Kosterfjorden	> 100		2
	– Swimming directly over sea-bed, caught by squat lobster *Munida sarsi*	Shetland coast	400	Summer	3
	– Swimming near basket star *Gorgonocephalus caputmedusae*	Oslofjorden	~ 100	Winter	4
Stomach content of benthic predators	– Caught by basket stars *Gorgonocephalus arcticus*	Bay of Fundy	14	Summer to Winter	5
Hyperbenthic sledge	– Caught along the seabed, (96 and 333 ind. per 100 m^3 of filtered water)	Olsofjorden	80–100	Spring, Autumn	6
Video Camera	– High numbers of kill a few meters above the seabed	Olsofjorden	~ 100	Autumn	6
Acoustics	– Krill SSLs in direct contact with the seabed	Øksfjorden	240	Winter	7
	– High krill concentrations a few meters above seabed	St. Lawrence Estuary	100 – 150	Summer	8
	– During tidal upwelling, krill get advected onto shallow areas and stay close to the bottom	St. Lawrence Estuary	50 – 150	Summer	9
					10

(*continued*)

Table 5.4 (continued)

Method	Observations	Location	Depth (m)	Season	Ref.
Stomach content of *M. norvegica*	- Inorganic detritus and parts of red, green and brown algae in stomachs of *M. norvegica*	Clyde Sea	170	Year-round	11
	- Filamentous algae and parts of polychaetes often associated with mud particles in stomach	Clyde Sea	170	Summer, Winter	12
	- Mud in 12 % of *M. norvegica* and 69% of *Thysanoessa raschii* stomachs	St. Lawrence Gulf	180	Winter	13
	- Filamentous algae common in *M. norvegica*, but absent in *T. raschii* & *T. inermis* stomachs	St. Lawrence Gulf	180	Year-round	14
	- Filamentous algae common year-round in *M. norvegica*	Oslofjorden	~100	Year-round	15
	- Detritus with benthic elements in stomach	Olsofjorden	100–160	Summer	6
	- Filamentous algae account for ~8 % of the identifiable stomach content	Gullmarsfjorden	100	Autumn	16

References: 1- Youngbluth *et al.*, 1998, 2- Jonsson *et al.* (2001), 3- Hudson and Wigham (2003), 4- Rosenberg *et al.* (2005), 5- Emson *et al.* (1991), 6- Beyer (1992), 7- Falk-Petersen and Hopkins (1981), 8 – Simmard *et al.* (1986), 9-Cotté and Simard (2005), 10 – Lavoie *et al.* (2000), 11- Fisher and Goldie (1959), 12-Mauchline (1960), 13- Berkes (1973), 14- Sameoto (1980), 15- Fevolden (1982), 16 – Schmidt, unpublished.

fjord bottom in Øksfjorden/Norway (Falk-Petersen and Hopkins, 1981). In the Oslofjord, the major daytime aggregation of krill was 10–30 m above the seabed, with demersal fish approaching them from below (Onsrud *et al.*, 2004). The authors suggested that krill was not descending right down to the seabed due to the high risk of predation by a large-eyed fish species associated with the seabed (Onsrud *et al.*, 2004). Thus, daytime distribution of *M. norvegica* might be a trade-off between predator avoidance and searching for food (Onsrud *et al.*, 2004). Alternatively, final migration from the krill layer to the seabed might have been on an individual basis only and of short duration, which is harder to detect with acoustics.

Indications from stomach content analysis of *M. norvegica* are vaguer. If, for instance, mud (= lithogenic particles) or filamentous algae are found in stomach samples (Berkes, 1973; Beyer, 1992; Fevolden, 1982; Fisher and Goldie, 1959; Mauchline, 1960; Sameoto, 1980), they might have been ingested at the seabed or were resuspended in the water column. Comparisons across species or developmental stages might give a clue. Feeding on resuspended material is likely if all species contain benthic material in their stomachs, for instance after a storm. However, it can be assumed that there were interactions with the seabed, if benthic items are only found in one species. In the Gullmarsfjord, for instance, filamentous algae and mud were found in all of the 20 analysed adult krill, but rarely seen in young stages (Schmidt, unpublished data), which suggests that the adult krill were foraging at the seabed. Likewise, filamentous algae are more common in stomachs of *M. norvegica* than in simultaneously sampled *Thysanoessa raschii* or *Thysanoessa inermis* (Fevolden, 1982; Sameoto, 1980), while *T. raschii* seems to feed readily on mud (Berkes, 1973).

In either case, whether obtained from the seabed or water column, these studies show that krill can supplement their diet with food of benthic origin (Fig. 5.9). The importance of benthic feeding for the nutrition of *M. norvegica* has never been quantified, but might be significant. After copepods, filamentous algae of benthic origin were the second-most abundant food item in the stomach content of *M. norvegica* from the Gullmarsfjorden (Schmidt, unpublished data). They accounted on average for 8% (maximum 15%) of the total identified volume, exceeding the contribution of pelagic sources such as tintinnids (5%), dinoflagellates (4.5%) and diatoms (0.3%). The volume of the lithogenic particles was equivalent to about one quarter of the identified food (Schmidt, unpublished data). Thus, further benefit might have come from bacteria, athecate protozoan or detritus, which are often abundant within seabed sediment of productive fjords (Sundbäck *et al.*, 2004; Wulff *et al.*, 2005). Fjordic populations of *M. norvegica* can attain a larger body size than their counterparts in the open ocean, which might at least partly be related to richer feeding conditions (Boysen and Buchholz, 1984; Cuzin-Roudy *et al.*, 2004; Fisher and Goldie, 1959; Mauchline, 1960).

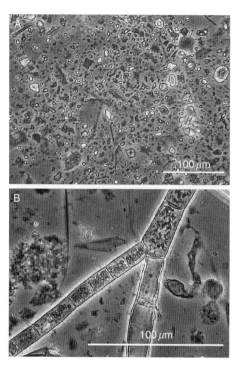

Figure 5.9 Food items of benthic origin as found in the stomach content of *M. norvegica*. (A) Lithogenic sediment, 3–15 μm ('mud'). These particles resist combustion (550°C, 12 h) and alkaline digestion (1.5 M NaOH, 100°C, 30 min), which indicates that they are neither detritus nor frustules of small diatoms. (B) Brown filamentous algae (identified by K. Eriksson, personal communication).

4.7. Comparison with other euphausiid species

In the northern part of their distribution, *M. norvegica* often co-occurs with three other euphausiid species of the genus *Thysanoessa*: *T. raschii*, *T. inermis* and *T. longicaudata*. These *Thysanoessa* species all have a boreal to low Arctic distribution, with *T. raschii* preferring shallow coastal and fjord areas, *T. inermis* being mainly a shelf species and *T. longicaudata* an oceanic species (Dalpadado and Skjoldal, 1996; Siegel, 2000). Comparative studies on the degree of stomach fullness of freshly caught specimens have suggested that *M. norvegica* has a higher and more constant feeding activity than the *Thysanoessa* species (Berkes, 1973; Fevolden, 1982; Mauchline and Fisher, 1969; Sameoto, 1980). In contrast to the other species, *M. norvegica* feeds more frequently at its daytime depth and most of the stomachs are at least half full even in winter (Berkes, 1973; Fevolden, 1982; Sameoto, 1980). When offered *Artemia salina* nauplii under laboratory conditions, *M. norvegica* larvae showed higher feeding rates at low food concentration than did

Thysanoessa spp. larvae, while both species fed equally well at high food concentration (Schmidt *et al.*, 2004b).

Evidence for species-specific differences in the diet comes from the morphology of the feeding appendages, the body composition and the stomach content (Table 5.5). Most studies agree that *M. norvegica* has a broader food spectrum and is more carnivorous than the *Thysanoessa* species (e.g. Mauchline, 1966; Båmstedt and Karlson, 1998; Berkes, 1973; Dalpadado *et al.*, 2008; Fevolden, 1982; Sameoto, 1980). However, the smallest of the species, *T. longicaudata*, is also considered to feed frequently on copepods (Båmstedt and Karlson, 1998; Falk-Petersen *et al.*, 2000; Dalpadado *et al.*, 2008). Like *M. norvegica*, this species has high amounts of the fatty acids 20:1($n − 9$) and 22:1($n − 11$) in their storage lipids, which indicate dietary input from wax ester-rich calanoid copepods (Falk-Petersen *et al.*, 2000). *T. raschii* seems to feed more detritivorously than the other species. Notable are the high amounts of phytol in their wax esters, which suggest feeding on decaying plant material, either from bottom sediment or suspended in the water column (Falk-Petersen *et al.*, 2000). The stomach content of *T. raschii* is usually dominated by 'amorphous detritus' (Fevolden, 1982), while ingestion of copepods is rare (Båmstedt and Karlson, 1998; Sameoto, 1980). *T. inermis* is considered the most herbivorous of the four species (Sargent and Falk-Petersen, 1981; Dalpadado *et al.*, 2008), but predation on copepods has also been observed (Båmstedt and Karlson, 1998; Sameoto, 1980; Falk-Petersen *et al.*, 2000).

Berkes (1976) studied the ecology of *M. norvegica*, *T. raschii*, *T. inermis* and *T. longicaudata* in the Gulf of St. Lawrence. The resource partitioning between the species was analysed on the basis of their distribution, stomach content and timing of the breeding season. *M. norvegica* and *T. longicaudata* had a very similar diet year-round with food overlap coefficients between 0.78 and 0.98 (1.00 indicating maximal overlap). However, competition between the two species would have been low due to a different geographical and vertical distribution. In contrast, *M. norvegica* closely coincided with *T. raschii* and *T. inermis*, but used different food sources in winter and spring (food overlap coefficients between 0.14 and 0.68). A later breeding season further differentiates *M. norvegica* from the *Thysanoessa* species (Berkes, 1976; Dalpadado, 2006). Competition between *T. raschii* and *T. inermis* was more likely as these species overlapped in distribution as well as in diet (food overlap coefficients of 0.97 in May and August). Berkes (1976) suggested that slight differences in the spacing of their feeding appendages permit preference for different food size-spectra and therefore niche diversification between *T. raschii* and *T. inermis*. In January, the diet of *T. raschii* did not overlap with that of the other species as their stomach content was dominated by bottom mud (Berkes, 1976). In contrast, in August, the food of all four euphausiid species was very similar (overlap coefficient 0.63–0.97) due to the higher frequency of copepod remains in the stomachs of *T. raschii* and *T. inermis* (Berkes, 1973; Sameoto, 1980). The feeding on

Table 5.5 Different degrees of predatory feeding in 4 North-Atlantic euphausiid species as indicated by their morphology, biochemical tissue composition and frequency of copepod remains in their stomachs

Characteristics		M. n.	T. l.	T. r.	T. i.	Ref.
Morphology	Body size (mm)	45	20	25	30	1
	Coarseness of feeding basket (μm)	25	14 – 20	6.5 – 8.5	8.5 – 9	2
	Mandible grinding : cutting region	0.74	0.85	0.97	no data	3
Fatty acid indicators in triacyglycerols	Copepods: 20:1 + 22:1 isomers (%)	18	10	1.3	0.9	1;4
	Algae: 16:1(n-7) (%)	4.5	12	11.9	15.3	1;4
	Carnivory: 18:1(n-9)/18:1(n-7)	1.9	2	1.1	0.6	1;4
Fatty alcohol indicator in wax esters	Decaying phytoplankton: Phytol (%)	none	none	93	10	4
Stomach content	Stomachs with copepod mandibles (%)	93	35	6	69	5
	Contribution of copepods to diet (%)	23	19	low	15	5

Note: Krill for fatty acid and fatty alcohol analysis were sampled in winter (except T.l., where only summer data were available). Those for stomach content analysis were sampled in spring. M.n.– *Meganyctiphanes norvegica*; T.l. – *Thysanoessa longicaudata*; T.r. – *T. raschii*, T.i.– *T. inermis*.

References: 1–Falk-Petersen et al. (2000), 2– Berkes (1973), 3– Mauchline (1980), 4– Sargent and Falk-Petersen (1981), 5– Bämstedt and Karlson (1998).

copepods coincides with the annual peak in copepod biomass (Lindahl and Hernroth, 1988; Titelman and Fiksen, 2004). Thus, seasonal differences in food availability were the key to diversification, with high dietary overlap between species when resources are plenty (autumn) and low overlap when food is scarce in winter (Berkes, 1976).

5. FEEDING RATES AND DAILY RATION

Feeding rates give valuable information on mortality of food organisms and the energy intake of the target species. However, accurate estimates of *in situ* feeding rates are hard to acquire and therefore attempts are rare. One approach is to conduct feeding incubations, where krill are enclosed in relatively small volumes and the disappearance of food is monitored. Problems are that feeding behaviour and food encounter rate might be different from those *in situ*. Alternatively, feeding rates can be estimated from the stomach content of freshly caught krill and their digestion rate. The digestion rate varies with food availability and diurnal feeding rhythm and should ideally be measured at the time of sampling (Båmstedt *et al.*, 2000). However, such *in situ* data are missing for *M. norvegica*. In feeding incubations with copepod prey, the digestion time of *M. norvegica* was about 3.0 h (Båmstedt and Karlson, 1998).

Four studies have quantified the predation rate of *M. norvegica* on copepods under laboratory conditions (Table 5.6). McClatchie (1985) observed a tenfold rise in predation rate (0–9 copepods per krill h^{-1}) when copepod concentrations in the container increased from 8 to 100 copepods l^{-1}. Other authors used concentration of 1–2 copepods l^{-1} and found predation rates < 3 copepods per krill h^{-1} (Båmstedt and Karlson, 1998; Torgersen, 2001). For adult krill (> 30 mg dry weight), the copepod predation rate was unrelated to krill body size (Båmstedt and Karlson, 1998), but some increase was noticed under illumination (Torgersen, 2001).

The energy intake by *M. norvegica* is usually expressed as daily ration (i.e. prey intake over 24 h, expressed as the percentage of the predator body mass). This ration reflects both the predation rate and the body size of the ingested copepods. According to calculations of McClatchie (1985), *M. norvegica* has to ingest an equivalent of at least 4.6% of its body calories per day to meet its metabolic demands. In two of the studies, *M. norvegica* did either not ingest a sufficient daily ration (Torgersen, 2001) or only with an exceptionally high prey abundance (McClatchie, 1985), whereas in experiments of Båmstedt and Karlson (1998) daily rations of $> 10\%$ were achieved with typical copepod concentrations. Two factors might have caused these differences: first, McClatchie (1985) used relatively small containers for his experiments, which might have suppressed feeding rates (Båmstedt and

Table 5.6 *Meganyctiphanes norvegica* feeding on copepods: Mean and range of predation rate, stomach content and daily ration

Feeding incubations

Krill size or dry weight	Prey species	Container size (L)	Prey conc. (ind. L^{-1})	Predation rate(ind. h^{-1})	Daily ration (%)	S-Daily ration (%)	Ref.
No info	M. lucens	0.6	17	0.5 (0.4 – 0.6)	No info	–	1
29 mm, 30 mg	C. t., P. spp. C.f., A. spp	4	8–100	3.5 (0.7 – 9.0)	3 (0.8-6.4)	3.0	2
10 – 45 mg	C.f.	45, 90	1-2	1.2 (0 – 3.2)	12.7 (0 – 33)	15.0	3
30 mm	C.f., M. longa	43	2	0.1(in light: 0.4)	–	3.8	4

In situ stomach content

Krill size or dry weight	Prey species	Variable	Copepods stomach^{-1}	Daily ration (%)	S-Daily ration (%)	Ref.
35 – 80 mg	C.f.	Region	6.3 (2-34)	5.6 (1 – 20)	15.9	3
> 30 mm	No info	Season	2.1 (1 – 3)	–	–	5
30 – 40 mm	T. l., P. spp., A. spp, C. spp.	Season and region	4.0 (0 – 16)	–	–	6
> 30 mm	T. l., P. spp., A. spp, C. spp.	Season	3.0 (0 – 18)	–	3.7	7
17 – 41 mm	C.f., M. longa C. h., C. g.	-	0.7 (0 – 2)	–	1.4	8
60 -100 mg	O. spp, A. spp P. spp., C. spp.	-	24 (2 – 75)	2.0 (0 – 4.3)	5.3	9

Note: *M. lucens* – *Metridia lucens*; *C. t.* – *Centropages typicus*; *P.* spp. – *Pseudocalanus* spp.; *C.f.* – *Calanus finmarchicus*; *A.* spp. – *Acartia* spp.; *M. longa* – *Metridia longa*; *T. l.* – *Temora longicornis*; *C.* spp.- *Calanus* spp.; *C. h.* – *Calanus hyperboreus*; *C. g.* – *Calanus glacialis*; *O.* spp. – *Oithona* spp.

References: 1- David and Conover (1961), 2- McClatchie, (1985). 3- Bämstedt and Karlson (1998), 4-Torgersen, (2001); 5- Onsrud and Kaartvedt (1998); 6- Lass et al. (2001); 7- Kaartvedt et al. (2002), 8- Dalpadado et al. (2008), 9- Schmidt, unpublished data, Gullmarsfjorden, September 2003

Data variability within the same study has been related to factors such as number of prey, size of prey, region or season. Feeding incubations were run for 8-24 h in darkness. McClatchie (1985) and Bämstedt & Karlson (1998) gave data on daily rations (% of krill dry weight). For other studies, which included information on copepod prosome length or carbon equivalents (Torgersen 2001, Kaartvedt *et al.* 2002, Dalpadado 2008), the mean copepod dry weight was calculated according to Karlson & Bämstedt (1994). Two assumptions were made to estimate a daily ration on copepods, (1) the copepod dry weight in the stomach reflects the feeding of the last 3 hours (Bämstedt & Karlson 1998) and (2) this feeding rate is not significantly different from the average value over 24 h. To allow comparison between studies, a standardised daily ration (S-Daily Ration) was calculated from the mean copepod dry weight per krill stomach and a standard krill dry weight of 30 mg. This approach is verified by the observation that neither the number of copepods per stomach nor the copepod predation rates showed a clear relationship with body size for adult krill (Bämstedt & Karlson 1998 Dalpadado *et al.* 2008, Schmidt unpublished data.).

Karlson, 1998); second, seasonal differences in temperature and/or feeding activity might have caused lower values in winter (Torgersen, 2001), but high values in spring and summer (Båmstedt and Karlson, 1998).

Peak numbers of copepod mandibles recorded in stomachs of freshly caught *M. norvegica* were 70 mandibles of *C. finmarchicus* (mainly copepodid stage III Båmstedt and Karlson, 1998), 150 mandibles of a copepod mixture dominated by *Oithona* spp. (Schmidt, unpublished data) or over 30 mandibles mainly from *T. longicornis* or *Pseudocalanus* spp. (Kaartvedt *et al.*, 2002; Lass *et al.*, 2001). Assuming a digestion time of 3 h (Båmstedt and Karlson, 1998), the stomach content of copepods translates into daily rations between 0 and 20% (Table 5.6). In the study of Båmstedt and Karlson (1998), daily rations were higher than in the other three studies. A number of factors might have contributed to this difference, for example, the benefit of feeding on large copepods (McClatchie, 1985), the peak feeding activity during spring (Fisher and Goldie, 1959) and the high copepod biomass in the Skagerrak and northwest coast of Norway (Båmstedt, 1988). The last point becomes most obvious when comparing krill from the NE Atlantic (Båmstedt and Karlson, 1998) with those from the Barents Sea/Arctic (Dalpadado *et al.*, 2008). Both studies were carried out in spring/summer and in both cases *M. norvegica* were mainly feeding on *Calanus* spp., but the copepod concentrations in the water column were 2–3 orders of magnitude lower in Arctic waters (Båmstedt and Karlson, 1998; Dalpadado *et al.*, 2008). As a consequence, the daily ration of *M. norvegica* feeding on copepods was about 10 times lower in the Barents Sea than in the Atlantic.

Except for the study of Båmstedt and Karlson (1998), most daily rations estimated from the stomach content were lower than 4.6%, especially for larger adult krill. This suggests that the energy intake was insufficient to cover the basic metabolic requirements (McClatchie, 1985). However, the data are based only on the copepod part of the diet. Significant feeding on phytoplankton or benthic food sources will have increased the daily ration. In the study of Kaartvedt *et al.* (2002), for instance, phytoplankton accounted for ~ 60% of the diet in spring and summer, which would mean that the daily ration based on the total carbon intake was at least twice as high as the ration based merely on copepod ingestion. The calculations were also based on the assumption that the digestion time was about 3 h, which is considered a conservative estimate (Båmstedt and Karlson, 1998). With a shorter digestion time, daily rations would be higher. Finally, further studies are necessary to verify the metabolic demand of *M. norvegica* within different seasons. A daily ration of 4.6% body energy per day given by McClatchie (1985) is clearly higher than estimates for *E. superba* of ~ 1% body energy per day (Atkinson *et al.*, 2002; Huntley *et al.*, 1994). Considering that overall growth rates of the two species are similar (Tarling, this volume), basic metabolic requirements and food intake of *M. norvegica* should not differ greatly from that of *E. superba*.

6. Future Work

There exists good background knowledge on the morphology and function of the feeding appendages of *M. norvegica*, and their food has been studied across a range of environments and seasons. However, we still lack basic understanding of the link between environmental conditions and the diet of *M. norvegica* on one hand and the relation between nutrition and production on the other hand. How are diet and daily ration influenced by the availability of various food sources (copepods, phytoplankton, benthic items) and by the differing vertical distributions of developmental stages and genders? What affect does the presence of predators or competitors have on food uptake and subsequently in measures of production? It is important to understand, which food sources are most beneficial for *M. norvegica* and which food concentrations are sufficient to allow growth, reproduction or accumulation of lipid stores. To address these questions in future studies, the food environment, the diet, daily ration and production parameters should be studied in concert (Fig. 5.10). Basic mechanisms and consequences will be easier to identify if the studies allow comparisons, for example, between onshore and offshore habitats, juvenile and adult krill or co-occurring species.

To achieve quantitative estimates on the role of different food sources, two approaches are suggested. Either to use a specific method for each food source: the paired mandible method for copepods, gut fluorescence for algae and microscope-based approaches for benthic food, with carbon equivalents, dry mass or volume as a common unit. Alternatively, a DNA analysis of the stomach content will potentially cover all of the food sources with one method, including soft items such as athecate protozoan. Ideally, the same method would be applied across a range of seasons and regions to gain an overview about the feeding behaviour of *M. norvegica* in relation to food abundance and seasonal nutritional requirements.

So far, little attention has been paid to ingestion rates of *M. norvegica* and their predation pressure on other species, for example, copepods or euphausiid larvae (Båmstedt & Karlson, 1998). Such studies can be based on results from stomach content analysis, but need additional information on digestion rates. Most likely the digestion time varies with seasonal and diurnal feeding behaviour. Therefore, it is more valuable to run direct estimates at the time of sampling (Båmstedt *et al.*, 2000), than to apply a fixed value. Finally, cross-species comparison of morphological, biochemical or behavioural characteristics (Berkes, 1973) might give clues on how food is acquired or processed. This not only helps in the understanding of individual species, but also provides insights into the wider food web.

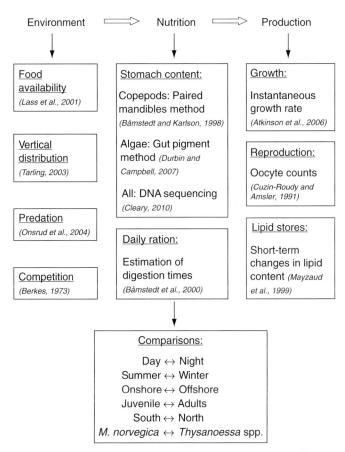

Figure 5.10 Schematic illustration of the link between food environment, nutrition and production of *M. norvegica*, with crucial parameters, study examples and references to be considered in future feeding studies. The food environment of *M. norvegica* is first of all determined by the abundance of different food sources. However, due to differences in vertical distribution, predation pressure or competition with other species, not all of this food might be available. To characterise the nutrition, both food composition and ingested quantities have to be estimated based on stomach content and daily ration. Quantity and quality of the food intake will be reflected in production parameters such as daily growth rate, number of oocytes or changes in lipid content.

Our knowledge has significantly increased since the last review on *M. norvegica* (Mauchline, 1980), but there are still uncertainties about food web relationships, environmental constrain and the link between food quantity/quality and production in *M. norvegica*. These questions have to be tackled through a combination of good sampling design, modern and established methods of sample analyses and hypothesis-based numerical models.

ACKNOWLEDGEMENTS

Special thanks go to H.-L. Suh for contributing an original drawing of the stomach of *M. norvegica* and photos on structures of the gastric mill published in his Ph.D. thesis in 1988. L. De Jong-Moreau kindly supplied scanning electron micrographs of mandibles, previously published in the Journal of Marine Biology Association, United Kingdom. G. Tarling encouraged the stomach content analysis on *M. norvegica* and supplied specimens from the Gullmarsfjord. K. Karlson and H. Saito helped with the identification of copepod mandibles, and K. Eriksson distinguished various types of filamentous algae. C. Östman identified the cnidarian nematocysts. I thank S. Kaartvedt, Y. Simard, R. Rosenberg, I. Hudson and D. Pond for discussing krill-seabed interactions and A. Atkinson for comments on this chapter.

REFERENCES

Abrahamsen, M. B., Browman, H. I., Fields, D. M., and Skiftesvik, A. B. (2010). The three-dimensional prey field of the northern krill, *Meganyctiphanes norvegica*, and the escape response of their copepod prey. *Mar. Biol.* **157,** 1251–1258.

Artiges, M., Pagano, M., and Thriot, A. (1978). Morphologie fonctionnelle des appendices nutritionnels de *Meganyctiphanes norvegica* (M. Sars, 1856) et *Euphausia krohnii* (Brandt, 1851) (Crustacea Euphausiacea). *Arch. Zool. Exp. Gen.* **119,** 95–106.

Atkinson, A., Meyer, B., Stübing, D., Hagen, W., Schmidt, K., and Bathmann, U. V. (2002). Feeding and energy budgets of Antarctic krill *Euphausia superba* at the onset of winter- II Juveniles and adults.. *Limnol. Oceanogr.* **47,** 953–966.

Atkinson, A., Shreeve, R. S., Hirst, A. G., Rothery, P., Tarling, G. A., Ond, D. W., Korb, R. E., Murphy, E. J., and Watkins, J. L. (2006). Natural growth rates in Antarctic krill (*Euphausia superba*): II. Predictive models baed on food, temperature, boy length, sex, and maturity stage. *Limnol. Oceanogr.* **51,** 973–987.

Båmstedt, U. (1988). The macrozooplankton community of Kosterfjorden, western Norway. Changes in the size and biochemical composition of *Meganyctiphanes norvegica* (Euphausiacea) in relation to its life cycle. *Sarsia* **61,** 15–30.

Båmstedt, U., and Karlson, K. (1998). Euphausiid predation on copepods in coastal waters of the Northeast Atlantic. *Mar. Ecol. Prog. Ser.* **172,** 149–168.

Båmstedt, U., Gifford, D. J., Irigoien, X., Atkinson, A., and Roman, M. (2000). Feeding. *In* "ICES zooplankton methodological manual" (R. Harris, P. Wiebe, J. Lenz, H. R. Skjødal and M. Huntley, eds.), pp. 297–399. Academic press, London.

Barange, M., Gibbons, M. J., and Carola, M. (1991). Diet and feeding of *Euphausia hanseni* and *Nematoscelis megalops* (Euphausiacea) in the northern Benguela Current: ecological significance of vertical space partitioning. *Mar. Ecol. Prog. Ser.* **73,** 173–181.

Berkes, F. (1973). Production and comparative ecology of euphausiids in the Gulf of St. Lawrence Ph.D. thesis. University of Montreal.

Berkes, F. (1976). Ecology of Euphausiids in the Gulf of St. Lawrence. *J. Fish. Res. Board Can.* **33,** 1894–1905.

Beyer, F. (1992). *Meganyctiphanes norvegica* (M. Sars) (Euphausiacea) a voracious predator on *Calanus*, other copepods, and ctenophores, in Oslofjorden, southern Norway. *Sarsia* **77,** 189–206.

Boyd, C. M., Heyraud, M., and Boyd, C. N. (1984). Feeding of the Antarctic krill *Euphausia superba. J. Crust. Biol.* **4,** 23–141.

Boysen, E., and Buchholz, F. (1984). *Meganyctiphanes norvegica* in the Kattegat: studies on the annual development of pelagic population. *Mar. Biol.* **79,** 195–207.

Browman, H. I. (2005). Applications of sensory biology in marine ecology and aquaculture. *Mar. Ecol. Prog. Ser.* **287,** 263–307.

Buchholz, F., Buchholz, C., Reppin, J., and Fischer, J. (1995). Diel vertical migration of *Meganyctiphanes norvegica* in the Kattegat: Comparison of net catches and measurements with Acoustic Doppler Current Profilers. *Helgoländer Meeresunters* **49,** 849–866.

Buchholz, C., Buchholz, F., and Tarling, G. (2006). On the timing of moulting processes in reproductively active Northern krill Meganyctiphanes norvegica. *Mar. Biol.* **149,** 1443–1452.

Clarke, A., and Morris, D. J. (1983). Towards and energy budget for krill: The physiology and biochemistry of Euphausia superba Dana. *Polar Biol.* **2,** 69–86.

Clarke, A., and Tyler, P. A. (2008). Adult Antarctic krill feeding at abyssal depths. *Curr. Biol.* **18,** 282–285.

Cleary, A. (2010). *In situ* feeding in the Northern krill, *Meganyctiphanes norvegica*: a DNA analysis of gut contents Ph. D. thesis, University of Rhode Island (in press).

Cotté, C., and Simard, Y. (2005). Formation of dense krill patches under tidal forcing at whale feeding hot spots in the St. Lawrence Estuary. *Mar. Ecol. Prog. Ser.* **288,** 199–210.

Cuzin-Roudy, J., and Amsler, M. (1991). Ovarian development and sexual maturity staging in Antarctic krill, *Euphausia superba* (Euphausiacea). *J. Crust. Biol.* **11,** 236–249.

Cuzin-Roudy, J., Albessard, E., Virtue, P., and Mayzaud, P. (1999). The scheduling of spawning with the moult cycle in Northern krill (Crustacea: Euphausiacea): a strategy for allocating lipids to reproduction. *Invertebrate Reproduction and Development* **36,** 163–170.

Cuzin-Roudy, J., Tarling, G. A., and Strömberg, J. O. (2004). Life cycle strategies of Northern krill (*Meganyctiphanes norvegica*) for regulating growth, moult, and reproductive activity in various environments: the case of fjordic populations. *ICES J. Mar. Sci.* **61,** 721–737.

Dalpadado, P. (2006). Distribution and reproduction strategies of krill (Euphausiacea) on the Norwegian shelf. *Polar Biol.* **29,** 849–859.

Dalpadado, P., and Skjoldal, H. R. (1996). Abundance, maturity and growth of the krill species *Thysanoessa inermis* and *T. longicaudata* in the Barents Sea. *Mar. Ecol. Prog. Ser.* **144,** 175–183.

Dalpadado, P., Yamaguchi, A., Ellertsen, B., and Johanessen, S. (2008). Trophic interactions of macro-zooplankton (krill and amphipods) in the Marginal Ice Zone of the Barents Sea. *Deep Sea ResII* **55,** 2266–2274.

David, C. N., and Conover, R. J. (1961). Preliminary investigation on the physiology and ecology of luminescence in the copepod, *Metridia lucens*. *Biol. Bull.* **121,** 2–107.

De Jong-Moreau, L., Casanova, B., and Casanova, J. P. (2001). Detailed comparative morphology of the peri-oral structures of the Mysidacea and Euphausiacea (Crustacea): An indication for the food preference. *J. Mar. Biol. Ass. UK* **81,** 235–241.

Durbin, E. G., and Campbell, R. G. (2007). Reassessment of the gut pigment method for estimating in situ zooplankton ingestion. *Mar. Ecol. Prog. Ser.* **331,** 305–307.

Einarsson, H. (1945). Euphausiacea. 1. Northern Atlantic species Dana Rep. No. 27, 185 pp.

Emson, R. H., Mladenov, P. V., and Barrow, K. (1991). The feeding mechanism of the basket star *Gorgonocephalus arcticus*. *Can. J. Zool.* **69,** 449–455.

Falk-Petersen, S., and Hopkins, C. C. E. (1981). Zooplankton sound scattering layers in North Norwegian fjords: Interactions between fish and krill shoals in a winter situation in Ullsfjorden and Øksjorden. *Kieler Meeresforsch* **5,** 191–201.

Falk-Petersen, S., Hagen, W., Kattner, G., Clarke, A., and Sargent, J. (2000). Lipids, trophic relationships, and biodiversity in Arctic and Antarctic krill. *Can. J. Fish. Aquat. Sci.* **57,** 178–191.

Fevolden, S. E. (1982). Feeding habits and enzyme polymorphism in *Thysanoessa raschii* and *Meganyctiphanes norvegica* (Crustacea, Euphausiacea). *Sarsia* **67,** 1–10.

Fisher, L. R., and Goldie, E. H. (1959). The food of *Meganyctiphanes norvegica* (M. Sars) with an assessment of the contributions of its components to the Vitamin A reserves of the animal. *J. Mar. Biol. Assoc. UK* **38**, 291–312.

Gibbons, M. J., Pillar, S. C., and Stuart, V. (1991). Selective carnivory by *Euphausia lucens*. *Cont. Shelf Res.* **11**, 625–640.

Hamm, E., Merkel, R., Springer, O., Jurkojc, P., Maier, Prechtel K., and Smetacek, V. (2003). Architecture and material properties of diatom shells provide effective mechanical protection. *Nature* **421**, 841–843.

Hamner, W. M. (1988). Biomechanics of filter feeding in the Antarctic krill *Euphausia superba*: review of past work and new observations. *J. Crustcean. Biol.* **8**, 149–163.

Hamner, W. M., and Hamner, P. P. (2000). Behaviour of Atartic krill (Euphausia supeba): Schooling, foraging, and antipredatory behaviour. *Can. J. Fish. Aquat. Sci.* **57**, 192–202.

Hirano, Y., Matsuda, T., and Kawaguchi, S. (2003). Breeding Antarctic krill in captivity. *Mar. Fresh. Behaviour Physiology* **36**, 259–269.

Holt, E. W. L., and Tattersall, W. M. (1905). Schizopodous Crustacea from the northeast Atlantic slope. *Rep. Sea Inl. Fish. Ire., Sci. Invest.* 99–152Year 1902–3, App. 4.

Hudson, I. R., and Wigham, B. D. (2003). *In situ* observations of predatory feeding behaviour of the galatheid squat lobster *Munida sarsi* (Huus, 1935) using a remotely operated vehicle. *J. Mar. Biol. Ass. UK* **83**, 463–464.

Huntley, M. E., Nordhausen, W., and Lopez, M. D. G. (1994). Elemental composition, metabolic activity and growth of Antarctic krill Euphausia superba during winter. *Mar. Ecol. Prog. Ser.* **107**, 23–40.

Jonsson, L. G., Lundälv, T., and Johannesson, K. (2001). Symbiotic associations between anthozoans and crustaceans in a temperate coastal area. *Mar. Ecol. Prog. Ser.* **209**, 189–195.

Kaartvedt, S., Larsen, T., Hjelmseth, K., and Onsrud, M. S. R. (2002). Is the omnivorous krill *Meganyctiphanes norvegica* primarily a selectively feeding carnivore? *Mar. Ecol. Prog. Ser.* **228**, 193–204.

Karlson, K., and Båmstedt, U. (1994). Planktivorous predation on copepods. Evaluation of mandible remains in predator guts as a quantitative estimate of predation.. *Mar. Ecol. Prog. Ser.* **108**, 79–89.

Kawaguchi, S., Finley, L. A., Jarman, S., Candy, S., Ross, R. M., Quetin, L. B., Siegel, V., Trivelpiece, W., Naganobu, M., and Nicol, S. (2007). Male krill grow fast and die young. *Mar. Eol. Prog. Ser.* **345**, 199–210.

Kreibich, T., Hagen, W., and Saborowski, R. (2010). Food utilization of two pelagic crustaceans from Greenland Sea waters: *Meganyctiphanes norvegica* (Euphausiacea) and *Hymenodora glacialis* (Decapoda, Caridea). *Mar. Ecol. Prog. Ser.* in press.

Lass, S., Tarling, G. A., Virtue, P., Matthews, J. B. L., Mayzaud, P., and Buchholz, F. (2001). On the food of northern krill *Meganyctiphanes norvegica* in relation to its vertical distribution. *Mar. Ecol. Prog. Ser.* **214**, 177–200.

Lavoie, D., Simard, Y., and Saucier, F. J. (2000). Aggregation and dispersion of krill at channel heads and shelf edges: the dynamics in the Saguenay – St. Lawrence Marine Park. *Can. J. Fish Aquat. Sci.* **57**, 1853–1869.

Liljebladh, B., and Thomasson, M. A. (2001). Krill behaviour as recorded by acoustic Doppler current profiler in the Gullmarsfjord. *J. Mar. Syst.* **27**, 301–313.

Lindahl, O., and Hernroth, L. (1988). Large-scale and long-term variations in the zooplankton Community of the Gullmar fjord, Sweden, in relation to advective processes. *Mar. Ecol. Prog. Ser.* **43**, 161–171.

Mauchline, J. (1960). The biology of the euphausiid crustacean, *Meganyctiphanes norvegica* (M. Sars). *Proc. Roy. Soc. Edinb. (B)* **67**, 141–179.

Mauchline, J. (1966). The biology of *Thysanoessa raschii* (M. Sars), with a comparison of its diet with that of *Meganyctiphanes norvegica* (M. Sars). *In* "Some contemporary studies in marine science" (H. Barnes, ed.), pp. 493–510. George Allen and Unwin, London.

Mauchline, J. (1980). The biology of mysids and euphausiids. *Adv. Mar. Biol.* **18**, 1–681.

Mauchline, J. (1989). Functional morphology and feeding of euphausiids. *In* "Functional Morphology of Feeding and Grooming in Crustacea" (B. E. Felgehauer, L. Watling and A. B. Thistle, eds.), pp. 173–184. A.A. Balkema, Rotterdam.

Mauchline, J. (1998). The biology of calanoid copepods. *Adv. Mar. Biol.* **33**, 1–710.

Mauchline, J., and Fisher, L. R. (1969). The biology of euphausiids. *Adv. Mar. Biol.* **7**, 1–454.

Mayzaud, P., Virtue, P., and Albessard, E. (1999). Seasonal variations in the lipid and fatty acid composition of the euphausiid *Meganyctiphanes norvegica* from the Ligurian Sea. *Mar. Ecol. Prog. Ser.* **186**, 199–210.

McClatchie, S. (1985). Feeding behaviour in *Meganyctiphanes norvegica* (M. Sars) (Crustacea: Euphausiacea). *J. Exp. Mar. Biol. Ecol.* **86**, 271–284.

McClatchie, S., and Boyd, C. M. (1983). Morphological study of sieve efficiencies and mandibular surfaces in the Antarctic krill, *Euphausia superba*. *Can. J. Fish Aquat. Sc.* **40**, 955–967.

Nemoto, T. (1977). Food and feeding structures of deep sea *Thysanopoda* euphausiids. *In* "Oceanic Sound Scattering Prediction" (N. R. Andersen and B. J. Zahuranec, eds.), pp. 457–480. Plenum Press, New York and London.

Nilsson, D. E. (1996). Eye design, vision and invisibility in plankton invertebrates. *In* "Zooplankton: Sensory ecology and physiology" (P. H. Lenz, D. K. Hartline, J. E. Purcell and D. L. Macmillan, eds.), pp. 149–162. Gordon and Breach Publisher, Amsterdam.

Onsrud, M. S. R., and Kaartvedt, S. (1998). Diel vertical migration of the krill *Meganyctiphanes norvegica* in relation to the physical environment, food and predators. *Mar. Ecol. Prog. Ser.* **171**, 209–219.

Onsrud, M. S. R., Kaartvedt, S., Røstad, A., and Klevjer, T. A. (2004). Vertical distribution and feeding patterns in fish foraging on the krill *Meganyctiphanes norvegica*. *ICES J. Mar. Sci.* **61**, 1278–1290.

Paulsen, O. (1909). Plankton investigations in the waters around Iceland and in the North Atlantic in 1904. *Medd Komm Havundersøg, Kbh. Ser. Plankton Bd.* **8**, 57.

Petursdottir, H., Gislason, A., Falk-Petersen, S., Hop, H., and Svavarsson, J. (2008). Trophic interactions of the pelagic ecosystem over the Reykjanes Ridge as evaluated by fatty acids and stable isotope analyses. *Deep Sea Res. II* **55**, 3–93.

Pond, D., Watkins, J., Priddle, J., and Sargent, J. (1995). Variation in the lipid content and composition of Antarctic krill *Euphausia superba* at South Georgia. *Mar. Ecol. Prog. Ser.* **117**, 49–57.

Ponomareva, L. A. (1954). Copepods in the diet of euphausiids in the Sea of Japan. *Dokl. Akad. Nauk SSSR* **98**, 153–154.

Price, H. J. (1989). Swimming behaviour of krill in response to algal patches: A mesocosm study. *Limnol. Oceanogr.* **34**, 649–659.

Priddle, J., Watkins, J., Morris, D., Ricketts, C., and Buchholz, F. (1990). Variation of feeding by krill in swarms. *J. Plankton Res.* **12**, 1189–1205.

Rosenberg, R., Dupont, S., Lundälv, T., Sköld, H. N., Norkko, A., Roth, J., Stach, T., and Thorndyke, M. (2005). Biology of the basket star *Gorgonocephalus caputmedusae* (L.). *Mar. Biol.* **148**, 43–50.

Saborowski, R., Bröhl, S., Tarling, G. A., and Buchholz, F. (2002). Metabolic properties of Northern krill, Meganyctiphanes norvegica, from different climatic zones. I. *Respiration and excretion*. *Mar. Biol.* **140**, 547–556.

Saether, O., Ellingsen, T. E., and Mohr, V. (1986). Lipids of North Atlantic krill. *J. Lipid Res.* **27,** 274–285.

Sameoto, D. D. (1980). Relationships between stomach contents and vertical migration in *Meganyctiphanes norvegica, Thysanoessa raschii* and *T. inermis* (Crustacea Euphausiacea). *J. Plankton Res.* **2,** 129–143.

Sargent, J. R., and Falk-Petersen, S. (1981). Ecological investigations on the zooplankton community of Balsfjorden, northern Norway: Lipids and fatty acids in *Meganyctiphanes norvegica Thysanoessa raschii* and *T. inermis* during midwinter. *Mar. Biol.* **62,** 131–137.

Schmidt, K., McClelland, J. W., Mente, E., Montoya, J. P., Atkinson, A., and Voss, M. (2004a). Trophic-level interpretations based on δ^{15}N values: Implications of tissue-specific fractionation and amino acid composition. *Mar. Ecol. Prog. Ser.* **266,** 43–58.

Schmidt, K., Tarling, G. A., Plathner, N., and Atkinson, A. (2004b). Moult cycle-related changes in feeding rates of larval krill *Meganyctiphanes norvegica* and *Thysanoessa* spp. *Mar. Ecol. Prog. Ser.* **281,** 131–143.

Schmidt, K., Atkinson, A., Petzke, K.-J., Voss, M., and Pond, D. W. (2006). Protozoans as a food source for Antarctic krill, *Euphausia superba*: Complementary insights from stomach content, fatty acids and stable isotopes. *Limnol. Oceanogr.* **51,** 2409–2427.

Siegel, V. (2000). Krill (Euphausiacea) life history and aspects of population dynamics. *Can. J. Fish Aquat. Sci.* **57,** 130–150.

Simmard, Y., Lacroix, G., and Legendre, L. (1986). Diel vertical migrantion and nocturnal feeding of a dense coastal krill scattering layer (*Thysanoessa raschii* and Meganyctiphanes norvegica) in stratified surface waters. *Mar. Biol.* **91,** 93–105.

Sourisseau, M., Simard, Y., and Saucier, F. J. (2008). Krill diel vertical migration fine dynamics, nocturnal overturns, and their roles for aggregation in stratified flows. *Can. J. Fish Aquat. Sci.* **65,** 574–587.

Stuart, V., and Pillar, S. C. (1990). Diel grazing patterns of all ontogenetic stages of Euphausia lucens and in situ predation rates on copepods in southern Benguela upwelling region. *Mar. Ecol. Prog. Ser.* **64,** 227–241.

Stübing, D., Hagen, W., and Schmidt, K. (2003). On the use of lipid biomarkers in marine food web analyse: An experimental case study on the Antarctic krill, Eupausia superba. *Limnol. Oceanogr.* **48,** 1685–1700.

Suh, H. L. (1988). Ecology of euphausiids (Crustacea) with special reference to the functional morphology of feeding apparatus. Ph.D. thesis, University of Tokyo,160pp.

Suh, H. L. (1996). The gastric mill in euphausiid crustaceans: A comparison of eleven species. *Hydrobiologia* **321,** 235–244.

Suh, H. L., and Choi, S. D. (1998). Comparative morphology of the feeding basket of five species of *Euphausia* (Crustacea, Euphausiacea) in the North Pacific, with some ecological considerations. *Hydrobiologia* **385,** 107–112.

Suh, H. L., and Nemoto, T. (1987). Comparative morphology of filtering structure of five species of *Euphausia* (Euphausiacea, Crustacea) from the Antarctic Ocean. *Proc. NIPR. Symp. Polar Biol.* **1,** 72–83.

Suh, H. L., and Nemoto, T. (1988). Morphology of the gastric mill in ten species of euphausiids. *Mar. Biol.* **97,** 79–85.

Sundbäck, K., Linares, F., Larson, F., and Wulff, A. (2004). Benthic nitrogen fluxes along a depth gradient in a microtidal fjord: The role of denitrification and microphytobenthos. *Limnol. Oceanogr.* **49,** 1095–1107.

Tarling, G. (2003). Sex-dependent diel vertical migration in northern krill *Meganyctiphanes norvegica* and its consequences for population dynamics. *Mar. Ecol. Prog. Ser.* **260,** 173–188.

Tarling, G. A., Matthews, J. B. L., Burrows, M., Saborowski, R., Buchholz, F., Bedo, A., and Mayzaud, P. (2000). An optimisation model of the diel vertical migration of 'Northern krill'(*Meganyctiphanes norvegica*) in the Clyde Sea and Kattegat. *Can. J. Fish. Aquat. Sci.* **57**(Suppl 3), 38–50.

Titelman, J., and Fiksen, Ø. (2004). Ontogenetic vertical distribution patterns in small copepods: Field observations and model predictions. *Mar. Ecol. Prog. Ser.* **284**, 49–63.

Torgersen, T. (2001). Visual predation by the euphausiid *Meganyctiphanes norvegica*. *Mar. Ecol. Pog. Ser.* **209**, 295–299.

Viitasalo, M., Kiørboe, T., Flinkman, J., Pedersen, L. W., and Visser, A. W. (1998). Predation vulnerability of planktonic copepods: Consequences of predator foraging strategies and prey sensory abilities. *Mar. Ecol. Prog. Ser.* **15**, 129–142.

Virtue, P., Nichols, P. D., Nicol, S., and Hosie, G. (1996). Reproductive trade-off in male Antarctic krill, *Euphausia superba*. *Mar. Biol.* **126**, 521–527.

Virtue, P., Mayzaud, P., Albessard, E., and Nichols, P. (2000). Use of fatty acids as dietary indicators in northern krill, *Meganyctiphanes norvegica*, from northeastern Atlantic, Kattegat, and Mediterranean waters. *Can. J. Fish Aquat. Sci.* **57**(Suppl 3), 104–114.

Wulff, A., Vilbaste, S., and Truu, J. (2005). Depth distribution of photosynthetic pigments and diatoms in the sediments of a microtidal fjord. *Hydrobiologia* **534**, 117–130.

Youngbluth, M. J., Bailey, T. G., Davoll, P. J., Jacoby, C. A., Blades-Eckelbarger, P. I., and Griswold, C. A. (1989). Fecal pellet production and diel migratory behaviour by the euphausiid *Meganyctiphanes norvegica* effect benthic-pelagic coupling. *Deep Sea Res.* **36**, 1491–1501.

CHAPTER SIX

GROWTH AND MOULTING IN NORTHERN KRILL (*MEGANYCTIPHANES NORVEGICA* SARS)

Friedrich Buchholz *and* Cornelia Buchholz

Contents

Abstract

Moulting and growth as a key aspect of the life-history of crustaceans has been reviewed here for *Meganyctiphanes norvegica*. Moulting is a cyclical process with relatively constrained, uniform phases that have been well documented. The crustacean moult cycle has a large influence on growth-rates, reproduction and metabolism. Moult and growth are under hormonal control with further environmental influences. A pre-requisite for intensive studies is a detailed moult staging system. Here, a further refinement and temporal phasing is presented for *M. norvegica*. On such a basis, the dynamics of cuticle synthesis and degradation are shown. Moult and reproductive cycles are interlinked and krill is able to combine

Alfred Wegener Institute for Polar and Marine Research, Bremerhaven, Germany

Advances in Marine Biology, Volume 57
ISSN 0065-2881, DOI: 10.1016/S0065-2881(10)57006-8

growth and reproduction in a way that allows females to achieve similar net growth rates to males. A synchronisation of physiological processes and behaviour related to growth and reproduction enhances environmental success. Moult staging can also be used to assess growth rates in the field. Some further technical approaches are reviewed. Laboratory maintenance and field data are combined to determine growth rates under various environmental conditions. These are related to life-growth assessments from population studies comparing krill along a latitudinal gradient. Life-cycle and physiological data indicate that Northern krill are able to cope with both warm and cold environments and have highly adaptable phenotypes. The species may serve as useful indicator of environmental change. Its potential proliferation in new environments may also have implications to regional food webs, given the krill's high level of growth and productivity and their pivotal trophic role.

1. Introduction: Linking Moult and Growth

Crustacean growth depends on a series of moults, exchanging the rigid exoskeleton for a larger or differently equipped one, a property shared with all arthropods. At a superficial level, growth in crustaceans occurs discontinuously given the saltatory nature of an abrupt growth increment at each ecdysis. It can be characterised by two parameters: the intermoult period (IMP) and the growth increment at moult (INC).

In fact, moulting is a continuous process where one moult cycle is immediately linked to the next under favourable trophic and environmental conditions. As soon as the new exoskeleton is completed the water, which has been taken up to burst-open the old shell at predetermined seams and extend the new shell, is replaced continuously by tissue and lipid stores. Apolysis, the retraction of the epidermis from the exoskeleton during the premoult phase, is the first sign of preparation for the next moult. The epidermis then starts to build new exoskeletal layers until water is pumped in to initiate ecdysis and define the size of the next growth increment.

These processes are controlled by hormones which determine the timing of the moult rhythm and are the functional interface to environmental signals, such as temperature, photoperiod and trophic input.

In insects, the last metamorphic moult is the one which determines the final size of the individual. In some crustacean species, an analogous terminal anecdysis prevails where the moulting gland that produces the indispensable moult hormone simply degenerates. However, crustacean growth is mostly indeterminate—leading to, for example giant lobsters. Moulting is cyclical and relatively uniform in its phases, and much study has been devoted to determining the constraints on these moult phases. In turn, the moult cycle has a large influence on other rates, such as growth, reproduction and physiology and so is a key aspect of the life-history of crustaceans.

2. The Basic Course of the Moult Cycle

The organic matrix of the crustacean cuticle is a complex structure composed mainly of α-chitin microfibrils embedded in a protein matrix. These are mainly stacked in chitin/protein lamellae, called laminae. Morphologically, there are three different layers in the cuticle of krill: an outermost epicuticle followed by an exocuticle and an endocuticle (Fig. 6.1A; Buchholz and Buchholz, 1989; Buchholz et al., 1989). The epicuticle is the only layer to contain glycoproteins in addition to the chitin/protein matrix but makes up only 1.6% of total cuticular thickness. The exocuticle makes up 20% of the cuticle. Both the epi- and exocuticle are deposited before the krill sheds its old shell in the moult. The remainder of the cuticle is endocuticle, which is quickly built up in the postmoult period. The epidermis lies beneath these layers and is responsible for forming the new cuticle and secreting enzymes that digest the old one. Cell processes of the epidermis travel through pore canals in the new cuticle to reabsorb digested old cuticle components (Buchholz and Buchholz,

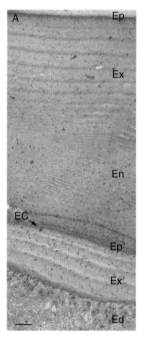

Figure 6.1 (A) Ultrastructure of the fully developed krill cuticle (*Meganyctiphanes norvegica*) at moult stage D2 with new epi- (Ep′) and exocuticle (Ex′) present underneath the old one. Ep epicuticle, Ex exocuticle, En endocuticle, EC exuvial cleft with the proximal layers of the endocuticle partly dissolved, Ed epidermis ('tendinal' cell). Scale bar 1 μm.

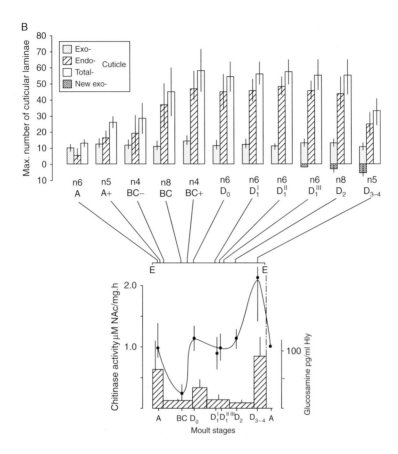

Figure 6.1 (B) *Euphausia superba*. Means and standard deviations of the max. cuticular laminae counted in the sixth pleomer at detailed moult stages. The lower part of the diagram depicts chitinase activity in the integument and N-acetyl-β-D-glucosamine content in the haemolymph (Hly) during the moult cycle (graph from Buchholz and Buchholz, 1989, with permission).

1989). The dynamic aspect of moult related cuticle build-up and break-down is illustrated in Fig. 6.1B showing increases and decreases in the number of cuticle laminae (note formation of the new endocuticle during late premoult). High activity of the chitinase coincides with elevated levels of glucosamine in the haemolymph indicating re-absorption of the amino-sugars from the cuticle. The early mobilisation of such energy rich components through chitinolysis at apolysis (stage D0) was recently confirmed by gene expression studies (Tarling, personal communication).

2.1. Hormonal and environmental control

The uniformity of the adult moult cycle allows the definition of finely graded moult stages (see below). The physiological relevance of such stages was first established by Adelung (1971) in the green crab, *Carcinus maenas* by determining the temporal course of the titre of the moult hormone 20-OH-ecdysone in the haemolymph over a complete moult cycle. This general pattern of moult stages is still valid today throughout most crustacean taxa (Covi *et al.*, 2009). Typically, directly after an ecdysis, an elevated level of 20-OH-ecdysone is still present declining to minimum levels during the postmoult phase. With apolysis, 20-OH-ecdysone concentration rises steeply, reaches a maximum a few days before the moult and falls abruptly again, well before ecdysis, to the low postmoult level. In *Euphausia superba*, this typical pattern was also found, but with a separate peak coinciding with apolysis (Buchholz, 1991).

The steroid moult hormone is produced in the thoracical moulting gland, the Y–organ, and is itself controlled by the moult inhibiting hormone MIH which is secreted by the X-organ. The X-organ–sinus-gland complex is a typical neuro-haemal organ located in the eye stalks, where a series of neuropeptides is produced, including some moult stimulating factors, which are shed into the haemolymph. The mode of action is apparently dependent on a reverse titre of MIH and ecdysone (Nakatsuji *et al.*, 2009). Specific steroid levels control protein and lipid synthesis in relation to complex developmental and internal growth processes. Organ specificity is achieved by receptors on cells and secondary messengers within the cells. In addition, ecdysones serve other functions in the control of reproduction and embryogenesis (Subramoniam, 2000).

The ecdysone titre controls the course of the regular moult cycle. Environmental cues modify the cycle via hormonal action, most possibly through variation of the MIH (Nakatsuji *et al.*, 2009): crabs can halt ongoing moult processes when predators are present (Adelung, 1971) and low ecdysone levels are associated with diapause in copepods (Johnson, 2003). The seasonal intensity of metabolic functions in Antarctic krill appear to be controlled by the photoperiod (Teschke *et al.*, 2008) and a light signal may also trigger apolysis, to set off the next pre-moult phase (Seear *et al.*, 2009). The MIH–ecdysone system is probably the mediator, adjusting the timing of seasonal development and growth.

The complex suite of processes directly linked to the shedding of the shell is triggered by the pre-moult peak and thereafter runs independently from ecdysone control, following a fixed temporal schedule in, for instance, final resorption of the old cuticle, water uptake, ecdysis and the consolidation and hardening of the new shell (Fig. 6.1B; Buchholz *et al.*, 1989). This process takes a fixed amount of time in the moult cycle and can be used to assess its overall duration in krill swarms in field studies (see below).

2.2. Categorisation and timing of stages in the moult cycle

The moult cycle can be divided into two main phases of postmoult and premoult. As a rough guide, 30% of the cycle is spent in the postmoult stage and the remainder is divided evenly between early and late premoult. The moult cycle is nevertheless comprised of a large number of incremental stages that can be distinguished morphologically or histologically. The definitive reference in this regard was published by Drach (1939) for crab and shrimp but has been found applicable and valid for a wide range of crustacean taxa since. Accordingly, it has been the most widely applied system until today and it is wise to use Drach's nomenclature of moult stages as a permanent reference. It was adopted for Antarctic krill by Buchholz (1982) and extended to include moult stage durations in Buchholz (1991). Cuzin-Roudy and Buchholz (1999) adapted the scheme for *Meganyctiphanes norvegica*. Table 6.1 is an overview of the main phases based on Buchholz (1991) and applies to both *M. norvegica* and *E. superba*, including an update to the estimated duration of each of the moult stages in *M. norvegica* (Buchholz *et al.*, 2006). The stage nomenclature was devised to match the Drach system as closely as possible, this applies also for the detailed version in Table 6.2.

A difficulty with the above scheme is that it is a labour intensive system to apply at the population level and also demands that all specimens be handled fresh, which is often not possible in the field. Therefore, other schemes of moult-stage categorisation have been devised which rely on features that are more readily distinguishable. For instance, Tarling *et al.* (1999a) considered the state of cuticle rigidity of specimens in freshly caught specimens over diurnal cycles to examine the interaction between moulting

Table 6.1 Overview table of moult phases, stages and updated durations in *Meganyctiphanes norvegica*

Phase	Process	Cuticle rigidity	Stage	Duration in % of moult cycle (previous value)	
Postmoult	Solidification of cuticle	Soft	**A**	2.0	(6.3)
		Hardening	**B**	7.4	(8.6)
	Reserve accumulation	Hard	**C**	21.1	(15.6)
Premoult	Secretion of new cuticle, setogenesis	Hard	**D0**	15.6	(15.6)
		Hard	**D1**	23.4	(23.4)
	New endocuticle	Hard	**D2**	18.2	(19.5)
	Cuticle resorption	Soft	**D3**	12.7	(10.9)
Ecdysis	Actual moult		**E**	< 1	

The original moult stage durations (brackets) were first published in Cuzin-Roudy and Buchholz (1999, with permission).

Table 6.2 Moult staging criteria for *Meganyctiphanes norvegica* (*M.n.*) with reference to *Euphausia superba* (*E.s*; Buchholz, 1991, with permission)

Stage			% IMP
M. n	*E.s*	Description	*M.n*
A	A−	Cuticle (Cut) very thin and soft, specimen is unsupported outside water (flaccid). Uropod tissue (Upt) not structured.	2.0
B$_{early}$	A	Cut: increasingly rigid. Upt: not structured. Contents of setae (Set.) is coarsely granular.	7.4
B	A+	Cut: increasingly rigid. Upt: stripe pattern starts to develop. Set: increasingly finely granular.	
C$_{early}$	BC−	Cut: still flexible at the rims of the carapace and lateral sclerites of abdomen. Upt: stripe pattern fully developed. Set: proximally still granular, otherwise clear.	21.1
C	BC	Cut: completely hard. Set: contents clear, no epidermal detachments.	
C$_{late}$	BC+	Epidermis (Epd) begins to retract at bases of setae.	
D$_0$	D$_0$	Epd completely detached from cuticle: apolysis	15.6
D$_{0\ late}$	D$_0$+	Epd begins to turn inside (invagination) at bases of setae.	
D$_{1\ early}$	D$_1'$	Annular fold of Epd reaches to less than 1/3 of the length of the old setae into the Upt.	23.4
D$_1$	D$_1''$	Fold of invagination has reached its farthest point.	
D$_{1\ late}$	D$_1'''$	Secondary bristles start to form on Set. Epd still lacks a cuticle.	
D$_2$	D$_2$	New cuticle is visible at bases to tips of setae and sec. bristles.	18.2
D$_{2\ late}$	D$_2$+	Full thickness attained by new cuticle. Krill is ready for moult.	
D$_{3\ early}$	D$_{3-4}$ −	Krill starts to become soft shelled (like C$_{early}$)	12.7
D$_3$	D$_{3-4}$	Cuticle completely softened	
D$_{3\ late}$	D$_{3-4}$ +	Specimen is unsupported outside water (flaccid). New cuticle detachable from old one by forceps.	
E	E	Ecdysis: krill moults within 10–20 s	

Our criteria and descriptions were cross-validated with J. Cuzin-Roudy (Villefranche-sur-Mer) through comparing categorisations of the same specimens.

and vertical migration behaviour. Specimens were categorised as 'moulting' when they were 'flaccid' (soft shelled and unsupported outside the water), corresponding to both moult stages A and D3, including the very short stage E of the actual ecdysis. Further moult stages were not differentiated in that study. Additional moult-time-series have been run on *M. norvegica* in the laboratory (at Marine Research Station Kristineberg, Gullmarsfjord in 1999 and 2000, Buchholz unpublished) and these new observations on moult-stage timings have now been integrated into the updated moult stage durations given here. Table 6.2 gives an extended version of the moult cycle for more practical application in the context of field and population studies on *M. norvegica*. Previous nomenclature used for *E. superba* is also included for comparison (Table 6.2; Buchholz, 1991).

Combining morphological moult staging (Fig. 6.2) with gene expression studies may give considerable further insights in the moult cycle of krill and

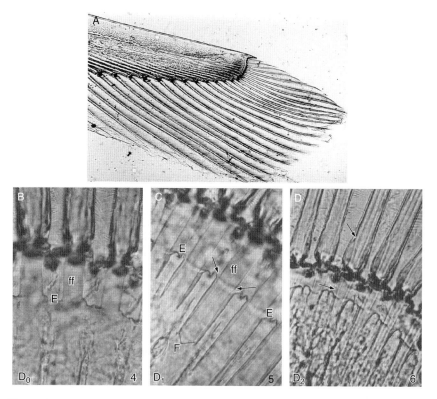

Figure 6.2   (A) Uropod of *Meganyctiphanes norvegica*, ~200×; (B)–(D) tip of the uropod in *Euphausia superba* in different premoult stages, ~400×. From Buchholz (1982), with permission; (B) Stage D0 = Apolysis, the epidermis (E) detaches from the cuticle; (C) Stage D1: Annular invaginations (arrows) start to form; (D) Stage D2: ready to moult specimen: cuticle doubled. The new seta bears a cuticle and secondary bristles (arrows).

other crustaceans, for example such as in the elucidation of chitin turnover- or cuticle-digestion processes (Seear *et al.*, 2009; Shechter *et al.*, 2007). In turn, the morphological stage definitions may be validated and refined further in this way (see also Fig.6.1B).

2.3. Timing of IMP and the dimension of INC

The interrelationship between IMP and INC depends on the exact timing of the moult phases. These, in turn, may be affected by environmental cues. In establishing any scheme on the physiological basis of growth, the adjustment of the moult cycle according to seasonal environmental signals must be considered.

IMP has been found to be invariably around 13 d in adult krill of ca 30 mm total length in a number of separate studies carried out at 10 °C (Buchholz *et al.*, 2006; Cuzin-Roudy and Buchholz, 1999 and unpublished data). This value also falls close to a regression line of a combination of laboratory experiments and field assessments of IMP (Fig. 1 from Cuzin-Roudy and Buchholz, 1999). Accordingly, we used this IMP value at 10 °C as an experimental standard. Furthermore, the regression parameters may be used to account for the effect of different environmental temperatures on IMP.

A set of maintenance experiments used this IMP-temperature relationship to test for the regularity of moulting and growth (Cuzin-Roudy *et al.*, 2004). Again, the majority of IMPs, either determined by counting the number of days between consecutive moults, or by extrapolation of the numbers of moulted specimens per day, confirmed the relationship. However, at a temperature of 8 °C, longer than expected IMPs were observed (see below).

Temporal variation of the IMP follows a fixed mechanism while the rhythmicity of the moult cycle appears to have a certain degree of autonomy independent of environmental factors. However, there are two 'windows' in the moult cycle where the timing can be adjusted. These are the so-called 'resting stages' C and D2. The last sub-stage within stage C is C4, according to Drach (1939): at this stage tissue and reserves are deposited until the next premoult phase can be initiated. If trophic input does not suffice, C4 is prolonged. This stage is sometimes referred to as 'intermoult', a term that should be avoided to be not confused with the 'IMP'. Additionally, in stage D2, a crustacean is ready to moult. The actual moult may then be triggered by the ecdysone peak and continue without further hormonal modulation. Apparently, a negative signal, be it the presence of predators, low temperature or lack of food may depress the hormonal trigger and delay the moult until the situation improves. This mechanism may be the cause of the outliers in the experiments and the field data above (Fig. 6.3).

The black rhombus in Fig. 6.3 represents non-reproductive krill in a wintery Clyde Sea situation and the round symbols, non-reproductive

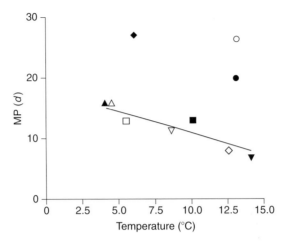

Figure 6.3 Relationship between intermoult period and temperature (graph from Cuzin-Roudy and Buchholz, 1999; with permission): white symbols: lab; black symbols: field, samples from Ligurian sea, Clyde sea and Kattegat. IMP = 19.11–0.843 Temp, $r = 0.94$; $p < 0.002$. The three outliers denote non-reproductive krill.

Ligurian sea krill. The trophic input in the former apparently does not suffice to sustain moulting: in the Ligurian Sea krill, growth and reproduction coincides with the very short spring phytoplankton bloom. The Mediterranean krill analysed was either from before this phase or well after (September) when krill is considered to be in a quiescent 'oversummering' mode, typical for the oligotrophic Mediterranean (Saborowski and Buchholz, 2002). Also typical is the stable temperature, close to 13 °C all year in the Ligurian sea. In such thermally stable conditions, the influence of the trophic environment overrides any temperature effect on moulting and growth.

The observation of prolonged IMP in krill in the Gullmarsfjord (Cuzin-Roudy et al., 2004) may also be associated with such a trophic effect. Cuzin-Roudy et al. (2004) estimated IMP in autumn in three consecutive years and found it to fit the temperature-IMP regression in one year, but to be much longer in the two other years. Cuzin-Roudy et al. (2004) considered that the amount of suitable available food was much less in the autumns of the two long IMP years, which was also reflected in the fact that females in the population were in an advanced stage of ovarian regression.

In these cases, the width of the C/D2—window may be the point of adjustment in the timing of moulting, depending on trophic input. However, IMP may be varied by temporal variation of most major moult stages at the same time in response to a suite of other cues, particularly reproductive processes (Buchholz et al., 2006).

Laboratory maintenance is the only way to differentiate the effect of the two parameters IMP and INC on growth rate: an adjustment of growth to environmental cues can be brought about by variation of either or both of these parameters. In Antarctic krill, the typical plasticity in growth is achieved by simultaneous variation of both parameters, including the occurrence of negative growth increments under very unfavourable conditions (Buchholz, 1991).

INC was calculated from the size increase between the cast exoskeleton and the postmoult specimen. Usually, carapace, uropod or antennal scale length are measured in each and compared to determine proportional increase or decrease in length. This can be calibrated against total body length or even body weight to estimate growth rate per moult. Knowledge of the moult rate (or IMP) then allows estimates of absolute growth rates (mm d^{-1}, mg d^{-1}).

A wide variation in INC has been noted in the maintenance experiments of *M. norvegica* cited above. On krill from Gullmarsfjord, our own unpublished measurements showed that INC was very variable, and ranged between -6.0% and $+6.0\%$ per moult with a mean of 0.72 (SD: 3.72; *n*: 25). Cuzin-Roudy *et al.* (2004) state that INC was not statistically different from zero and variable (-0.1% to 0.1%; *n*: 72). Accordingly, these measurements cannot be taken to differentiate relationships in the co-variation of IMP and INC. Maintenance procedures may contribute to these observed variations (see below).

The interdependence of the two moult parameters has not been well studied. However, the common hypothesis (Adelung, 1971; Hartnoll, 1982) is that crustaceans start to accumulate body reserves immediately after the postmoult period. If these suffice, the next pre-moult phase is initiated without pause, eventually leading to a further moult. In most cases, this mechanism implies that the size increase at moult stays constant because it is associated with the status of the body mass present. In contrast, IMP may be adjusted according to the feeding conditions encountered while passing through one moult cycle. The typical INC of brachyuran crabs is near 27%, which corresponds roughly to a doubling in volume of the crab (Adelung, 1971; Hartnoll, 1982). However, in euphausiids, INC is between 4% and 10% and another mechanism of moult prevails (Buchholz, 2003). INC and IMP can be varied at the same time, giving the opportunity to flexibly co-adjust growth alongside energetic demands and energy intake. In this way, moulting may be tuned to the temporally variable and typically patchy trophic environment of the open sea. Under favourable trophic conditions, fast growth rates can be achieved in this way.

2.4. Coordination of moult and reproductive cycles

Cuzin-Roudy and Buchholz (1999) illustrated that moulting and spawning in female *M. norvegica* are interlinked processes. A model was presented in which eggs were released during the early premoult phase in two to three consecutive spawning events. Then moult occurs three days after spawning has been completed. After moult, females enter a 'vitellogenic moult cycle' (VMC) where a new egg batch is developed but not released. After a further moult, they enter a 'spawning moult cycle' (SMC) where the eggs are released in the early premoult stage and the cycle is repeated. Both the VMC and SMC are of equal duration (see Chapter 7).

The bi-phasic spawning-moult cycle appears to be linked to behavioural traits. According to Tarling et al. (1999a), ready to spawn females are almost exclusively found in the surface layer at night, generally closer to the surface than males. Meanwhile, moulting specimens occur deep in the water column. As a consequence, in females, moulting as well as spawning are directly linked to a highly coordinated and cyclical vertical migratory behaviour with a time scale of 24 h which, in turn, depends on a complex set of controlling factors, for example light, predators (avoidance), trophic-, temperature-, and salinity-regimes. Underlying this is an endogenous rhythm synchronised by external factors or 'Zeitgebers', most notably light (see Gaten et al., 2008; Tarling et al., 1999b for further discussion). Accordingly, moulting in Northern krill is closely coordinated with cycles of vertical migration and spawning in order to optimise overall physiological performance. Spawns however occur at a predetermined time window of the SMC, temporally well separated from ecdysis, while vitellogenesis is coupled to the following moult cycle, the VMC. Vitellogenesis, as well as egg maturation is under the control of neuro-hormones, produced in the eyestalks of the krill, which most likely coordinate the whole set of complex cycles (Covi et al., 2009; Subramoniam, 2000).

In summary, vertical migration behaviour, moult and reproductive processes are functionally linked. The temporal coordination of these complex functions, with different time scales, depends on the synchronisation of physiological processes within the individuals. Such synchronisation may also occur at the level of the swarm (Tarling et al., 1999b). However, synchrony in swarms does not seem to persist over longer, annual timescales (see below). Presumably, the situation met in the Tarling et al. (1999b) study may have been typical for a summer situation when external conditions were optimal in terms of temperature and the prevailing feeding conditions.

2.5. Growth versus reproduction

Crustaceans display a certain 'periodicity and phasing' in their growth that is set by their pattern of moulting. Regional as well as seasonal differences have been observed between and within species. The variations are directly or indirectly influenced by temperature regimes and other environmental conditions like photoperiod or food availability (Conan, 1985 and references therein). Animal size, be it species specific or age dependent, is also always inversely correlated with IMPs, as shown for euphausiids by Fowler *et al.* (1971), Mauchline (1980) and Buchholz (1991) and compiled for other crustaceans by Hartnoll (1982). Evolution of genetic traits phasing moult and reproduction seems indispensable since, in most crustaceans, mating occurs between a hard (late postmoult) males and a soft (just postmoult) females. Moreover, the common crustacean trait of brood carrying, where eggs are attached to integumental setae, in some cases only present in a specific 'breeding dress' (e.g. *Palaemonetes varians*, Jefferies, 1964; for other examples see Nelson, 1991), often requires the extension of IMPs because the eggs would otherwise be lost with the exuvia during ecdysis (Conan, 1985; Hartnoll, 1985; Nelson, 1991). For the great majority of euphausiids that spawn freely, this can be discounted. Nevertheless, Hartnoll's (1985) general conclusion that "reproduction can never enhance [somatic] growth but must always restrict it" seemed until recently to be true for euphausiids as well. In a short-term (13 d), small scale (30 × 30 nm) study on 38 swarms (Buchholz *et al.*, 1996) *E. superba* continued moulting during the breeding period, while the familiar pattern of gravid females having a longer IMP (13 d) than mature males (8.3 d) of a similar body length was found. The same swarm study indicated that, after production and spawning of eggs had ceased, females moulted more frequently again. From an extensive study by Tarling *et al.* (2006) it must be concluded that, more commonly, mature female *E. superba* moult more often than mature males. Moreover, female *E. superba* maintained in the laboratory after the final spawning of the year continued moulting, but regressed in sexual maturity and size while extending their IMP (Thomas and Ikeda, 1987). The authors also reported that, after rematuration of external sexual characteristics and concurrent with maturing ovaries, a reduced IMP and positive growth increment lead to an overall size increase compared to the start of the experiment (provided food was sufficient).

In the course of our studies on Northern Krill, sex dependent differences in moulting became conspicuous at three locations and at different time scales (Buchholz *et al.*, 2006). Not only did sexual maturation result in a synchronisation effect on females at the population level (Tarling and Cuzin-Roudy, 2003) and modulate female vertical migration patterns (Tarling *et al.*, 1999b), but it also increased their moult frequency compared to males, at least temporarily. This pattern was observed at both a seasonal scale (in the Clyde population, Scotland) and at a weekly scale (in the Gullmarsfjord population, Sweden and the Kattegat, near Denmark) and through the use of different techniques: the assessment of the ratio of moulting females/moulting males in field samples and the daily checking for moulting in incubated specimens. Overall, female body length was equal to that of males, an indication that the overall net growth was similar, despite the fact that females moulted more often. Furthermore, the shortened IMPs in females were not a result of the abbreviation of specific moult stages, but of an equiproportional shortening across all moult stages, except those in 'moult'. Accordingly, reproductive activity in females did not alter the course of the normal moult cycle.

Thomas and Ikeda (1987) similarly found that moulting and somatic growth continued during egg development in *E. superba*. In both *E. superba* and *M. norvegica*, it is apparent that despite the amount of resource invested in the ovary, there is still spare energy to channel into somatic growth. An increase in size is important since it, in turn, enhances further fecundity given the strong positive correlation between body size and brood size (Cuzin-Roudy, 2000; Mauchline, 1980; Mauchline and Fischer, 1969). Larger individuals also have a higher swimming capacity, which increases their chances of finding new plankton patches and insuring continued energy intake (Buchholz, 1991). Alternatively, large ready to spawn females may have a greater swimming capacity to migrate into the warmer surface layers of the water column which may be beneficial to the development of their brood (Tarling *et al.*, 1999b).

It seems that female Northern krill in a number of different populations are able to combine growth and reproduction in a way that they achieve similar net growth rates to males. This feature is also common to growth patterns in Antarctic krill. It reflects that energy intake in both species must be high to fuel both processes simultaneously, a fact that may also be a reason for the ecological success of this group.

3. PRACTICAL APPROACHES

3.1. Laboratory maintenance and catch of specimens

Experience in laboratory maintenance of euphausiids has shown that *E. superba* is a hardy animal that can be incubated for months in both through-flow and fixed volume aquaria (Buchholz, 1991). Unfortunately, individual juvenile and adult *M. norvegica* do not survive well in aquaria for longer than a maximum of 3–4 weeks. While it is difficult to isolate a single reason, the most probable cause is bacterial infections of the integument. Coryne-type bacteria were isolated and identified from infections in the cuticle of moribund specimens. These bacteria belong to a group with a chitinolytic capacity which may affect the survival of specimens (Buchholz, unpublished). Experiments applying antibiotics were not successful for as yet unknown reasons. Frequent water exchange and/or better flow through systems can alleviate the problem but not solve it completely. A prerequisite is the careful capture of specimens, avoiding damage to the integument, particularly to the long and fragile antennae, filtering basket and pleopods. This begins with the method of catch. Opening/closing nets (Tucker Trawl, MOCNESS) allow catches to be aimed at krill marks on echosounder traces, avoiding lengthy tows, particularly through warmer surface layers. Soft and large mesh netting (min. 2 mm), short trawling times (10 min) and slow ship speed (2 kn) is recommended. After the catch, specimens should immediately be transferred to cooled seawater. After 10 min, they should be further sorted individually into a large bin for acclimation. Prior to physiological experiments, acclimation time in test chambers etc. should last for at least 6 h. Containers for maintenance should not be too small, that is at least 1 l of seawater for a single specimen. Through flow systems, preferably with a current induced in the containers, is advantageous (Fig. 6.4; Buchholz, 1991).

Artemia or natural zooplankton collected with a small net, preferably in combination with cultured phytoplankton, may serve for food. Systems in which the krill exhibit zero or negative growth increments may indicate that the holding conditions are inadequate.

3.2. Moult staging

Details on the categorisation of moult stages are given above. The staging method is best applied to live krill. First, cuticle hardness is tested with fine forceps, particularly at the (anterior) rims of the carapace. Subsequently one or both outer uropods are detached, placed on a slide with cover-glass and analysed under a microscope with 100–400× magnification (Fig. 6.2).

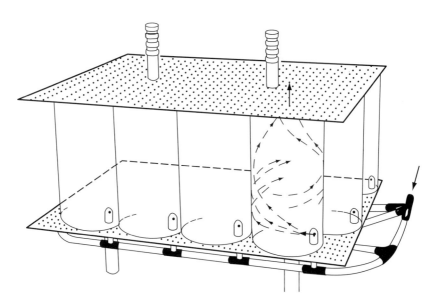

Figure 6.4 Inset for maintenance tanks from Buchholz (1991) (with permission), consisting of 18 Perspex tubes for individual krill. The flow of seawater and resulting laminar current inside a chamber is shown by arrows. The PVC-top is removable.

A phase contrast facility is helpful. Antennal scales may be used instead of uropods with no detectable bias in results.

The technique is partly applicable to defrosted krill, on condition that they are rapidly frozen post-capture and analysed immediately after defrosting (Buchholz, 1991). Krill preserved in formalin is more difficult to categorise in terms of moult stage as the cuticle rigidity is modified through the process of fixation. It may nevertheless be performed successfully given a certain level of experience built up on analysing fresh samples. It is likely that only the major moult stages (post-, premoult) are distinguishable in such preserved specimens.

3.3. Field assessment of IMP and moult activity

Field data attempting to describe the frequency of moulting over the course of the life history of krill are sparse, although there has been some attempts, at a range of time scales. One method is to assess the proportion of 'moulting' or 'non-moulting' specimens at various time points in the year, assuming that moulting is random in the population and each time point provides a good representation of the general level of moulting activity, of which the inverse is a means of estimating IMP (see below). Alternatively, a more sophisticated staging system can be applied, where individual moult stages are identified in laboratory incubated specimens (Buchholz, 1991; Cuzin-Roudy and Buchholz, 1999). Combining both approaches provides insights into moulting patterns over short and long timescales and between different populations, as demonstrated in Buchholz et al. (2006). The approach has been successfully adapted for mysids (Gorokhova, 2002).

The 'moulting' or 'moult active' stages comprise stages A and D3, which last for 14.7%, of the moult cycle, or 1.9 d at a standard temperature of 10 °C. This period is believed to be independent of temperature and nutrition and can thus be considered temporally fixed. Using this value, and having determined the percentage of krill found in stages D3 and A, the duration of the IMP can be assessed in a field population as follows:

$$\text{IMP}(d) = \frac{1.9}{\%D3 + A} \times 100. \tag{1}$$

However, it is not always feasible to distinguish moult stage A from B, since cuticle hardening is still on-going as endo-cuticle laminae are added and sclerotization continues throughout the new shell layers in stage B (Buchholz and Buchholz, 1989). Therefore, an alternative is to include all krill that have not attained complete cuticle rigidity, so adding the duration of stage B (7.4%) to Eq. (1), corresponding to a new total for stages D3, A and B of 2.9 d. The corresponding algorithm is therefore as follows:

$$\text{IMP}(d) = \frac{2.9}{\%\text{D3} + A + B} \times 100. \tag{2}$$

To assess and compare the potential for growth between locations or seasons, the relationship between the numbers of 'moult active' to 'moult inactive' specimens can be determined. The moult active fraction consists of krill actually involved in the immediate preparation for moult, ecdysis itself or the early consolidation of the cuticle after moult. This phase (D3 A B) has been found to comprise 22% of a moult cycle in an actively moulting population, meaning that if random moulting is going on in a population, 22% of the krill can be expected to be found in that phase (see above; Buchholz et al., 2006). Accordingly, comparison of the observed percentage against this number is an indicator of whether a population is active with regards moult and growth or whether it is in a quiescent state and spending proportionally larger amounts of time in, for instance, stages, C and D2 (Buchholz et al., 2010), so decreasing the proportion of moult active specimens.

3.3.1. Moulting frequency
Cuzin-Roudy and Buchholz (1999) assessed 'moulting frequency' in large samples of individually maintained krill. A simple extrapolation to 100% of the cumulative percentage of moulted krill at the end of the maintenance period was used to estimate IMP, given that the time available during the sampling cruise, and thus maintenance period, was shorter than a complete IMP.

3.3.2. Moulting frequency through instantaneous growth rates
In E. superba, the first moult under maintenance conditions is always the moult with the largest INC (Buchholz, 1991). Apparently, the growth increment at moult is more sensitive to maintenance induced effects than the IMP. As a consequence, incubations should be kept as short as possible to avoid such artefacts. This led to development of the 'instantaneous growth rate' method for assessing growth in Antarctic krill (IGRs; Quetin and Ross, 1991) which has since been widely used and improved (Tarling et al., 2006; see Chapter 3). The IGR technique is fully applicable to M. norvegica and other euphausiids (Pinchuk and Hopcroft, 2007).

3.3.3. Moult synchrony
A pre-requisite for all the methods mentioned above is that moulting in the field is random. If this were not the case, data could be strongly biased and growth rate either under- or overestimated, depending on the phase of the moult cycle in which capture and incubations were made. Mass moults and peaks in certain stages of sexual development have been observed indicating

a synchronisation of the moult and reproductive activity in some circumstances (Buchholz et al., 1996; Tarling and Cuzin-Roudy, 2003). However, using the results from an extensive study on 38 swarms of Antarctic krill (Buchholz et al., 1996; Watkins et al., 1992), it was concluded that synchrony in moulting is more the exception than the rule. Furthermore, it may be restricted to single swarms or, smaller aggregations, and probably only lasts for relatively short periods. However, as a precaution, IGR or field moult assessments should not rely on a single sample only. To represent a larger area, at least three different samples taken some distance apart should be analysed independently. It would be desirable also to pay attention to the moult synchronisation of individuals in future swarming studies to be able to improve insights into the extent of this phenomenon.

4. Growth in the Field Along a Climatic Gradient

4.1. Seasonal timing, synchronisation and autonomy of moulting

M. norvegica is mainly found in—often very dense—layers or swarms and is an extremely prolific diurnal vertical migrator. Moult and spawning rhythms are coordinated with vertical migration (Tarling et al., 1999b), where spawning happens in the warm surface layers and moulting in the cool and safer depths. This short-term rhythmicity is linked to a synchronisation of the egg maturation cycle with the moult cycle, which has a period of ~2 weeks (Cuzin-Roudy and Buchholz, 1999). Patterns are also apparent at the seasonal scale: in *M. norvegica*, pulses of enhanced plankton production elicit a synchronisation of both moulting and spawning in a swarm for a certain period of time, after which such coordination slowly vanishes (Tarling and Cuzin-Roudy, 2003). A subsequent pulse of productivity may induce synchrony again. Linked to a flexible omnivorous diet (Lass et al., 2001), such synchronisation may be a plastic physiological reaction to temporal oscillations in the trophic environment. Synchronisation was also found in Antarctic krill (Buchholz, 1985) and a recent study in *Euphausi hanseni* indicated that moult and spawning may be closely controlled by up-welling events which induce sudden changes in trophic conditions (Buchholz, unpublished). Further comparative work on the cues for physiological rhythmicity and synchronisation in krill swarms would be highly desirable in pelagic species.

On an individual basis, a further interesting trait of krill is its tendency to continue moulting even in unfavourable conditions, for example extended periods of starvation, when they may even shrink at moult (Buchholz, 2003). Apparently, krill moult autonomously, that is de-coupled from the necessity to grow, although the moult rate may vary to some degree in line with conditions. It indicates that the process is controlled by a strong endogenous rhythm. Ultimately, the trait may provide some selective advantage in ensuring a clean and fresh epicuticle, free of drag-inducing epibionts, which may otherwise hinder its fast swimming-capability (Buchholz and Buchholz, 1989; Buchholz et al., 2006).

4.2. Life-growth assessments

A way to illustrate growth over time is to plot body length versus month of the year over the life time of the species. Einarsson (1945) provided one of the first such plots using data collected over a wide geographic range, from Iceland to the NE Atlantic. Accordingly, a seasonal growth pattern was described. This was used by Boysen and Buchholz (1984) as a reference for Northern krill in the Danish Kattegat (Fig. 6.5).The curve shows that growth slows with increasing age. Furthermore, the growth period is closely correlated to seasonal food availability, being most pronounced from April to October in the 0-group. During the winter months, when the abundance of phyto- and zooplankton is scarce, there is little increase in size. In the II-group, growth re-gains momentum towards summer. In the case of the 0-group, a wide range in lengths are apparent. This phenomenon is due to the long and irregular spawning time, from April to September. The spread

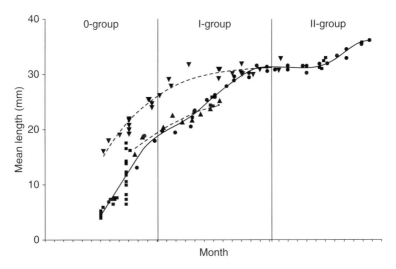

Figure 6.5 Hypothetical Life-Growth curves of *Meganyctiphanes norvegica* in NE Atlantic waters and the Mediterranean. Dots: Boysen and Buchholz (1984), Kattegat; squares: Einarsson (1945), NE Atlantic; triangles: Falk-Petersen and Hopkins (1981), Balsfjorden; triangles (pointing down): Labat and Cuzin-Roudy (1996), Ligurian sea. Vertical lines: annual limits, December–January. Reproduced with permission.

of the data diminishes considerably with increasing age. Within this data, the life span of Northern krill is 2.5 years during which it attains a maximal length of 36 mm. Other datasets presented by Tarling (Chapter 3) show that Northern krill may reach beyond 3 years of age in some more northerly locations.

In the Mediterranean Northern krill population (Labat and Cuzin-Roudy, 1996), recruitment starts earlier in the year so the 0-group krill are generally more advanced and larger than other populations relative to calendar date. However, the actual trajectory of growth in 0-group individuals is similar to other populations and they converge at the I-group stage (~30 mm body length). Throughout the summer months, growth is relatively low in I-group individuals from this population, which probably reflects the poor trophic conditions in summer in the Mediterranean. A true II-group is missing in this population, probably due to predation mortality.

Data points from Falk-Petersen and Hopkins (1981) from Balsfjorden (69°21'N; 19°06'E) are included in Fig. 6.5 to represent high latitude populations. These generally fit the NE Atlantic population trajectory (Einarsson, 1945). The krill are presumably advected from the open Norwegian sea and may reflect its environmental condition rather than the fjord situation. Norwegian sea krill are believed to contain the northernmost self sustaining population of *M. norvegica* (Dalpadado *et al.*, 2008; Buchholz *et al.*, 2010).

'Fjord-krill' grow faster and more continuously and attain a larger size than open-ocean populations. This is true for western Norwegian fjords (Wiborg, 1971; see Boysen and Buchholz, 1984, also for more comparisons) as well as the western Swedish Gullmarsfjord and the Scottish Clyde sea: the growth curves (Cuzin-Roudy *et al.*, 2004) are similar to the NE Atlantic curve during the first year (0-group), but show much faster and continued growth through the second year (I-group). Accordingly, end sizes of up to 43 mm are reached. This probably reflects krill profiting from the enhanced plankton production through nutrient fertilisation from the land surrounding fjords (Buchholz *et al.*, 2010).

Interestingly, Siegel (2000) states that the life-growth curve in Antarctic krill has a similar trajectory to the Northern krill curve, with the difference being that Antarctic krill continues to grow over 5 years to an end-size of 60mm—whereas the Northern krill stops growing at around 2.5 years at a size of around 35–40 mm. It suggests that the physiological performance of the two species is comparable, at least in the first two years of life.

As a conclusion, *M. norvegica* in the open waters of the North Atlantic and fringe seas show a comparable seasonal growth pattern. Apparently, growth is much dependent on the trophic situation and seasonal pattern of nutrition rather than the latitudinal position, that is climatic influences. Variation in local environmental and trophic regimes affects other body dimensions also. For instance, differences in weight are noted which run on a different time course to increases in body length. According to the nutritional situation, krill of the same length may have a different weight—with the ratio of tissue to water altering between specimens (see discussion in Boysen and Buchholz, 1984). Much of the difference is seen in the protein and lipid fractions, which constitute the reserves used for ovary maturation and overwintering (Falk-Petersen *et al.*, 2000).

5. An Outlook: Growth, Productivity and Environmental Change

M. norvegica must be considered an eurythermal species since it occurs from the Arctic to the Mediterranean, over which area, mean sea-surface temperature changes by over 13 °C. Indeed, Northern krill has been used as a model organism to consider adaptation in the marine pelagic environment within a cooperative European project entitled "Impact of a climatic gradient on the Physiological Ecology of a Pelagic crustacean (PEP)". Populations were studied from the thermally variable Danish Kattegat, the cool and stable Atlantic Clyde Sea and the warm and stable Ligurian sea.

From the above project and other studies on growth, reproduction and physiological performance, it has been shown that nutritional flexibility and high-energy expenditure are characteristic features of *M. norvegica* and that fitness appears to be maximised through seasonal shifts in behavioural and physiological traits. Molecular work has shown some unique features in Northern krill, helping to explain functional diversity in the species (Zane *et al.*, 2000). Physiological plasticity appears to be the key to the euphausiid's exceptional ecological success in various environments, brought about by fast growth and reproduction (Buchholz and Saborowski, 2000).

These studies enable us to come to some preliminary conclusions with regards the resilience of this species to climate variability and change. The Northern krill populations spread around the various margins of the NE Atlantic differ in certain traits but are not genetically separated from each other (see Chapter 2), even if their local climates and thermal regimes are distinctly different. A further unexpected finding of the genetic analysis was that an Atlantic population off Cadiz, S. Spain, differed substantially from all the other populations investigated. The swarm sampled may have been advected from warmer Atlantic regions, presumably from near the Canary Islands. This finding indicates that the species may tolerate even warmer temperatures than experienced in the Mediterranean. Although the physiological properties of this southern population are not known, it illustrates that *M. norvegica* can in fact cope with an extremely wide range of (changing) environmental temperatures, demonstrating a high degree of phenotypic plasticity with regards eurythermy. Accordingly, krill may easily adjust to the higher temperatures, predicted by many future climate scenarios. However, metabolic energy expenditure increases strongly with rising temperature, potentially leading to the loss of energy needed for maintenance of growth and reproduction. As a vertically migrating species, Northern krill may adapt through altering its vertical distribution or the amount of time spent in surface waters.

In general, Northern krill swarms or populations are found in areas where feeding is enhanced by environmental conditions, particularly linked to the existence of frontal systems, for example at the continental slopes or generally, areas of enhanced advection. Increase in sea temperatures may cause changes in the species composition of phyto-, and zooplankton communities associated with these food-rich areas. Due to the versatility in food choice in Northern krill, this may not necessarily create a major problem and the species may even benefit from certain changes. The well developed vertical and horizontal swimming capacity of adults make it capable of searching out alternative food patches or relocating to other frontal systems.

The northernmost record of Northern krill populations had been the Arctic Barents Sea which are, so far, considered non-reproductive (Dalpadado et al., 2008). However, there are increasing numbers of reports from the west-Spitsbergen area to which M. norvegica seems to be regularly advected. In the Kongsfjord, there are even the first indications of reproductive processes in krill (Buchholz et al., 2010). These observations may be taken as a further indication of a warming effect or a change in food web composition which may be advantageous to Northern krill, enabling it to expand its distributional range northwards. Accordingly, Northern krill may serve as a useful indicator of change. Their proliferation in new environments may also have implication to those food webs, given their high levels of productivity and their pivotal trophic role.

ACKNOWLEDGEMENTS

Part of the work was funded by the German Science Foundation (DFG) grants Bu 548 1-2 and EU-MAST III (Marine Science and Technology) grant MAS3-CT95-0013, The PEP Programme: Impact of a climatic gradient on the Physiological Ecology of a Pelagic crustacean. On behalf of the 'PEP-community' and their followers Geraint Tarling is thanked for his lasting and consistent enthusiasm and interest in its target: the Northern krill, which in the end led to the collation of the current review. Last not least we are grateful to Geraint and Paul Seear for considerable improvements of the text. Figures and tables from Mar Biol and MEPS were reproduced with kind permission of Springer Science+Business Media, those from J Plankton Res of Oxford University Press.

REFERENCES

Adelung, D. (1971). Untersuchungen zur Häutungsphysiologie der dekapoden Krebse am Beispiel der Strandkrabbe Carcinus maenas L. Helgoländer wiss. Meeresunters. 22, 66–119.

Boysen, E., and Buchholz, F. (1984). *Meganyctiphanes norvegica* in the Kattegat. Studies on the development of a pelagic population. *Mar. Biol.* **79**, 195–207.

Buchholz, F. (1982). Drach's molt staging system adapted for euphausiids. *Mar. Biol.* **66**, 301–305.

Buchholz, F. (1985). Moult and growth in euphausiids. *In* "Antarctic Nutrient Cycles and Food Webs, Proceedings of the 4th SCAR Symposium on Antarctic Biology" (W. R. Siegfried, P. Condy and R. M. Laws, eds.), pp. 339–345. New York, Springer.

Buchholz, F. (1991). Moult cycle and growth of Antarctic krill, *Euphausia superba*, in the laboratory. *Mar. Ecol. Prog. Ser.* **69**, 217–229.

Buchholz, F. (2003). Experiments on the physiology of Southern and Northern krill, *Euphausia superba* and *Meganyctiphanes norvegica*, with emphasis on moult and growth— A review. *Mar. Freshw. Behav. Physiol.* **36**, 229–247.

Buchholz, C., and Buchholz, F. (1989). The ultrastructure of the Integument of a pelagic crustacean: Moult cycle related studies on *Euphausia superba*. *Mar. Biol.* **101**, 355–365.

Buchholz, F., and Saborowski, R. (2000). Metabolic and enzymatic adaptations in the Northern and Southern krill, *Meganyctiphanes norvegica* and *Euphausia superba*. *Can. J. Fish. Aquat. Sci.* **57**(Suppl. 3), 115–129.

Buchholz, C. M., Pehlemann, F. W., and Sprang, R. R. (1989). The cuticle of krill (*Euphausia superba*) in comparison to that of other crustaceans. *Pesq. Antart. Bras.* **1**, 103–111.

Buchholz, F., Watkins, J. L., Priddle, J., Morris, D. J., and Ricketts, C. (1996). Moult in relation to some aspects of reproduction and growth in swarms of Antarctic krill, *Euphausia superba*. *Mar. Biol.* **127**, 201–208.

Buchholz, C., Buchholz, F., and Tarling, G. A. (2006). On the timing of moulting processes in reproductively active Northern krill *Meganyctiphanes norvegica*. *Mar. Biol.* **149**, 1443–1452.

Buchholz, F., Buchholz, C., and Weslawski, J. M. (2010). Ten years after: Krill as indicator of changes in the macro-zooplankton communities of two Arctic fjords. *Polar Biol.* **33**(1), 101–113.

Conan, G. Y. (1985). Periodicity and phasing of molting. *In* "Crustacean Issues 3" (F. A. Schram and A. M. Wenner, eds.), pp. 73–99. A. A. Balkema, Rotterdam, Boston.

Covi, J. A., Chang, E. S., and Mykles, D. L. (2009). Conserved role of cyclic nucleotides in the regulation of ecdysteroidogenesis by the crustacean molting gland. *Comp. Biochem. Physiol. A* **152**, 470–477.

Cuzin-Roudy, J. (2000). Seasonal reproduction, multiple spawning, and fecundity in northern krill, *Meganyctiphanes norvegica*, and Antarctic krill, *Euphausia superba*. *Can. J. Fish. Aquat. Sci.* **57**, 6–15.

Cuzin-Roudy, J., and Buchholz, F. (1999). Ovarian development and spawning in relation to the moult cycle in Northern krill, *Meganyctiphanes norvegica* (Crustacea: Euphausiacea), along a climatic gradient. *Mar. Biol.* **133**, 267–281.

Cuzin-Roudy, J., Tarling, G. A., and Strömberg, J.-O. (2004). Life cycle strategies of Northern krill (*Meganyctiphanes norvegica*) for regulating growth, moult, and reproductive activity in various environments: The case of fjordic populations. *ICES J. Mar. Sci.* **61**, 721–737.

Dalpadado, P., Ellertsen, B., and Johannessen, S. (2008). Inter-specific variations in distribution, abundance and reproduction strategies of krill and amphipods in the Marginal Ice Zone of the Barents Sea. *Deep Sea Res.* **55**, 2257–2265.

Drach, P. (1939). Mue et cycle d'intermue chez les Crustaces decapodes. *Ann. Inst. Oceanogr (Monaco)* **19**, 103–391.

Einarsson, H. (1945). Euphausiacea. I. Northern Atlantic species. *Dana-Rep. Carlsberg Found* **27**, 1–184.

Falk-Petersen, S., and Hopkins, C. C. E. (1981). Ecological investigations on the zooplankton community of Balsfjorden, northern Norway: Population dynamics of the euphausiids Thysanoessa inermis (Kroyer), Thysanoessa raschii (M. Sars) and *Meganyctiphanes norvegica* (M. Sars) in 1976 and 1977. *J. Plankton Res.* **3**, 177–192.

Falk-Petersen, S., Hagen, W., Kattner, G., Clarke, A., and Sargent, J. (2000). Lipids, trophic relationships, and biodiversity in Arctic and Antarctic krill. *Can. J. Fish. Aquat. Sci.* **57** (Suppl. 3), 178–191.

Fowler, S. W., Small, L. F., and Keckeš, S. (1971). Effects of temperature and size on molting of euphausiid crustaceans. *Mar. Biol.* **11**, 45–51.

Gaten, E., Tarling, G., Dowse, H., Kyriacou, C., and Rosato, E. (2008). Is vertical migration in Antarctic krill (*Euphausia superba*) influenced by an underlying circadian rhythm? *J. Genet.* **87**, 473–483.

Gorokhova, E. (2002). Moult cycle and its chronology in *Mysis mixta* and *Neomysis integer* (Crustacea, Mysidacea): implications for growth assessment. *J. Exp. Mar. Biol. Ecol.* **278**, 179–194.

Hartnoll, R. G. (1982). Growth. *In* "The Biology of Crustacea 2" (D. E. Bliss and L. G. Abele, eds.), pp. 111–196. Academic Press, New York, NY.

Hartnoll, R. G. (1985). Growth, sexual maturity and reproductive output. *In* "Crustacean Issues 3" (F. A. Schram and A. M. Wenner, eds.), pp. 101–128. A.A. Balkema, Rotterdam.

Jefferies, D. J. (1964). The moulting behaviour of *Palaemonetes varians* (Leach) (Decapoda; Palaemonidae). *Hydrobiologia* **24**, 457–488.

Johnson, C. L. (2003). Ecdysteroids in the oceanic copepod *Calanus pacificus*: Variation during molt cycle and change associated with diapause. *Mar. Ecol. Prog. Ser.* **257**, 159–165.

Labat, J. P., and Cuzin-Roudy, J. (1996). Population dynamics of the krill *Meganyctiphanes norvegica* (M. Sars, 1857) (Crustacea: Euphausiacea) in the Ligurian Sea (NW Mediterranean Sea), Size structure, growth and mortality modelling. *J. Plankton Res.* **18**(12), 2295–2312.

Lass, S., Tarling, G. A., Virtue, P., Matthews, J. B. L., Mayzaud, P., and Buchholz, F. (2001). On the food of northern krill *Meganyctiphanes norvegica* in relation to its vertical distribution. *Mar. Ecol. Prog. Ser.* **214**, 177–200.

Mauchline, J. (1980). The Biology of Mysids and Euphausiids. Academic Press, London \New York\Toronto\Sydney\San Francisco.

Mauchline, J., and Fischer, L. R. (1969). The Biology of Euphausiids. Academic Press, London\New York.

Nakatsuji, T., Lee, C. Y., and Watson, R. D. (2009). Review: Crustacean molt-inhibiting hormone: Structure, function, and cellular mode of action. *Comp. Biochem. Physiol. A* **152**(2), 139–148.

Nelson, K. (1991). Scheduling of reproduction in relation to molting and growth in malacostracan crustaceans. *In* "Crustacean Issues 7" (F. A. Schram, A. W. Wenner and A. Kuris, eds.), pp. 77–113. A. A. Balkema, Rotterdam, Brookfield.

Pinchuk, A. I., and Hopcroft, R. R. (2007). Seasonal variations in the growth rates of euphausiids (Thysanoessa inermis, T-spinifera, and Euphausia pacifica) from the northern Gulf of Alaska. *Mar. Biol.* **151**, 257–269.

Quetin, L. B., and Ross, R. M. (1991). Behavioral and physiological characteristics of the Antarctic krill *Euphausia superba*. *Am. Zool* **31**, 49–63.

Saborowski, R., and Buchholz, F. (2002). Metabolic properties in Northern Krill, *Meganyctiphanes norvegica*, from different climatic zones: Enzyme characteristics and activities. *Mar. Biol.* **140**, 557–565.

Seear, P., Tarling, G. A., Teschke, M., Meyer, B., Thorne, M. A. S., Clark, M. S., Gaten, E., and Rosato, E. (2009). Effects of simulated light regimes on gene expression in Antarctic krill (*Euphausia superba* Dana). *J. Exp. Mar. Biol. Ecol.* **381,** 57–64.

Shechter, A., Tom, M., Yudkovski, Y., Weil, S., Chang, S. A., Chang, E. S., Chalifa-Caspi, V., Berman, A., and Sagi, A. (2007). Search for hepatopancreatic ecdysteroid-responsive genes during the crayfish molt cycle: From a single gene to multigenicity. *J. Exp. Biol.* **210,** 3525–3537.

Siegel, V. (2000). Krill (Euphausiacea) demography and variability in abundance and distribution. *Can. J. Fish. Aquat. Sci.* **57**(Suppl.), 151–167. (Proceedings of the Second International Symposium on Krill).

Subramoniam, T. (2000). Crustacean ecdysteriods in reproduction and embryogenesis. *Comp. Biochem. Physiol.* **125**(2), 135–156.

Tarling, G. A., and Cuzin-Roudy, J. (2003). Synchronization in the molting and spawning activity of northern krill (*Meganyctiphanes norvegica*) and its effect on recruitment. *Limnol. Oceanogr.* **48,** 2020–2033.

Tarling, G. A., Matthews, J. B. L., and Buchholz, F. (1999a). The effect of a lunar eclipse on the vertical migration behaviour of *Meganyctiphanes norvegica* (Crustacea: Euphausiacea) in the Ligurian Sea. *J. Plankton Res.* **21,** 1475–1488.

Tarling, G. A., Cuzin-Roudy, J., and Buchholz, F. (1999b). Vertical migration behaviour in the northern krill *Meganyctiphanes norvegica* is influenced by moult and reproductive processes. *Mar. Ecol. Prog. Ser.* **190,** 253–262.

Tarling, G. A., Shreeve, R. S., Hirst, Atkinson A., Pond, D. W., Murphy, E. J., and Watkins, J. L. (2006). Natural growth rates in Antarctic krill (*Euphausia superba*): I. Improving methodology and predicting intermolt period. *Limnol. Oceanogr.* **51,** 959–972.

Teschke, M., Kawaguchi, S., and Meyer, B. (2008). Effects of simulated light regimes on maturity and body composition of Antarctic krill. *Euphausia superba. Mar. Biol.* **154,** 315–324.

Thomas, P. G., and Ikeda, T. (1987). Sexual regression, shrinkage, re-maturation and growth of spent female *Euphausia superba* in the laboratory. *Mar. Biol.* **95,** 357–363.

Watkins, J. L., Buchholz, F., Priddle, J., Morris, D. J., and Ricketts, C. (1992). Variation in reproductive status of Antarctic krill swarms; evidence for a size-related sorting mechanism? *Mar. Ecol. Prog. Ser.* **82,** 163–174.

Wiborg, K. F. (1971). Investigations of the euphausiids in some fjords on the west coast of Norway in 1966–1969. *Fisk. Dir Skr. Ser. HavUnders.* **16,** 10–35.

Zane, L., Ostellari, L., Maccatrozzo, L., Bargelloni, L., Cuzin-Roudy, J., Buchholz, F., and Patarnello, T. (2000). Genetic differentiation in a pelagic crustacean (*Meganyctiphanes norvegica*, Euphausiacea), from the North East Atlantic and the Mediterranean Sea. *Mar. Biol.* **136,** 191–199.

CHAPTER SEVEN

REPRODUCTION IN NORTHERN KRILL (*MEGANYCTIPHANES NORVEGICA* SARS)

Janine Cuzin-Roudy

Contents

Abstract

This review presents the current state of knowledge with regard to the reproductive biology of Northern krill (*Meganyctiphanes norvegica*). Reproduction is limited to a distinct period of the year. First development of the ovary occurs at the

Observatoire Océanologique, Université Pierre et Marie Curie (Paris)-CNRS, Villefranche-sur-Mer, France

Advances in Marine Biology, Volume 57

ISSN 0065-2881, DOI: 10.1016/S0065-2881(10)57007-X

onset of the season, when the stock of primary oocytes issued from the germinal zone starts to accumulate glycoproteic yolk. Previtellogenesis continues through-out the entire reproductive season, but oosorption (the retrieval by the ovary of the yolk constituents from the growing oocytes) may occur in unfavourable conditions and represents an important metabolic process for sustaining females during such periods. Oosorption also occurs at the onset of the resting season.

It has been established that individual females may perform several cycles of reproduction each year. Each reproductive cycle spans two moult cycles, one in which lipid yolk is accumulated (vitellogenesis) and another when spawning occurs. The time of spawning does not coincide with the moult (*ecdysis*), but with the onset of moult preparation (C-D0 moult stages). The complete egg-batch is spawned well before the moult.

Storage lipids are accumulated preferentially in the ovary with distinctly high levels of ω-3 polyunsaturated fatty acids in the polar lipid fraction as well as phosphatidylcholine, a key component in the development of the embryo. There is no difference concerning lipid storage between resting females, males and juvenile krill.

Beside the ovary, the fat body is an important organ involved in the metabo-lism and storage of the glycoproteins and lipids that will be transformed into the lipoglycoproteins of the yolk platelets in the ovary.

M. norvegica produce large egg batches with the number of mature oocytes in one batch being proportional to the size of the female, with a mean number of 1000–1200 eggs per batch. The number of reproductive cycles per year is a function of the trophic conditions, with the first reproductive cycle being trig-gered by the first phytoplankton bloom.

Other reproductive features reflect specific adaptations of krill to a pelagic life, like swarming and vertical migration behaviour. *M. norvegica* segregate at night for moulting and mating or spawning, while swimming constantly during their diel vertical migration (DVM).

Key questions concerning krill reproduction remain, particularly in identifying the cues that switch krill in and out reproductive development, or between egg-building and oosorption. New molecular tools are now available to tackle such questions.

1. INTRODUCTION

The first major consideration of sexual features in *Meganyctiphanes. norvegica* (M.Sars) was carried out by Einarsson (1945), in his extensive report about northern euphausiids species. The adult gonad structure was first described by Raab (1915) using classical histological techniques and serial sections, for *M. norvegica* and *Euphausia krohni*. Many features of reproduction in *M. norvegica* resemble those in *E. superba*, as first described by Bargmann (1937, 1945) and then by Ross and Quetin (1983, 1986, 2000). Another euphausiid species that has been studied at a similar level of detail is *E. pacifica* (Feinberg *et al.*, 2007; Gomez-Gutierrez *et al.*, 2006; Ross *et al.*, 1982).

Most studies have considered reproduction in *M. norvegica* from the point of view of the population dynamics of the species (Lindley, 1982;

Mauchline, 1985) and focused on estimating growth, production, fecundity, recruitment, longevity and mortality. The morphology of external (secondary) sexual characters, the size and age of sexual maturity and spawning were established using morphological observations of krill from field preserved samples (Mauchline, 1980; Mauchline and Fisher, 1969).

Sexual differentiation and maturity from the juvenile to adult state were assigned according to the development of secondary sexual characters: in males, the *petasmata* (s. *petasma*) are *appendix masculina* situated on the internal *ramus* of the first pleopods, and *ampullae*, situated on the 7th thoracic sternite, are for the storage of spermatophores. In females the *thelycum*, a receptacle for the sperm, is situated on the 6th sternite (Mauchline, 1981; Mauchline and Fisher, 1969). A scoring system was developed, based on external morphological characteristics of *E. superba* (Makarov and Denys, 1980; Mauchline, 1981), which has since been adapted for *M. norvegica* (Buchholz and Prado-Fiedler, 1987), and is still in general use for population studies.

Nevertheless, simply examining the external appearance of a krill with regard to secondary sexual characters does not provide any information on the development of the ovary for egg production (or the testis spermatozoa), or on the timing of the physiological processes associated with reproductive development. For instance, the live *M. norvegica* reproductive female presented in Fig. 7.1 would be scored as 'ready-to-spawn' (III D, Makarov and Denys, 1980), but the ovary is still in the process of yolk accumulation and not ready for egg release. Gaining a fuller understanding of *M. norvegica* reproductive cycles, their seasonal timing and their links with environmental variations has required a number of detailed studies that will be described further in this chapter.

Due to their entirely pelagic life cycle, their swarming and migrating behaviour, krill and especially *M. norvegica*, are not expected to develop normally when confined to laboratory conditions for long durations. However, live observations of freshly captured krill samples or of krill kept in controlled laboratory conditions for short durations (Buchholz *et al.*, 2006) afford new insights into the functioning of the gonads, and other physiological

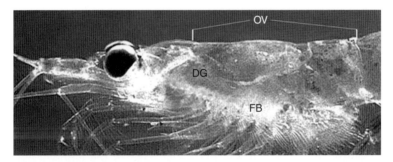

Figure 7.1   *M. norvegica.* Live female developing an egg batch. Note the extension of the ovary inside the thoracic cavity from the stomach to the 2nd abdominal segment. The large nuclei of the future eggs indicate that the oocytes were still in the process of lipid yolk accumulation (SDS 3). DG, digestive gland; FB, fat body; OV, ovary.

processes involved in reproduction. Preserved krill samples are also a source of critical information in this regard, particularly when related back to the environmental context within which they were captured.

In Malacostraca crustaceans, the gonad is not differentiated during the larval and early juvenile stages (references in Charniaux-Cotton and Payen, 1985). The gonads of both genetic males and females contain only undifferentiated sexual cells, or *gonia* (s. *gonium*), associated with mesodermal cells that will form the gonoducts. In genetic females, the ovary, the oviduct and the female secondary sexual characters (which is the *thelycum* for krill), differentiate spontaneously and are maintained during the whole life cycle. In genetic males, the transformation of *gonia* into *spermatogonia* starts only after the formation of the androgenic gland (Payen, 1980). The androgenic hormone produced by the gland is also responsible for the development of male primary (*spermatogonia*) and secondary sexual characters (*petasmata* and *ampullae* in krill) and also for the conservation of testis activity during the whole life cycle. In the absence of the androgenic gland, all the male characteristics, including the *spermatogonia*, disappear and the testis goes back to an ovarian structure (Meusy and Payen, 1988).

It is likely that *M. norvegica* follows the same sequence of development as other malacostracans with regard to the differentiation of the gonads in juveniles and the acquisition of sexual maturity by adults. Consequently, an individual's sex cannot be determined from external examination of morphological characteristics during larval and juvenile phases of krill, before the appearance of secondary sexual characters in sub-adults and gonad activity in adults, producing sperm or eggs (Mauchline, 1980; Mauchline and Fisher, 1969). For population studies and the analysis of individual krill in preserved random samples, adults are usually considered to be sexually mature when developed *petasmata* and *ampullae* filled with spermatophores are present in males, and a red *thelycum* bearing a sperm plug present in females, indicating that mating has already occurred, that is, using the same characteristics as the ones used for Antarctic krill (Makarov and Denys, 1980; Quetin and Ross, 1991). However, the following section describes detailed examinations that have revealed the physiological underpinning of the cycle of reproduction in *M. norvegica*.

2. From Gonad Differentiation to Adult Maturity

2.1. Differentiation of the gonad in larvae and early juveniles

In Euphausiacea, the germ cells originate, together with endoderm, from a single blastomere (called DII) of the 4-cell stage of the embryo and they separate from the endoderm at the 32-cell stage (Charniaux-Cotton and Payen, 1985) to form an undifferentiated gonad, like in other malacostracans.

The differentiation of the gonads has not been studied in detail for *M. norvegica*, but was described for *E. superba* (Cuzin-Roudy, 1987a), using live observations and histological sections of successive developmental stages including larvae (*furciliae*), juvenile (0-group, see Chapter 3) and young mature adult krill. Furciliae and early juvenile stages, up to a total body length of 23 mm (Standard length 2, Mauchline, 1981) all had undifferentiated gonads containing only *gonia* 1 and 2, which appeared as very small ovaries (see Cuzin-Roudy, 1987a for further details). Ovaries with *oogonia* 1 and 2 and primary young oocytes (*yoc*) were found in 0-group krill. These persist as the principal germinal zone (GZ) in the developed ovary (the GZ is indicated in Fig. 7.3). Full development of the *thelycum* bearing a sperm plug occurred in females of I-group or older (I-group+). In males, the gonad was not differentiated until the appearance of the first budding of the *petasmata* digits, in 0-group krill. Fully active testes, in spermatogenesis, with spermatocytes 1 and 2, spermatids and *spermatozoa*, were found in I-group+ males, bearing also fully developed *petasmata* and spermatophores in their *ampullae*. Despite a different control of gonad differentiation, both male and female *E. superba* attained full sexual maturity at similar ages and body sizes: 24–33 mm for females and 24–36 mm for males as confirmed by Siegel and Loeb (1994).

This same process of gonad differentiation is likely to occur in *M. norvegica* as confirmed by the size structure and cohort analysis of various populations (see Chapter 3).

2.2. Sexual development in sub-adults

2.2.1. Sub-adult stages

Euphausiids moult and grow over their entire life cycle. There is no 'adult moult' that signifies the arrest of postembryonic development and the initiation of reproduction, as occurs in insects and some decapods. Consequently, there is no evidence in krill for a control of adult sexual development by a juvenile hormone-like substance, like the methyl farnesoate secreted by the mandibular organ in the crab *Libinia emarginata* (Laufer *et al.*, 1987; Meusy and Payen, 1988).

Secondary sexual characters of euphausiids (Mauchline and Fisher, 1969) develop progressively in late juvenile krill. The 'sub-adults' are essentially the transitional stage between juveniles and adults. The development of secondary sexual characters in *M. norvegica* proceeds over a number of successive moults. This development is presumably under the control of hormones produced by the ovary in females, and by the androgenic gland in males, as experimentally demonstrated in some peracarids and decapods Penaeidea or Caridea (Charniaux-Cotton and Payen, 1985). The males of the later taxa develop an '*appendix masculina*' on the first pleopods equivalent to the krill petasmata, which suggest a similar endocrine control by the androgenic hormone for *M. norvegica*. A pair of glandular sacs situated on the sperm ducts of *M. norvegica*, described by Raab (1915), have been recognised as the androgenic glands by Zerbib (1967). They are easily visible from both sides through the transparent carapace of live krill (personal observation), as a pair of irregular pinkish expansions on the distal part of each ejaculatory canal on its way to the ventral *ampullae* (7th thoracic sternite).

2.2.2. Male sexual maturity

Mature adult males of *M. norvegica* have, besides a developed testis, complete *petasmata* on the first pair of pleopods, one pair of *ampullae* on the 7th thoracic sternite, paired spermatophore sacs (Bargmann, 1937) on the ejaculatory canal and one pair of androgenic glands.

The adult *petasma* of *M. norvegica* was described by Einarsson (1945) with four long processes and some small additional ones. The development of this complex structure has been described in detail in other euphausiids species as it is one of the main means of species identification (Mauchline and Fisher, 1969). Various aspects of the growth and development of the *petasmata* were observed (personal observations) in a cohort of live male sub-adults of *M. norvegica* captured during a campaign in the Ligurian Sea in May and maintained on board ship for moulting. Three successive sub-adult stages were evident from examination of the *exuvia* (the old cuticle casts) of the newly moulted krill. They were characterized by: (1) the appearance of a pair of small buds on the anterior surface of the internal *ramus* of the first pleopods in krill that were around 24 mm total body length; (2) short fingers replacing the buds after the next moult in krill of 26 mm; (3) long and thin fingers approaching the final shape of the adult *petasma* in krill of 32 mm. Fully developed petasmata, with all the digits and spines of the adult shape (Einarsson, 1945) were present in young adults beyond 32 mm. Over the next adult moults, the *petamata* grow in proportion to the size of the krill, without changing their final adult shape.

2.2.3. Female sexual maturity

Mature adult females have a red *thelycum* on the 6th thoracic sternite (Einarsson, 1945), generally filled with a white sperm plug and one or two empty spermatophores attached to the thelycum (Mauchline, 1980).

As with the males, the development of the *thelycum* also occurs over several moult cycles, but it is difficult to follow the progressive growth of the two expansions of epiderm that comprise it: the expansions originate from the paired bases of the 6th thoracic appendages and merge together ventrally to form a cup that receives and retains the sperm plug (Einarsson, 1945). The *thelycum* becomes red when it is ready to receive the plug. During the reproductive season, mature females with an empty red thelycum are rare in field samples, because, although the plug is lost each time the female moults, it is normally refilled at mating immediately afterwards.

2.3. Mating

In field samples, as soon as the sexual secondary characters are fully developed, females with a complete *thelycum*, generally red in colour, are mated and bear a sperm plug in the *thelycum* until their next moult. As in many crustaceans (Adiyodi, 1985), adult females are receptive immediately after moulting once the old cuticle has been shed and the new cuticle is still soft. Like for other crustaceans, it is believed that males are attracted by pheromones liberated by the female during moulting, and various substances have been suggested but never proved to be involved. A likely candidate is β-hydroxyecdysone, the moulting hormone of crustaceans that is released at moulting with the exuvial fluid.

Quetin and Ross (2001) report that male *E. superba* exhibit a specific courtship behaviour and chase the female while swimming fast around her. Mating in *M. norvegica* has been achieved in captivity when ready to moult females were incubated with one or two adult males (personal observation). The females bore one or two pairs of spermatophores after the mating.

The actual process of mating is difficult to observe in *M. norvegica*, possibly due to the extreme sensitivity of this species to light, which may disturb a process that normally occurs at night. The transfer of the spermatophores by the male, from the male ampullae to the female thelycum, where they are glued precisely, is so rapid that no recorded images are yet available. Putatively, the *petasmata* are likely to play an important role in this complex process.

2.4. Seasonal reproduction

Like for most malacostracans (Charniaux-Cotton and Payen, 1985), once the ovary and the thelycum have developed in *M. norvegica* (during the sub-adult stages), they remain for the rest of the lifespan. In *E. superba*, by contrast, these structures regress during the non-reproductive season. Common to both these species is the fact that the activity of the ovary for producing eggs occurs during only a limited period of the year, the reproductive season, when successive batches of eggs are produced cyclically (Cuzin-Roudy, 1993).

The lapse of time between the acquisition of sexual maturity and the first spawn of the season is variable between populations, depending on their adaptation to local environmental conditions. The first sign of activity in the ovary is the development of the *yoc*, that have been blocked in meiotic prophase (diacynesis) since their formation in the differentiated ovary and now start to increase in size by accumulating *vitellus* (or yolk). The process culminates in the formation of a batch of large mature oocytes ready for egg release. Meiosis progresses in these oocytes until it is blocked at the first meiotic metaphase, where it remains until spawning. During spawning, the polar globules are expelled and then the secondary oocytes are fertilised as they come into contact with the sperm plug during their exit from the oviducts, as reported in *E. superba* (Tarling *et al.*, 2009). Consequently, female sexual maturity should not be confused with oocyte maturity, since oocytes within the ovary of mature females may be found in various states of development.

E. superba is known to reproduce during a limited part of the year. *E. pacifica* breeds all year round in the North Pacific (Ross *et al.*, 1982), but seasonally in the Northern Sea of Japan (Mauchline, 1980). The timing of the reproductive cycles in *M. norvegica* varies geographically as *M. norvegica* populations are adapted to various climatic conditions of the North Atlantic Ocean, from Iceland (Astthorsson, 1990) and Norway (Falk-Petersen and Hopkins, 1981), and even the Barents Sea (Dalpadado and Skjoldal, 1991), to the Mediterranean Sea (Labat and Cuzin-Roudy, 1996) and the Mauritanian coast of Morocco (Mauchline, 1980; Mauchline and Fisher, 1969). In the North Atlantic, reproduction and recruitment essentially take place during summer (Boysen and Buchholz, 1984; Lindley, 1982; Mauchline, 1980; Mauchline and Fisher, 1969), but reproduction may occur in spring for some Atlantic populations (Astthorsson, 1990; Mauchline, 1985). By contrast, the Ligurian Sea population (Mediterranean Sea) reproduces in winter and early spring (January–May) and the resting season is longer than elsewhere, from May to the next January (Cuzin-Roudy, 1993; Labat and Cuzin-Roudy, 1996). Life cycle variations between different *M. norvegica* populations in various environments are discussed further in Chapter 3.

3. GONAD DEVELOPMENT DURING THE REPRODUCTIVE SEASON

3.1. Spermatogenesis

Activity in the testis is permanent in adult males of both *M. norvegica* and *E. superba*. The mitotic activity of *spermatogonia* produces, in different testicular cysts (as illustrated for both krill in Cuzin-Roudy, 1993; Cuzin-Roudy, 1987b), *spermatocysts* 1 and 2, that undergo meiosis to become *spermatids*. These later differentiate into ovoid non-flagellated *spermatozoa* (Raab, 1915). Non-flagellated *spermatozoa* are also found in some other crustaceans (Adiyodi, 1985) and are linked to specific adaptions for egg fertilization (Tarling *et al.*, 2009) due to their lack of mobility.

The spermatozoa leave the testis and are next packed into spermatophores in the spermatophore-glands of the sperm ducts. There they are surrounded by a rich seminal fluid for sustaining the spermatozoa until fertilization (Adiyodi, 1985), and by a cuticular membrane that packs them into a spermatophore. A pair of spermatophores is stored in the *ampullae* ready for release at mating. The *ampullae* are almost always full in adult males from field samples and, after mating, replacement spermatophores soon appear in the *ampullae*, even in laboratory maintained krill (personal observation).

3.2. Oogenesis

In contrast to the testis, the ovary of *M. norvegica* is active during only a limited part of the year, the reproductive season, when eggs are built from the stock of *yoc* in the sexually mature female. The development of the ovary follows the general pattern of other crustaceans, such as malacostracans and copepods, and has been described in many species (references in Adiyodi, 1985 and Charniaux-Cotton and Payen, 1985).

In both *M. norvegica* and *E. superba*, young primary oocytes (*yoc*) produced by the differentiated ovary in mature females stay 'dormant', blocked in prophase of meiosis during the entire resting season. The saddle shaped ovary occupies then 1/4 of the height of the thoracic cavity (Cuzin-Roudy and Buchholz, 1999). An early indication that the reproductive season has started is the further development of the *yoc*, when they start to grow by accumulating yolk during two different phases of vitellogenesis (Cuzin-Roudy, 1987b; Cuzin-Roudy, 1993), in response to an unknown signal that is probably linked to environmental conditions (see below).

3.2.1. Ovarian activity from primary oocytes to 'eggs'

The structure of both male and female gonads of *M. norvegica* was described by Cuzin-Roudy (1993), using the same histological techniques as for *E. superba* (Cuzin-Roudy, 1987b). The different steps of ovarian development are similar in both species (Cuzin-Roudy and Amsler, 1991). The specific pattern of seasonal reproduction for *M. norvegica* has further been defined for various populations: for example, west Scotland and Clyde sea, Kattegat and Gullmarsfjord and Ligurian Sea (Cuzin-Roudy, 2000; Cuzin-Roudy and Buchholz, 1999; Cuzin-Roudy et al., 1999; Cuzin-Roudy et al., 2004; Tarling and Cuzin-Roudy, 2003).

Mitoses of primary oogonia (*og1*, diameter about 20 μm) that form the germinal band (GZ) in the ventral part of the ovarian lobes continue throughout the reproductive season, as well as the transformation of secondary oogonia (*og2*, diameter about 40 μm) into *yoc* (diameter about 60 μm), blocked at diakinesis of meiotic prophase. *Og1*, *og2* and *yoc* have a very large nucleus surrounded by a very thin cytoplasm (Cuzin-Roudy, 1993). The transformation of *og2* into *yoc* is easily distinguishable on histological sections (Cuzin-Roudy, 1993; Cuzin-Roudy and Buchholz, 1999; Cuzin-Roudy et al., 1999) as the conspicuous chromatin of *og2* is not visible anymore in the blocked nucleus of primary oocytes, from *yoc* to mature *oc4*, when meiosis resumes. During this transformation, the *og2* migrate out of the germinal band (*GZ* in Fig. 7.3), and travel to the dorsal and lateral secondary germinal zones, forming conspicuous bunches of *yoc* (Fig. 7.2), ready for yolk accumulation and growth.

3.2.2. Oocyte growth during yolk accumulation

Yolk accumulation in the oocytes occurs in two phases, following the same pattern as other malacostracans (Adiyodi, 1985; Adiyodi and Subramoniam, 1983). During the first phase, called previtellogenesis, the oocytes accumulate glyco-proteic yolk while during the second phase, vitellogenesis, lipids also are accumulated. On histological sections treated alternatively with different histochemical stains, Cuzin-Roudy (1993) found that the glycoproteic yolk accumulated in previtellogenic oocytes stained bright red with PAS (Periodic Acid Schiff) (Figs. 7.3 and 7.4) while the lipids of the vitellogenic oocytes stained black with Sudan Black (Fig. 7.2, right and Fig. 7.6) (see Everson Pearse, 1960, for the techniques). Towards the end of oocyte maturation, the previously large nucleus becomes no longer visible in live or preserved krill. The nucleus has lost its nuclear membrane, migrated into the peripheral cytoplasm of the oocyte and resumed meiosis (Cuzin-Roudy, 1993), but is still visible on histological sections (Fig. 7.2, right).

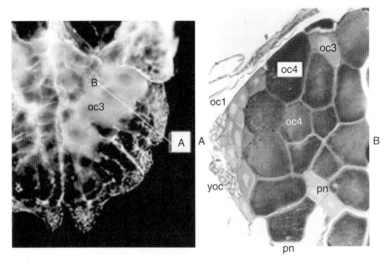

Figure 7.2 *M. norvegica.* Live and Sudan Black stained ovaries of reproductive females. Left: note the large nuclei of the vitellogenic oocytes (*oc3*) still accumulating lipid yolk. Right: note the lipid yolk filling the mature oocytes and the peripheral nuclei in meiosis (pn). In both views, oocytes at different stages of development are present along lines (A–B). *Oc1*, previtellogenic oocytes; *oc2*, early vitellogenic oocytes; *oc3*, late vitellogenic oocytes; oc4, mature oocytes (eggs); pn, peripheral nucleus in meiosis; *yoc*, primary oocytes.

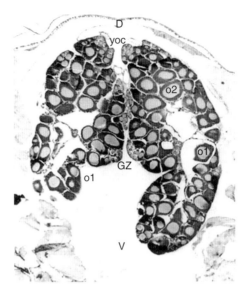

Figure 7.3 *M. norvegica.* Transverse section of the ovary of a post-spawned female. Note the large gaps left in the centre of the ovarian lobes after complete spawning of the mature egg batch and the presence of *o1* and *o2* ready for the next cycle of egg production. PAS, Groat's hematoxylin staining. 40×. D, dorsal; V, ventral side; GZ, principal germinal zones with oognia; *o1*, previtellogenic oocytes; *o2*, early vitellogenic oocytes; *yoc*, young oocytes.

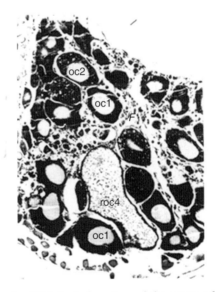

Figure 7.4 *M. norvegica.* Histological section of the ovary of a post-spawn female undergoing oosorption for the already developed oocytes. *oc1*, previtellogenic oocytes; *oc2*, early vitellogenic oocytes; roc4, residual mature oocytes.

Oocyte development has been categorised into four stages, *oc1* to *4* (Cuzin-Roudy, 1993), In *oc1*, the previtellogenic oocytes accumulate gly-coproteins; in *oc2*, the oocytes start vitellogenesis, with dispersed small lipid yolk platelets in their cytoplasm; *oc3* is the principal growth phase for lipid accumulation and synthesis of the lipoglycoproteins or 'vitellins' of the eggs; in *oc4*, the final step of maturity, the nucleus is blocked in first metaphase of meiosis.

Oocyte diameters, measured on histological sections, varied from 50 to 250 µm for *oc1*, 180 to 250 *µm* for *oc2*, 250 to 350 *µm* for *oc3*, and up to 450 µm for *oc4*.

In the ovary of a reproducing female of *M. norvegica*, getting ready for the first spawning event of the season, one can expect to find all of the following (Cuzin-Roudy and Buchholz, 1999): (1) a germinal band running ventrally in each half of the saddle shaped ovary (Fig. 7.3), with *og1* in mitotic activity producing abundant *og2*; (2) bunches of *yoc* forming the dorsal and lateral germinal zones of each ovarian lobe; (3) abundant *oc1* (next to the *yoc*), organised in rows and converging towards the centre of the ovarian lobes (Fig. 7.3); (4) a row of *oc2* (Fig. 7.3); (5) a large batch of oc3 that have attained their final size but have yet to mature (Fig. 7.2, left); (6) or alternatively, a batch of mature *oc4* that have resumed meiosis (Fig. 7.2, right) and attained their final maturation. The last step resembles the externally defined maturation state of III D (Makarov and Denys, 1980) as defined for *E. superba*, in which the female krill possesses a swollen ovary that overfills the thoracic cavity down to the 2nd abdominal segment, and with a conspicuous blue-grey aspect to its external appearance.

3.2.3. Staging female sexual development and the reproductive cycle

Given the greater insight granted from examining oocyte maturation within the ovary, Cuzin-Roudy (1993) established a new staging method to categorise stages of krill sexual development in large field-population sam-ples. Six sexual development stages (SDS 1–6) were defined for *M. norvegica*.

SDS 1 is defined as females entering their first reproductive season, with a small saddle shaped ovary that occupies about 1/4 of height of the thoracic cavity in lateral view and contains only the principal germinal zone (*og1* and *2*) and abundant *yoc* (Cuzin-Roudy et al., 1999).

SDS 2 is characterised by a larger ovary with lobes filled with abundant growing *oc1s* next to the germinal band and the abundant *yoc* of the dorsal and lateral germinal zones. Previtellogenesis has started and will be active during the whole reproductive season. The ovary then occupies up to 1/2 of the thoracic cavity.

At SDS 3, the ovarian lobes become opalescent, in live and preserved krill, due to the presence of a large batch of *oc3* that are accumulating lipid yolk.

Vitellogenesis is on-going. When the *oc3* have attained their full size, the ovary fills completely the thoracic cavity.

SDS 4 refers to females that are ready to spawn, with an ovary filled with a large batch of oc4, ready for release and fertilization. The ovarian lobes expand anteriorly up to the stomach, laterally and ventrally, and posteriorly inside the abdomen, up to the 2nd abdominal segment. The carapace in *M. norvegica* does not swell as in larger *E. superba* III D in Makarov and Denys, (1980). Nevertheless, the blue-grey aspect of the mature ovary has also been recognised in ready to spawn *E. pacifica* (Ross *et al.*, 1982) and *E. superba* (Hirano *et al.*, 2003; Ross and Quetin, 1983) and is good signal of imminent egg release in *M. norvegica* (Boysen and Buchholz, 1984; Cuzin-Roudy and Buchholz, 1999). The blue-grey coloration is not due to a pigment accumulated during vitellogenesis, as known for crabs (Adiyodi and Subramoniam, 1983), but to the disappearance of the large refractive nucleus of the previous oocyte phases at the resumption of meiosis (Fig. 7.2, right). The transformation can be observed through the transparent carapace in live krill maintained on board ship: a wave of blue-grey coloration proceeds within minutes along the whole batch of *oc3* when their conspicuous nuclei disappear (personal observation).

M. norvegica and *E. superba* females are usually qualified as 'gravid' when the ovary has filled the thoracic cavity. The term 'gravid' is confusing as it does not make a distinction between an ovary filled by a mature egg batch (SDS 4) and an ovary of similar size filled by a batch of immature oocytes (SDS 3). We will see below that vitellogenesis is a long process spanning more than one moult cycle. Consequently, spawning can only be predicted with confidence when SDS 4 has been attained. SDS 4 lasts for less than half a day, with eggs being released during the first part of the night at a temperature maintained as in the field (about 10 °C; Cuzin-Roudy *et al.*, 2004).

Two stages, SDS 5^1 and 5^2, were defined for the post-spawn phase, because of the partial spawning events observed in females maintained in the laboratory, with one, two or three successive spawning events occurring at 1 day intervals until the egg batch has been completely released, as shown in Fig. 7.3.

The release of the complete egg batch leaves a large gap in the centre of each lobe of the ovary (Cuzin-Roudy and Buchholz, 1999), which now resembles a less advanced stage of ovarian development (SDS 2). The abundance *oc1* and *oc2* in histology sections of recently spawned females clearly indicates that previtellogenesis remains active and that a new cycle of vitellogenesis is underway (Fig. 7.3). Under the Makarov and Denys (1980) scheme, such a krill would be categorised as III B (i.e. pre-spawn), since there was no recognition that a krill may undergo repeated cycles of vitellogenesis.

At the end of a reproductive cycle, the *oc1* and *oc2* left in the SDS 5^2 ovary may be instead arrested in their development and broken down by oosorption, as illustrated for the Clyde Sea krill in Cuzin-Roudy *et al.* (2004) and here by Fig. 7.4. Oosorption is a crustacean strategy (Adiyodi and Subramoniam, 1983) for the retrieval of resources to sustain basic demands for survival and activity besides reproduction. It is reversible and may occur several times during the reproductive season, in coincidence with periods when trophic conditions are insufficient (Tarling and Cuzin-Roudy, 2003; Chapter 3). This is different to the end of the reproductive season, when the females enter a stage of reproductive rest (SDS 6).

In SDS 6 the ovary is contracted and small (1/4 of carapace height), and could be confused with SDS 1. The ovary is still saddle shaped, but with irregular lobes that are deformed by large residual *oc4* from the last spawn of the year that will be re-absorbed later on. In contrast to SDS 5, the *oc1* produced continuously during the reproductive season have been arrested in their previtellogenic development and broken down by oosorption. The germinal band nevertheless remains and some mitotic activity continues through the whole resting season, for the restoration of a stock of *yoc* in readiness for the next reproductive season.

The use of the SDS scoring system, based on physiological changes in the reproducing ovary rather than a simple examination of external appearance, helped to reveal the cyclical nature of reproduction within the krill reproductive season. The system was first applied to *E. superba* (Cuzin-Roudy and Amsler, 1991) before its use on the Ligurian population of *M. norvegica* (Cuzin-Roudy, 1993) and then other North Atlantic populations of this species (Cuzin-Roudy and Buchholz, 1999, Table 2; Cuzin-Roudy *et al.*, 2004). The method can be used on live or formalin preserved specimens and uses a 'squash' technique to identify the different types of oocytes present simultaneously in the ovary. The degree of ovarian development attained by a population at a given time of year can be defined from the proportion of different sexual development stages (SDS 1–6) averaged across all sampled individual females. This metric is referred to as the Sexual Development Index (SDI) (Cuzin-Roudy and Labat, 1992; Cuzin-Roudy *et al.*, 2004) that ranges from 1 to 6, like the SDS system, and characterises the average degree of sexual development within the sampled population (Tarling and Cuzin-Roudy, 2003). This method can be applied to females of all sizes and ages, from live and preserved field samples, at any time of year.

This method was used to describe the seasonal reproductive pattern of *M. norvegica*, in which there were successive reproductive cycles of vitello-genesis and egg production during the reproductive season followed by a period of reproductive rest (Fig. 7.5).

3.2.4. Role of the 'fat-body' for yolk accumulation

A strongly PAS positive tissue has been described in histological sections of the cephalothorax of *M. norvegica* and *E. superba* (Cuzin-Roudy, 1993). This tissue, first described by Raab (1915), does not have a definite shape and

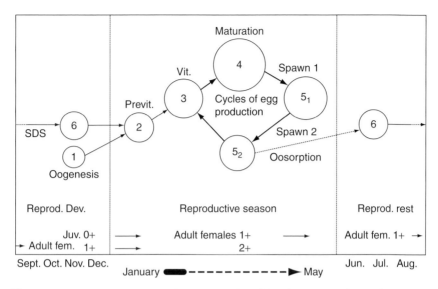

Figure 7.5 *M. norvegica*. Seasonal ovarian cycles of development and sexual rest. The succession of sexual development stages (SDS 2 to 5^2) corresponds to cycles of egg production during the reproductive season, followed by a period of sexual rest (SDS 1; SDS 6). Time scale is for the Ligurian krill; Previt, previtellogenesis; Vit, vitellogenesis; 0+, 1+, 2+, krill in their first, second or third year (from Cuzin-Roudy and Buchholz, 1999).

extends loosely inside the ventral part of the thoracic cavity, in males as well as females. It attains a large size in reproductive females, and is apparent in live or preserved krill as a whitish tissue that expands down into the bases of the last thoracic legs, as well as laterally and upwards, coming into close contact with the ovarian membrane of the pre- and vitellogenic ovary (Cuzin-Roudy, 1993). Due to its structure of glandular cells regrouped into *acini*, it was first referred to as the 'shell gland' (Bargmann, 1937; Raab, 1915). However, krill eggs are released without any 'shell' (Tarling *et al.*, 2009) and the embryonic envelopes are extruded from the eggs after fertilization. A more realistic interpretation was offered by Gabe (1962) in his extensive study of a similar tissue described as the fat body (FB) in insects and various crustaceans. Studies on crustaceans have shown that the fat body plays a role in the accumulation and synthesis of the precursors of the vitellins (lipoglycoproteins) accumulated in eggs for the development of the embryo (Adiyodi, 1985; Adiyodi and Subramoniam, 1983).

In krill, the strong positive reaction to PAS in the FB, as well as *oc1* in the ovary (Figs. 7.3 and 7.4), indicates the presence in both organs of glycoproteins, neutral mucopolysaccharides (like chitin), and possibly unsaturated lipids and phospholipids, during previtellogenesis and vitellogenesis (Everson Pearse, 1960).

The developed FB shows also a positive response to Sudan black (Fig. 7.6) that indicates the presence of lipids in the large blood lacunae of the FB, and in association with the cell membranes of the acini (Cuzin-Roudy et al., 1999). Total lipid content of haemolymph is 60% higher in reproductive females than in males, a confirmation of the role of the FB in vitellogenesis (see below). After the reproductive season, total lipids are still abundant in the FB of both sexes, but become localised in the large blood lacunae within the gland (Cuzin-Roudy et al., 1999).

The FB in male M. norvegica is less developed than in the females and also reacts differently to histochemical stains. The male FB can be slightly PAS+ in some specimens, marking the presence of neutral mucopolysaccharides or glycoproteins, or could react strongly to Alcian blue, indicating the presence of acid mucopolysaccharides. There is no clear seasonal pattern of activity in the male FB, which may reflect the fact that males do not show a clear regression of the testis and accessory structures during the resting season.

In the histological studies of the FB of M. norvegica cited above, the same stains (PAS and Alcian blue, or Sudan black) also revealed glandular cells grouped into acini, and surrounded by large blood lacunae (Fig. 7.6A) as

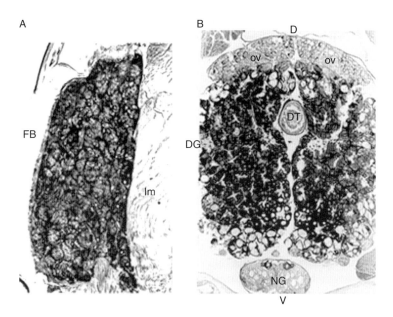

Figure 7.6 M. norvegica. Histological sections of a female. (A) Fat body of a reproducing female showing lipid accumulation in association with the cell membranes and the blood lacunae. 40×. (B) Transverse section of a female krill in sexual rest at the level of the digestive gland. Note the small dorsal ovary reduced to the germinal zones and the lipid accumulation in the rich digestive gland. 20×. FB, fat body; dg, digestive gland; dt, digestive tract; ng, nervous ganglia; og, oogonia (from Cuzin-Roudy et al., 1999).

described by Raab (1915). The structure (bouclier glandulaire) was also found by Gabe (1962) in the phyllocarid *Nebalia geoffroyi* in the same part of the body. These glandular cells produce alternatively two types of secretion in a cyclical fashion: (1) acid mucopolysaccharides that have a strong affinity for Alcian blue (at pH 3.5), and (2) neutral muco- and glycoproteins that are strongly PAS+.

The FB is maximally developed in reproductive females of both *M. norvegica* (Cuzin-Roudy, 1993; Cuzin-Roudy and Buchholz, 1999; Cuzin-Roudy *et al.*, 1999, 2004) and *E. superba* (Cuzin-Roudy, 1987b, 2000; Cuzin-Roudy and Amsler, 1991). In these specimens, this structure is strongly PAS +, staining the same bright red colour as in the cytoplasm of previtellogenic *oc1*. The development of the FB alongside the previtellogenic and vitellogenic ovaries suggests that the main role of the FB is in the synthesis and transformation of glyco- and lipoproteins into vitellins (glyco-lipoproteins), which form the yolk platelets of the mature oocytes, as occurs in the FB of insects and various other crustaceans (Adiyodi, 1985). Alternatively, the FB could have other functions, such as the metabolism of chitin (a neutral polysaccharide which is PAS+) in preparation for the next moult.

3.2.5. Regression of sexual maturity during the resting season

A regression of the ovary and of secondary sexual characteristics during the long sexual rest after the breeding season is well known in *E. superba*, together with a possible reduction in body size through negative growth (Ikeda and Dixon, 1982). For instance, Cuzin-Roudy and Labat (1992) captured large *E. superba* from the ice-zone in November in which secondary sexual characteristics were regressed or absent. In *M. norvegica*, a regression of the *petasmata* or *thelycum* during the resting season has not been observed. During the resting period of both the Atlantic and Mediterranean populations, the ovary regresses to a sub-adult size, with an active germinal band of *oogonia* and 'dormant' *yoc* (Fig. 7.6B) (Cuzin-Roudy *et al.*, 1999), as occurs in *E. superba* (Cuzin-Roudy, 1993; Cuzin-Roudy and Amsler, 1991).

For example, the termination of the reproductive season and ovary regression were rapid in the Gullmarsfjord population (Sweden), which consisted of 91% of females preparing a spawn or spawning (SDI = 4) in early September and 73% of females in spent and regressed condition (SDI = 2.9) in mid October (Cuzin-Roudy *et al.*, 2004).

The testis in males similarly reduces in size, but remains active during the resting season. Spermatophores are still visible in the ampullae, but are reduced in size with sometimes only a single spermatophore being present (Cuzin-Roudy, 1993; Mauchline and Fisher, 1969).

3.2.6. Allocation of resources for egg production

Krill continue to moult and grow throughout their whole life cycle which implies that resources must be allocated to a number of other processes besides reproduction during adulthood. *M. norvegica* mainly produce eggs during the most productive time of the year, thus insuring food for the survival of the offspring. Prior to the release of embryos, the eggs are supplied with glycoproteins and lipids during pre-vitellogenesis and vitellogenesis, which means that females also require optimal trophic conditions to perform this function.

Four types of storage lipids have been found in marine zooplankton (Lee *et al.* 2006): triacylglycerols (TAG), wax esters (WE) and diacylacylglycerol ethers (DAG) among the neutral lipids, and the phospholipids, which are polar lipids (PL). TAG are the most common storage neutral lipid class across the animal kingdom, including zooplankton. WE are major storage lipids in high latitude zooplankton, including euphausiids, but do not occur in *M. norvegica* nor *E. superba*. Both *M. norvegica* and *E. superba* have a lipoglycoproteic vitellus in yolk platelets instead of WE droplets (Cuzin-Roudy, 1987b; Cuzin-Roudy, 1993). Among polar lipids, phosphatidylcholine (PC), a key component of biomembranes in embryos, is an important component of the egg vitellins of high latitude euphausiids, including *M. norvegica* (Falk-Petersen *et al.*, 1981, 2000). The quantity and composition of stored lipids is likely to vary widely between females in reproductive or resting stages, as they have to convert a large portion of their lipid storage into the lipoglycovitellins which are synthesised and stored within the egg for the embryo development.

Lipid content and composition have been studied in the Clyde, Kattegat and Ligurian populations of *M. norvegica*, both in and out of the reproductive season (Albessard *et al.*, 2001; Cuzin-Roudy *et al.*, 1999; Mayzaud *et al.*, 1999; Virtue *et al.*, 2000). It terms of total lipid content, a higher level was found in Clyde Sea krill (4% wet weight) than in the Kattegat and Ligurian krill (2–3%). *M. norvegica* has distinctly high levels of ω-3 polyunsaturated fatty acids (PUFA) particularly in the PL fraction, which was comparatively higher in reproductive females than in males, in the three populations. There were no differences in PUFA levels between sexes outside the reproductive season.

In ready to spawn krill (SDS 4), the large *oc4s* showed Sudan black stained lipid yolk platelets (Fig. 7.2, right) (Cuzin-Roudy, 1993; Cuzin-Roudy *et al.*, 1999). In their developed FB, lipids appeared as thin granules associated with the membranes of the expanded acini cells, while the large blood lacunae were stained densely black, indicating the presence of another type of lipid in the haemolymph of this organ in reproductive krill (Fig. 7.6A). The digestive gland of both sexes occurred in various stages of development over the year, but oil droplets in vacuoles of the R cells (lipid storage cells, see Dall and Moriarty, 1983) were evident in all seasons, even during the oligotrophic summer of the Ligurian Sea (Fig. 7.6B).

Cuzin-Roudy *et al.* (1999) and Albessard *et al.* (2001) reported that there was no significant difference in the total body lipid content between mature *M. norvegica* females (SDS 4) and males or females with a regressed (spent) ovary. However, clear significant differences were obtained when the different organs involved in lipid metabolism were analysed separately.

Females with a mature egg batch in the ovary (SDS 4) had almost twice the amount of lipids in the ovary compared to regressed females (SDS 6). No differences were seen in the total lipid content of the digestive gland (DG) and the FB between sexes and reproductive stages. Concerning the lipid fractions, the proportion of TAG and PL was significantly different between SDS 4 (ready to spawn) females and spent (SDS 6) females. The amount of PL was higher than TAG in the ovary and FB of SDS 4 females, but no difference was found between these lipid classes in SDS 6 females. Haemolymph during the early reproductive season contained 50% more lipid in females than in males, with levels of PL markedly higher than TAG in both sexes. High levels of PUFA were noticed, especially in the PL fraction in reproductive females compared to males. PUFA levels in the TAG fraction were lower in females than males. There were no differences in PUFA levels between male and female krill out of the reproductive season. As the SDS 4 stage is of very short duration (less than 6 h at 12–13 °C in the Ligurian krill, less than 12 h at 4–5 °C in the Clyde, Kattegat, and Gullmarsfjord populations (Cuzin-Roudy and Buchholz, 1999), these results imply that there is a high level of circulating lipids in reproductive females, corresponding to a rapid accumulation of lipids in the egg batch during vitellogenesis.

By comparison, Albessard (2003) found that eggs recently released (mean wet weight = 36.2 μg) contained 6% of total lipids, 13% of proteins and a negligible amount of carbohydrates. Among lipids, DAG and TAG represented 9% of the total, and PL, 75%. This confirms the essential role of the PUFA from PL in embryonic development.

To summarise, the major site of lipid storage is the DG in males and spent (SDS 6) females, while it is the ovary for ready to spawn females. The FB serves as a secondary site for lipid deposition, and putative synthesis of the glyco- and lipoproteins that will be transformed into vitellins by the ovary. There is a reversal in the relative amounts of glycoproteins and lipid in the ovary and FB between SDS 4 and SDS 6 krill, indicating a close physiological link between the two organs for the synthesis in the ovary of the lipoglycoproteins of the egg yolk platelets. It is most likely therefore that the FB functions as the main source of easily mobilised glycoproteins and lipids for egg production in the ovary (Cuzin-Roudy et al., 1999).

4. Ovarian Development Linked to Moult Development

4.1. Apparent independence of spawning and moulting

In both *M. norvegica* and *E. superba*, the time of spawning does not coincide with the point of moult. Nicol (1989) reported that, even though laboratory maintained *E. superba* spawned up to three times over successive days, they did not moult. Such a pattern has been confirmed for laboratory maintained spawning *M. norvegica* (Cuzin-Roudy, 2000; Cuzin-Roudy and Buchholz, 1999; Cuzin-Roudy *et al.*, 2004). Furthermore, in preserved samples, females ready to spawn or with partially emptied ovaries were not close to *ecdysis*, the point of moulting in their moult cycle, but rather midway through the moult cycle (Labat and Cuzin-Roudy, 1996; Tarling and Cuzin-Roudy, 2003) as shown in Fig. 7.7. This pattern is unlike that of many other crustaceans, where egg release and ecdysis co-occur.

The relationship between ovarian development and moult development is very diverse in malacostracans (Adiyodi, 1985; Nelson, 1991). Compared to taxa that produce small egg batches every moult cycle (e.g. mysids, Cuzin-Roudy, 1985), taxa that produce large egg batches (e.g. brachyuran crabs) have a prolonged moult cycle which involves a pause during the intermoult stage (stage C). The prolonged intermoult period is called *anecdysis*, and moulting becomes rare in large adults, with growth reduced or even arrested. Another strategy is found in amphipods, isopods and some decapods (Penaeidea) that go through several successive moult cycles within each reproductive cycle. *M. norvegica* and *E. superba* adopt the latter strategy, undertaking two successive moult cycles for every one reproductive cycle, as explained below.

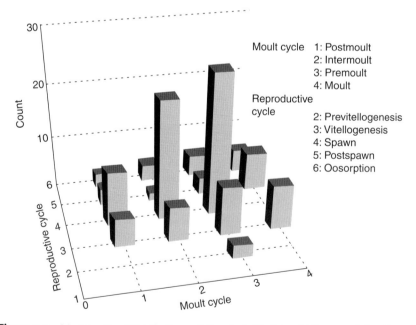

Figure 7.7 *M. norvegica.* Distribution of female krill ($N = 116$) scored for Sexual development Stages (SDS 2–6) and moult phases (1 to 4) during the reproductive season (Clyde Sea, March, May, June, August pooled samples).

4.2. Egg release in early premoult

A link between ovarian development and moult cycle development was first observed for *E. superba* (Cuzin-Roudy, 1987b) and defined more precisely in both *E. superba* and *M. norvegica* through applying the 'SDS staging method' on preserved and also live material (Cuzin-Roudy and Amsler, 1991) from different populations (Cuzin-Roudy, 2000). In live and preserved samples of *M. norvegica*, Cuzin-Roudy and Buchholz (1999) scored for total body length, moult stage (A, B, C, D0, D1, D2, D3-main and D3-late, see Chapter 6) and SDS (1–6), in order to study the relationship between ovarian development and moult preparation at the population level. Such a level of detail is unfeasible in preserved samples, for which simplified schemes have been devised by Cuzin-Roudy and Labat (1992) for *E. superba* and Tarling and Cuzin-Roudy (2003) and Cuzin-Roudy *et al.* (2004) for *M. norvegica* The simplification mainly involved combining certain moult stages into easily distinguishable phases, which were (1) 'Moult' (D3 -late + Ecdysis + A); (2) Postmoult (B); (3) Intermoult (C + D0); and (4) Premoult (D1+ D2 + D3main). Full details on the moult cycle are given in Chapter 6 so only a brief overview is given here.

The phase 'Moult' is easy to spot in preserved samples as live krill have a soft cuticle that lasts for a few hours before and after ecdysis and are flaccid when taken out of the water. They become distorted when dropped in a 10% formalin solution, and have a false unhealthy aspect compared to other krill in the sample. In postmoult (stage B), the soft epidermic tissue has restructured itself and become denser. The epidermic tissue resumes activity during intermoult (stage C) and the secretion of the endocuticle, arrested before the moult, is now completed. The long stage C is terminated by the retraction of the epidermis from the cuticle, termed *apolysis*. The formation of the new cuticle (epi-, exocuticle) secretion during premoult (stages D1 and D2) is easy to distinguish in microscope preparations of the antennal scale. By late premoult (D3- late), the segmentation of the long, curled, setae of the antennal scale becomes apparent. By D3-late, thinning and softening in the old cuticle is also evident.

With regard to the relationship between the moult and reproductive cycle, Cuzin-Roudy and Buchholz (1999) demonstrated that spawning in *M. norvegica* coincided with late stage C and stage D0, with one or two further spawning events occurring during stage D1. The whole egg batch was found to have been released several days before the moult.

4.3. The reproductive cycle and its periodicity

Cuzin-Roudy and Buchholz (1999) examined *M. norvegica* samples from the Ligurian Sea, the Clyde, the Kattegat and the Gullmarsfjord to establish the interaction between the timing of the reproductive and moult cycles.

From these examinations, a generic model of these interactions was built (Fig. 7.8), which is applicable to all the studied populations of *M. norvegica* (Cuzin-Roudy *et al.*, 2004; Tarling and Cuzin-Roudy, 2003).

According to the Cuzin-Roudy and Buchholz (1999) model, during the 1st moult cycle, females with a mature egg batch (SDS 4) release eggs at moult stage C late or D0, that is, at *apolysis*, and eventually a second and third batch during early premoult (D1). The first moult cycle is termed the 'spawning moult cycle' (Fig. 7.8). The *oc1* and the *oc2* oocytes that are already present in the ovary before the release of the mature egg batch, then start vitellogenesis to build-up the next egg batch. Vitellogenesis is not interrupted by the first moult and the next moult cycle, which is termed the 'vitellogenic moult cycle', is a cycle entirely devoted to the building of a new egg batch of *oc3*. After the moult of the 2nd moult cycle, the 3rd moult cycle will be a 'spawning moult cycle'. Due to the coincidence of spawning with *apolysis* and not with moult, a full reproductive cycle spans two moult cycles, starting at *apolysis* and not at *ecdysis*. *Apolysis* is also the first sign that cell multiplication is occurring in the epidermis (*mitosis*) and that the physiological activities within the epidermis for the next moult preparation are underway.

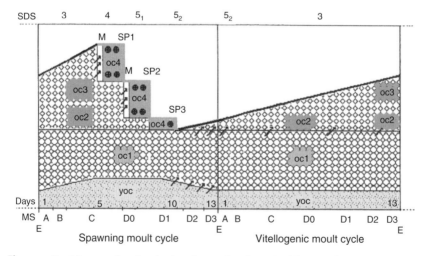

Figure 7.8 *M. norvegica*. Synthetic schema for the coincidence of one reproductive cycle with two successive moult cycles, with vitellogenesis re-starting right after spawning. E, ecdysis; M, meiosis; *oc1*, previtellogenic oocytes; *oc2–3*, vitellogenic oocytes; *oc4*, mature oocytes; SP 1,...3, spawns; *yoc*, young oocytes. Time scale in days, for a 13 day intermoult period (IMP) (from Cuzin-Roudy and Buchholz, 1999).

4.4. Spawning and moulting synchrony in krill populations

A noticeable feature in certain time-series studies is apparent peaks and troughs in moulting or spawning activity within *M. norvegica* populations (Tarling and Cuzin-Roudy, 2003; Tarling *et al.*, 1999a). One would expect the proportion of individual females performing these functions to be relatively constant over time if they occurred randomly within populations. The observed pattern otherwise suggests that there is a degree of synchrony between individuals in performing these functions. Such synchrony would imply that there should be spawning peaks, with a resulting pulsing of eggs in the water column. According to the above reproductive cycle model (Fig. 7.8), such pulses should occur at a periodicity of two moult cycles, which is temperature dependent and around 26 days in the Clyde Sea and Kattegat, and 14 days in the Ligurian Sea (Cuzin-Roudy and Buchholz, 1999).

The moulting, spawning and juvenile recruitment of *M. norvegica* were analysed over an annual cycle in the Clyde (Tarling and Cuzin-Roudy, 2003). According to the model proposed by Cuzin-Roudy and Buchholz (1999) (Fig. 7.8), the spawning period (SDS 4 and 5^1) should last for 31% of the reproductive cycle of an individual female. In the situation where females spawns every other moult cycle, only 15.5% should be in spawning condition at any one time in a randomly distributed population. It was observed that, during some sampling campaigns, the actual percentage went as high as 51%. These campaigns must have coincided with synchronised spawning events, which also infers that the moulting and reproductive cycles of females must have been synchronised at the population level in the Clyde.

Tarling and Cuzin-Roudy (2003) built a semi-empirical model (termed a reproductive-phase model), on the basis that seven possible successive reproductive cycles of 26 days occurred within a 5 month reproductive season (April–August). The model predicted that the first spawning period took place around 26 days after the first *Chl a* peak in February, which was in agreement with the observations of 51% of females spawning in March and suggested that the bloom triggered egg development in all adult females.

Three cohorts of juvenile krill were actually observed, in April, June and August rather than the seven that were theoretically possible. A back-calculation of the time of spawning of the three cohorts coincided well with the spawning periods predicted by the model and also with when peaks in *Chl a* occurred. Of the four cohorts that never appeared, the most likely explanation is that there were insufficient food resources for the larvae to survive since *Chl a* levels were low at the predicted dates that the cohorts reached the sensitive furcilia stage. Alternatively, the lack of food resources may have resulted in females entering temporary phases of oosorption, as shown in Fig. 7.4, instead of starting a new vitellogenic cycle, meaning the failure of the next reproductive cycle in such females. Overall, the good match between model predictions and observations in this study serves to reinforce the validity of the Cuzin-Roudy and Buchholz (1999) reproductive cycle model.

5. Fecundity and Predictive Models

5.1. Estimates of the size of the mature egg-batch

Fecundity is difficult to determine in krill that release their eggs freely in the water column in multiple spawns, like *M. norvegica* and *E. superba*. Fecundity has been estimated from counting the eggs released by ready to spawn females kept in the laboratory, in *E. pacifica* (Ross *et al.*, 1982) and *E. superba* (Harrington and Ikeda, 1986; Nicol, 1989; Ross and Quetin, 1983), or by counting 'ovarian eggs' in the ovary of females from preserved samples, in *M. norvegica* (Mauchline and Fisher, 1969). There is considerable variability between estimates due to uncertainties concerning krill egg production : (1) eggs from a mature ovary are released in partial spawns over a number of consecutive days in laboratory maintained krill; (2) after the completion of egg release, numerous oocytes at various degree of development are still present in the postspawn ovary during the reproductive season; (3) ready to spawn females are present in population samples throughout the reproductive season; (4) but moulting and spawning synchrony may produce an apparent bias in the samples from a population if they are not repeated at multiple occasions during the season.

Denys and McWhinnie (1982) first recognised that egg production in the ovary of *E. superba* was cyclical and also that eggs are released in two successive spawning events in laboratory maintained krill. Counting both eggs released by live females and 'eggs' present in the developed ovary of females from preserved samples, they estimated a yearly brood size of 2200–8800 eggs per female. They also observed the presence of 'eggs' at different stages of development or resorption in the ovary, but neither recognised a cycle of development for the egg batch, nor the possibility of more than one reproductive cycle per year.

During the reproductive season, all female krill are engaged in the development of a brood, but eggs are produced periodically and released in several spawning events over the season. Nevertheless, the development of the ovary over a complete reproductive cycle has never been achieved in the laboratory, at least for *M. norvegica*. Consequently, determining whether females release several successive broods during the season had to be answered indirectly through examining the proportions of spawning females in large field samples of *E. superba* brought back to the laboratory (Ross and Quetin, 1983), and also by staging the ovarian development in females from large random samples of *E. superba* and *M. norvegica* taken at various times of the reproductive season, (*E. superba*: Cuzin-Roudy and Amsler, 1991; Cuzin-Roudy and Labat, 1992; Cuzin-Roudy, 2000; Ross and Quetin, 2000; Tarling *et al.*, 2007; *M. norvegica*: Cuzin-Roudy and Buchholz, 1999; Cuzin-Roudy, 2000).

Estimates of fecundity are a function of both the number of spawns per year and the numbers of eggs produced and released. Cuzin-Roudy (2000) estimated the latter for both *E. superba* and *M. norvegica* through using oocyte count data from mature ovaries of preserved krill. A predictive model of multivariate allometry was then applied to ascertain the functional relationship between egg batch size and female total body length (see below for further details). An example where both spawning frequency and egg batch size information was combined to estimate (annual) fecundity is presented in Tarling *et al.* (2007) for *E. superba* around South Georgia.

5.2. Models for estimating and predicting fecundity

Cuzin-Roudy (2000) estimated the fecundity of *M. norvegica* through counting mature and previtellogenic oocytes in the ovary of a total of 277 ready to spawn females (SDS 4 with a blue-grey appearance, total body length 22.5–40.2 mm) from preserved samples collected from the Clyde Sea, Kattegat, Ligurian Sea. Principal Component Analysis (PCA) was used for the study of the co-variation in the population between quantitative morphometric descriptors, namely: total body length (BL), total body wet weight (WWT), ovary wet weight (WWO) and number of mature oocytes (NMO). The analysis identified BL and WWT to be the best predictors of NMO in the multivariate allometric model. The NMO ranged from 199 in the smallest females (22.5 mm) to 4 069 in the largest (40.2 mm). As the average NMO for an average female was around 1000–1200 eggs and the mean number of previtellogenic oocytes was 2136 *oc1* per ovary, without taking into account the innumerable 'dormant' *yoc*, it was expected that a minimum of three reproductive cycles could be achieved over the whole reproductive season in each of the three sampled regions. This approach allows size–structure data to be used in predicting the size of egg-batches and fecundity within a population. However, the fecundity estimate is still a function of the number of spawning episodes per season, which is under the ultimate control of seasonal and trophic conditions that may vary widely from year to year (see above).

6. Reproductive Behaviour

6.1. Mating behaviour

Mating in krill is a complex manoeuvre, as the spermatozoa are deprived of any mobility due to the absence of a flagellum and have to be deposited in close contact with the extruded eggs. A role for the male petasmata in helping the transfer of the spermatophores from the ampullae to the thelycum has been suggested (Mauchline and Fisher, 1969), but not yet documented in detail for *M. norvegica*. Another pair of appendages with potential involvement is the 7th pereiopods, which appear specialised in *E. superba* (Tarling *et al.*, 2009). Their structure is different in *M. norvegica* however and more like the other pereiopods that form the food basket.

During courtship, Ross and Quetin (2000) reported that male *E. superba* chase the females for around 5 s and then contact or approach closely for a further 10–30 s. Males appeared to perceive a chemical trail within 8–10 cm of the female and turned in pursuit.

In *M. norvegica*, mating behaviour begins up to 2 months prior to the first spawning event in the Kattegat (Boysen and Buchholz, 1984) and continues throughout the whole reproductive season, as indicated by the almost constant presence of the sperm plug in the female thelycum over this time. Females lose this plug each time they moult (every 7–20 days) so the presence of the plug indicates that mating must be ongoing. Some have considered the presence of the sperm plug as a sign of reproduction. However, this feature is not an indication of egg preparation or of a proximate egg release.

6.2. Spawning and vertical segregation during DVM at night

Daytime surface swarms that were dominated by ready to spawn or recently spent females, together with abundant eggs in the surface, have been reported in *M. norvegica* by Nicol (1984). A segregation of ready to spawn females closer to the surface than the rest of the population has also been reported in the Kattegat population in summer (Tarling *et al.*, 1999a). The phase of the moon may also influence how close to the surface the 'spawners' may go (Tarling *et al.*, 1999b). The scheme in Fig. 7.9 illustrates a typical vertical migration pattern of inshore *M. norvegica* that is modified by spawning and moulting: (1) during the daytime, krill aggregate in a swarm close to the bottom; (2) at night, segregation occurs, with moulting individuals (moulters) remaining deep while the rest of the population moves up the water column, 'spawners' closest to the surface; (3) in new moon phases or eclipse, or cloudy nights 'spawners' approach even closer to the surface.

This night time segregation may bias samples for population studies, especially concerning the sex ratio and the proportions of 'spawners' and 'moulters' in the samples. Consequently, sampling should be carried out in the deep during the day to avoid such bias (see Chapter 3).

6.3. Individual spawning behaviour

Tarling *et al.* (2009) showed experimental evidence in *E. superba* that spawning of the entire egg batch can occur over one long continuous period rather than in partial spawns on consecutive days. Egg release was performed at night while the female swam constantly. The swollen ovary was completely emptied within a 10 h period. More than 4000 eggs were

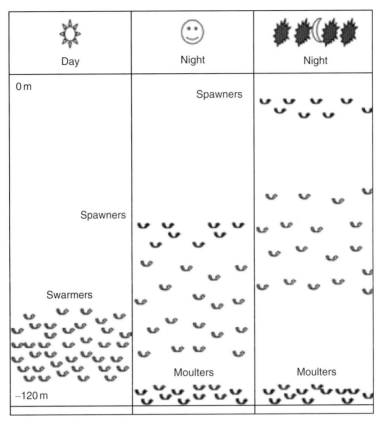

Figure 7.9 *M. norvegica*. Schema illustrating the diel vertical migration (DVM) and segregation at night of swarming krill, and the effect of moonlight.

released, singly or in clusters of 2, 3 or 4 at a time. This may be a pattern common to other free-spawning euphausiids such as *M. norvegica*, that are excellent swimmers and migrate into the surface layers at night from potentially very deep daytime depths (800 m in the Ligurian Sea). Krill appear to prefer to spawn in the surface layers, so the main constraint on complete expulsion of the entire egg batch in a single night would be the amount of time spent in these layers, which will be a function of the duration of the night time period (see Chapter 9).

Tarling *et al.* (2009) also documented that the seventh thoracic leg (pe7) in *E. superba* is involved in the process of egg expulsion, accelerating the egg downwards and away from the krill. Pe7 appears different to the other thoracic legs in *E. superba*. In *M. norvegica*, pe7 is reduced to two segments but has otherwise kept the morphology of the other thoracic legs and makes up part of the food basket (see Chapter 5).

7. CONCLUSION AND PERSPECTIVES

7.1. Concluding remarks

The regulation of the seasonal and cyclical reproduction of krill is complex as it does not only concern the timing in the year for sexual maturity, gametogenesis and egg production, but also the link with the moult cycle and the drastic metabolic changes necessary for egg production. For this reason, a transition from a dormant to active reproductive state is very much dependent on fluctuations in trophic conditions. Baker (1938) defined two types of factors that regulate reproductive activity: (1) *proximate factors* for the large annual cycles of photoperiod and temperature, that regulate the timing of the reproductive season each year by acting, directly or indirectly *via* hormones, on the mechanisms of sexual maturity, mating and gonad development, long before egg production. The *proximate factors* vary from year to year and also involve trophic conditions and food availability to krill; (2) *ultimate factors* which take account of the selective value of individuals in relation to when they produce offspring relative to the optimal time for larval survival. In krill, there may be certain advantages to individuals that spawn earlier in the season, or spawn on many different occasions so as to increase the chances of survival in their offspring. Individuals that move away from the swarm to spawn may also have a selective advantage through reducing the level of cannibalism. *M. norvegica* also demonstrate a trophic flexibility in which they can shift from feeding on spring diatom blooms which facilitate the start of previtellogenesis, on to dinoflagellates and lipid rich copepods (Virtue *et al.*, 2000) to fuel vitellogenesis later in the season. These adaptations ensure the permanence of the different populations of the species in environments as diverse as the cold and sub-neritic north Atlantic and sub-Arctic habitats and the warm, purely oceanic and very oligotrophic Mediterranean.

7.2. New perspectives

This review has described the current state of knowledge with regard to the morphological and physiological processes involved in the process of reproduction in *M. norvegica*. We now understand that the process is cyclical within the reproductive season and that there is interplay between moulting and spawning. Moreover, our understanding of what controls these processes at the molecular level is now starting to develop. Seear *et al.* (2009) provide a promising new approach through examining the genomic reaction of the physiology of Antarctic krill to an adjustment in photoperiod. Amongst the observed responses was a cessation in the molecular processes involved in the moult after just 7 days of exposure to continuous darkness (representing winter). Although a relatively limited study in terms of its experimental approach, the study highlights that we are now close to unravelling much of the molecular architecture behind the physiological cycles and traits observable at the whole organism level in krill. Some of the key questions that can be answered with these new techniques involve the process that initiate the switching on and off of the reproductive apparatus at the start and end of the reproductive season. Furthermore, *M. norvegica* is able to switch between egg production and oosorption midway through a season which is likely to involve some threshold response to resources or other environmental cues that may be addressed through experimental manipulation and subsequent molecular characterisation. The expression of genes during development in krill is an open question and one of great importance to the field of reproduction in terms of the development of the gonads and secondary sexual features. Molecular tools to tackle such questions are now developing at an almost exponential rate. With regard to krill reproduction, this review will hopefully provide the required insights necessary for asking the right questions.

ACKNOWLEDGEMENTS

The author wishes to thank Dr G. A. Tarling for the stimulating discussions and comments that were of great help to improve this manuscript, and also Dr S. Falk-Petersen for his helpful review. This work could not have been done without the support of the Director (Prof. L Legendre) and my colleagues from the Laboratoire d'Océanographie de l'Observatoire de Villefranche-sur-mer (UMR 7093).

REFERENCES

Adiyodi, R. G. (1985). Reproduction and its control. *In* "The Biology of Crustacea" (D. E. Bliss and H. Mantel, eds.), Vol.9, pp. 147–215. Academic Press, New York, London.

Adiyodi, R. G., and Subramoniam, T. (1983). Arthropoda-Crustacea. *In* "Reproductive Biology of Invertebrates. Vol.1. Oogenesis, Oviposition and Oosorption" (R. G. Adiyodi and K. G. Adiyodi, eds.), pp. 443–495. John Wiley and Sons, New York.

Albessard, E. (2003). Etude de la structure du métabolisme lipidique de l' euphausiacé *Meganyctiphnaes norvegica* : Influence d'un gradient tropho-climatique, Thèse de Doctorat de l'Université de Lyon 1, 122pp.

Albessard, E., Mayzaud, P., and Cuzin-Roudy, J. (2001). Variation of lipid classes among organs of the Northern krill *Meganyctiphanes norvegica*, with respect to reproduction. *Comp. Biochem. Physiol. Part A* **129**, 373–390.

Astthorsson, O. S. (1990). Ecology of the Euphausiids *Thysanoessa raschi*, *T. inermis* and *Meganyctiphanes norvegica* in Isafjord-Deep, northwest Iceland. *Mar. Biol.* **107**, 147–157.

Baker, J. R. (1938). The evolution of breeding season. *In* "Evolution; Essays on Aspects of Evolutionary Biology" (G. R. de Beer, ed.), pp. 161–177. Oxford University Press, London, Presented to E. Goodrich.

Bargmann, H. E. (1937). The reproductive system of *Euphausia superba*. *Discov. Rep.* **14**, 237–249.

Bargmann, H. E. (1945). The development and life history of adolescent krill *Euphausia superba*. *Discov. Rep.* **23**, 103–176.

Boysen, E., and Buchholz, F. (1984). *Meganyctiphanes norvegica* in the Kattegat. *Mar. Biol.* **79**, 195–207.

Buchholz, F., and Prado-Fiedler, R. (1987). Studies on the seasonal biochemistry of the Northern krill *Meganyctiphanes norvegica* in the Kattegat. *Helgoländ. Meeresunters* **41**, 443–452.

Buchholz, C. M., Buchholz, F., and Tarling, G. (2006). On the timing of moulting processes in reproductively active Northern krill *Meganyctiphanes norvegica*. *Mar. Biol.* **149**, 1443–1452.

Charniaux-Cotton, H., and Payen, G. (1985). Sexual differentiation. *In* "The Biology of Crustacea, Vol 9" (D. E. Bliss and L. H. Mantel, eds.), pp. 217–299. Academic Press, New York, London.

Cuzin-Roudy, J. (1985). Chronology of the female molt cycle in *Siriella armata* M. Edw. (Crustacea: Mysidacea) based on marsupial development. *J. Crust. Biol.* **5**, (1), 1–14.

Cuzin-Roudy, J. (1987a). Gonad history of the Antarctic krill *Euphausia superba* during its breeding season. *Polar Biol.* **7**, 237–244.

Cuzin-Roudy, J. (1987a) Sexual differentiation in the Antarctic krill *Euphausia superba* Dana (Crustacea; Euphausiacea). *J. Crust. Biol.* **7** (3), 518–524.

Cuzin-Roudy, J. (1993). Reproductive strategies of the Mediterranean krill, *Meganyctiphanes norvegica* and the Antarctic krill, *Euphausia superba* (Crustacea:Euphausiacea). *Invertebr. Reprod. Dev.* **23**, 105–114.

Cuzin-Roudy, J. (2000). Seasonal reproduction, multiple spawning, and fecundity in northern krill, *Meganyctiphanes norvegica* and Antarctic krill *Euphausia superba*. *Can. J. Fish. Aquat. Sci.* **57**, 6–15.

Cuzin-Roudy, J., and Amsler, M. O. (1991). Ovarian development and sexual maturity staging in Antarctic krill, *Euphausia superba* Dana (Euphausiacea). *J. Crustac. Biol.* **11**, 236–249.

Cuzin-Roudy, J., and Buchholz, F. (1999). Ovarian development and spawning in relation to the moult cycle in Northern krill *Meganyctiphanes norvegica* (Crustacea: Euphausiacea), along a climatic gradient. *Mar. Biol.* **133**, 267–281.

Cuzin-Roudy, J., and Labat, J. P. (1992). Early simmer distribution of *E. superba* sexual development in the Scotia-Weddell region: A multivariate approach. *Polar Biol.* **12**, 65–74.

Cuzin-Roudy, J., Albessard, E., Virtue, P., and Mayzaud, P. (1999). The scheduling of spawning with the moult cycle in Northern krill, (Crustacea: Euphausiacea): A strategy for allocating lipids to reproduction. *Invertebr. Reprod. Dev.* **36**, 163–170.

Cuzin-Roudy, J., Tarling, G. A., and Strömberg, J. O. (2004). Life cycle strategies of Northern krill (*Meganyctiphanes norvegica*) for regulating growth, moult and reproductive activity in various environments: The case of fjordic populations. *ICES J. Mar. Sci.* **61**, 721–737.

Dall, W., and Moriarty, D. J. W. (1983). Functional aspects of nutrition and digestion. *In* "The Biology of Crustacea. Vol. 5: Internal Anatomy and Physiological Regulation" (D. E. Bliss and Linda H. Mantel, eds.), pp. 215–251. Academic Press, NewYork, London.

Dalpadado, P., and Skjoldal, H. R. (1991). Distribution and life history of krill from the Barents Sea. *Polar Res.* **10**, 443–460.

Denys, C. F., and McWhinnie, M. A. (1982). Fecundity and ovarian cycles of the Antarctic krill *Euphausia superba*. *Can. J. Zool.* **60**, 2414–2423.

Einarsson, H. (1945). Euphausiacea I. Northern Atlantic species. *Dana Rep.* **27**, 1–184.

Everson Pearse, A. G. (1960). Histochemistry. Theoretical and applied (2nd edition, reprinted in 1961) J and A Churchill Ltd, London, pp. 998.

Falk-Petersen, S., and Hopkins, C. C. E. (1981). Ecological investigations on the zooplankton community of Balsfjorden northern Norway. Population dynamics of the euphausiids *Thysanoessa inermis*, *Thysanoessa raschi* and *Meganyctiphanes norvegica* in 1976 and 1977. *J. Plankton Res.* **3**, 177–192.

Falk-Petersen, S., Gatten, R. R., Sargent, R., and Hopkins, C. C. E. (1981). Ecological investigations on the zooplankton community of Balsfjorden, northern Norway: Seasonal changes in the lipid class composition of *Meganyctiphanes norvegica* (M Sars), *Thysanoessa raschii* (M.Sars) *Thysanoessa inermis* (Kröyer). *J. Exp. Mar. Biol. Ecol.* **54**, 209–224.

Falk-Petersen, S., Hagen, W., Katner, G., Clarke, A., and Sargent, J. (2000). Lipids, trophic relationships and biodiversity in Arctic and Antarctic krill. *Can. J. Aquat. Sci.* **57**(Suppl 3), 178–191.

Feinberg, L. R., Shaw, C. T., and Peterson, W. T. (2007). The timing and location of spawning in the euphausiids *Euphausia pacifica* and *Thysanoessa spinifera* off the Oregon coast. USA, Hiroshima, Japan.

Gabe, M. (1962). Résultats de l'histochimie des polysaccharides: invertébrés. *In* "Handbuch der Histochemie" (W. Graumann and K. Neumann, eds.), pp. 95–356. G. Fischer, Stuttgart.

Gomez-Gutierrez, J., Feinberg, L. R., Shaw, T., and Peterson, W. T. (2006). Variability in brood size and female length of *Euphausia pacifica* among three populations in the North Pacific. *Mar. Ecol. Prog. Ser.* **323**, 185–194.

Harrington, S. A., and Ikeda, T. (1986). Laboratory observation of spawning, brood size and egg hatchability of the Antarctic krill *Euphausia superba* from Pridz Bay, Antarctica. *Mar. Biol.* **92**, 231–235.

Hirano, Y., Matsuda, T., and Kawaguchi, S. (2003). Breeding Antarctic krill in captivity. *Mar. Freshw. Behav. Physiol.* **36**, (4), 259–269.

Ikeda, T., Dixon, P., and Kirkwood, J. (1982). Laboratory observations of moulting, growth and maturation in Antarctic krill Euphausia superba Dana. *Polar Biol.* **4**, 1–18.

Labat, J.-P., and Cuzin-Roudy, J. (1996). Population dynamics of the krill *Meganyctiphanes norvegica* (Crustacea: Euphausiacea) in the Ligurian Sea (N-W Mediterranean Sea). Size structure, growth and mortality modelling. *J. Plankton Res.* **18**, 2295–2312.

Laufer, H., Borst, D., Baker, F. C., Carrasco, C., Sinkus, M., Reuter, C. C., Tsai, L. W., and Schooley, D. A. (1987). Identification of a juvenile hormone-like compound in a crustacean. *Science* **235**, 202–205.

Lindley, J. A. (1982). Population dynamics and production of Euphausiids. III. *Meganyctiphanes norvegica* and *Nyctiphanes couchii* in the North Atlantic Ocean and the North Sea. *Mar. Biol.* **66**, 37–46.

Makarov, R. R., and Denys, C. J. (1980). Stages of sexual maturity of *Euphausia superba* Dana. *BIOMASS Handb.* **11**, 1–11.

Mauchline, J. (1980). The biology of mysids and euphausiids. *Adv. Mar. Biol.* **18**, (part II), 371–681.

Mauchline, J. (1981). Measurements of body-length of *Euphausia superba* Dana-. *BIOMASS Handb.* **4**, 1–9.

Mauchline, J. (1985). Growth and production of Euphausiacea (Crustacea) in the Rockall Trough. *Mar. Biol.* **90**, 19–26.

Mauchline, J., and Fisher, L. R. (1969). The biology of euphausiids. *Adv. Mar. Biol.* **7**, 1–454.

Mayzaud, P., Virtue, P., and Albessard, E. (1999). Seasonal variations in the lipid and fatty acid composition of the euphausiids *Meganyctiphanes norvegica* from the Ligurian Sea. *Mar. Ecol. Prog. Ser.* **186**, 199–210.

Meusy, J. J., and Payen, G. G. (1988). Female reproduction in Malacostracan Crustacea. *Zool. Sci.* **5**, 217–265.

Nelson, K. (1991). Scheduling of reproduction in relation to moult cycle and growth in malacostracans crustaceans. In "Crustacean Issues" (A. Wenner and A. Kuris, eds.), pp. 77–113. Balkema, Rotterdam.

Nicol, S. (1984). Population structure of daytime surface swarms of the euphausiids Meganyctiphanes norvegica in the bay of Fundy. Mar. Ecol. Progress. Ser. **18**, 241–251.

Nicol, S. (1989). Apparent independence of the spawning and moulting cycles in female Antarctic krill (Euphausia superba Dana). Polar Biol. **9**, 371–375.

Payen, G. G. (1980). Experimental studies in reproduction in Malacostraca crustaceans. Endocrine control of spermatogenic activity. In "Advances in Invertebrate Reproduction" (W. M. Clark and T. S. Adams, eds.), pp. 187–196. Elsevier, North Holland, NewYork.

Quetin, L. B., and Ross, R. R. (1991). Behavioural and physiological characteristics of the Antarctic krill. Am. Zool. **31**, 49–63.

Quetin, L., and Ross, R. M. (2001). Environmental variability and its impact on the reproductive cycle of Antarctic krill. Am. Zool. **41**, 74–89.

Raab, F. (1915). Beitrag zur Anatomie und Histologie der Euphausiidae. Arbeiten aus dem Zoloogischen Institution zu Wien, Bull. XX. Heft 2. Taf. X.

Ross, R. M., and Quetin, L.B. (1983). Spawning frequency and fecundity of the Antarctic krill. Euphausia superba. Mar. Biol. **77**, 201–205.

Ross, R. M., and Quetin, L. B. (2000). Reproduction in Euphausiacea. In "Krill: Biology, Ecology and Fisheries-" (I. Everson, ed.), pp. 150–181. Blackwell Science, Cambridge.

Ross, R. M., Daly, K. L., and English, T. S. (1982). Reproductive cycle and fecundity of Euphausia pacifica in Puget Sound, Washington. Limnol. Oceanogr. **27**, 304–314.

Seear, P., Tarling, G. A., Teschke, M., Meyer, B., Thorne, M. A. S., Clark, M. S., Gaten, E., and Rosato, E. (2009). Effects of simulated light regimes on gene expression in Antarctic krill. JEMBE **381**, 57–64.

Siegel, V., and Loeb, V. (1994). Length and age at maturity of Antarctic krill. Antarct. Sci. **6**, 479–482.

Tarling, G. A., and Cuzin-Roudy, J. (2003). Synchronization in the molting and spawning activity of northern krill (Meganyctiphanes norvegica) and its effect on recruitment. Limnol. Oceanogr. **48**, (5), 2020–2033.

Tarling, G. A., Cuzin-Roudy, J., and Buchholz, F. (1999a). Vertical migration behaviour in the northern krill Meganyctiphanes norvegica is influenced by moult and reproductive processes. Mar. Ecol. Progr. Ser. 253–262.

Tarling, G. A., Buchholz, F., and Matthews, J. B. L. (1999b). The effect of a lunar eclipse on the vertical migration of Meganyctiphanes norvegica (Crustacea: Euphausiacea) in the Ligurian Sea. J. Plankton Res. **21**, 1475–1488.

Tarling, G. A., Cuzin-Roudy, J., Thorpe, S. E., Shreeve, R. S., Ward, P., and Murphy, E. (2007). Recruitment of Antarctic krill Euphausia superba in the South Georgia region; adult fecundity and the fate of larvae. Mar. Ecol. Prog. Ser. **331**, 161–179.

Tarling, G. A., Cuzin-Roudy, J., Wooton, K., and Johnson, M. L. (2009). Egg release behaviour in Antarctic krill. Polar Biol. **32**, 1187–1194.

Virtue, P., Mayzaud, P., Albessard, E., and Nichols, P. (2000). Use of fatty acids as dietary indicators in northern krill Meganyctiphanes norvegica from north eastern Atlantic, Kattegat and Mediterranean waters. Can. J. Fish. Aquat. Sci. **57**, (Suppl 3), 104–114.

Zerbib, C. (1967). Première observation de la glande androgène chez un Crustacé Syncaride : Anaspides tasmaniae Thomson et un Crustacé Eucaride : Meganyctiphanes norvegica Sars. C. R. Acad. Sci. D **265**, 415–418.

CHAPTER EIGHT

Laboratory-Based Observations of Behaviour in Northern Krill (*Meganyctiphanes norvegica* Sars)

Edward Gaten,* Konrad Wiese,✠ *and* Magnus L. Johnson[†]

Contents

Abstract

The behaviour of planktonic animals remains poorly understood due to the difficulty of observing them *in situ* without influencing their behaviour. Here we review experiments on the behavioural responses of Northern krill, *Meganyctiphanes norvegica* (and related organisms), in isolation in laboratory-based aquaria. The value of this approach lies in the close observation that is possible; the downside is the uncertainty as to how well the observed behaviour relates to the natural behaviour of the subject animal. We discuss studies of swimming and swarming, and the responses of krill to light. We consider techniques involving automatic recordings that avoid, to some extent, making subjective decisions on behaviour. The effects of isolation of such a gregarious animal and of exposure to unnaturally high light levels are also considered. We conclude that such experiments can be of great value as long as these limiting factors are addressed.

* Department of Biology, University of Leicester, Leicester, United Kingdom
[†] Centre for Environmental and Marine Sciences, University of Hull, Scarborough, United Kingdom
✠ Deceased

Advances in Marine Biology, Volume 57   © 2010 Elsevier Ltd.
ISSN 0065-2881, DOI: 10.1016/S0065-2881(10)57008-1

1. INTRODUCTION

The history of research in Northern krill mirrors a dichotomy apparent in the wider field of research in biological oceanography that has been apparent since the 1880s (Kunzig, 2000). Back then, Victor Hensen thought that the oceans could be described in purely numeric terms and spent decades counting plankton. Ernst Haeckel was meanwhile scathing, saying "Mathematical treatment of these does more harm than good, because it gives a deceptive semblance of accuracy, which in fact is not attainable" (Mills, 1989). Because of the difficulties encountered in trying to observe plankton *in situ* (Hamner and Hamner, 2000), modern day oceanographers have been drawn towards Hensen's approach, something made obvious by the complete absence of behavioural observation as a technique in the zooplankton biologists' handbook by Harris *et al.* (2005). Kawaguchi *et al.* (2010) suggest that many studies view krill as passive particles drifting at the behest of physical processes, yet an understanding of the behaviour of individual animals is fundamental to oceanic ecology. Here we review the work that has been attempted on *Meganyctiphanes norvegica* and similar organisms in aquaria housed on land or at sea. Our hope is that we encourage a re-appraisal of the utility of observational science and promote the individual plankter to a position more than that of a mere integer.

Potentially, *M. norvegica* is an ideal model organism for laboratory-based behavioural studies of plankton—it is large, robust, locally abundant and widely distributed, and it is well-studied with an extensive literature base (Johnson and Tarling, 2011; Mauchline, 1980; Mauchline and Fisher, 1969). However, the major drawback to its routine use in the laboratory is the difficulty in keeping stocks alive for an extended period. *M. norvegica* rarely survives for more than 4 weeks in aquaria, probably due to injury during capture and bacterial infections of the integument (Buchholz, 2003). However, they have been kept alive in outside tanks during winter for up to 9 weeks, although only when maintained in virtual darkness; where the tanks were exposed to daylight, none of the krill survived beyond 15 days (Macdonald, 1927). In contrast to the long-term survival rates now achieved in Antarctic krill, *Euphausia superba* (Kawaguchi *et al.*, 2010), it appears that even flow-through systems are insufficient for lengthy experiments using *M. norvegica*.

A further problem with the use of *M. norvegica* as an experimental animal in a laboratory setting is that, like all krill, they are social animals invariably found in swarms (Aitken, 1960; Nicol, 1984). Isolation of krill in aquaria has undoubtedly led to data being produced that are difficult to interpret, for example in the study of swimming energetics (Ritz, 2000) and circadian migratory rhythms (Gaten *et al.*, 2008).

2. PELAGIC ANIMALS IN AQUARIA

A major drawback to making observations of the behaviour of any pelagic animals in the laboratory is that they would rarely, if ever, encounter the limitations imposed by the walls of an aquarium. To an animal that spends its entire life in the ocean without ever necessarily reaching either the surface or the sea-bed, the confines of an aquarium usually result in a modification of their behaviour (Strand and Hamner, 1990). After working with mesopelagic hyperiid amphipods, Land (personal communication) suggested that, based on observations of pelagic crustaceans in aquaria, one could come to the naive conclusion that they lie at the bottom of the tank on their sides swimming in circles repeatedly or are attracted to solid surfaces. Over the years there have been steady developments in techniques used to house krill in laboratory-based and ship-based aquaria that permit experiments on larval and adult growth, food preferences and assimilation, all carried out under varying environmental conditions (Ross and Quetin, 2003).

Feeding behaviour of krill is one area where researchers have attempted to use aquaria in order to develop an understanding of growth, consumption rates, selectivity and the role of krill as grazers and/or predators. Early qualitative observations suggested that krill were able to generate a feeding current with their pleopods that drew food towards the basket formed by the periopods (Berkes, 1975; Macdonald, 1927). More quantitative work carried out on *M. norvegica* in aquaria suggested that there was an ontogenetic shift from herbivory to carnivory with increasing size and that it selectively predated on larger copepod species (McClatchie, 1985). Torgersen (2001) carried out some work on feeding by *M. norvegica* in aquaria at different light intensities. He noted that predation rates on *Calanus* spp. and *Metridia longa* were approximately three times higher under light levels comparable to that found at the depths they inhabit during the day than they were in the dark. A difference in the relative proportion of the two species consumed under light and dark conditions was noted—under illuminated conditions more *Calanus* were consumed than in the dark, despite there being little apparent difference in the swimming speeds of the prey under either condition. Although Weissburg and Browman (2005) suggested that it would be inaccurate to characterise *M. norvegica* as a visual predator on the basis of this work, they later acknowledged that light intensity does seem to affect predation efficiency of this species (Abrahamsen et al., 2010).

Price (1989) was able to observe feeding behaviour of krill (*Thysanoessa rachii*) in a mesocosm and reported changes in behaviour when the krill encountered food. When they came across phytoplankton patches that had been introduced into their large aquarium, the krill that encountered the patch swam faster and exhibited less turning than they had in the absence of food. Individual krill were also reported to turn back when they reached the edge of the algal patch. More recently Kawaguchi *et al.* (2010) have managed to develop an aquarium set up that promotes schooling in *Euphausia superba* and can thus be used to develop a better understanding of complex social interactions amongst these animals. They report that schooling krill swim faster than non-schooling and that schools appear to disperse under conditions of low light.

One interesting problem with interpreting the behaviour of krill in aquaria is whether it is the same as would be found *in situ*. Macdonald (1927) and Mauchline (1989) both observed *M. norvegica* apparently feeding at the bottom of an aquarium by either suspending sediment through the action of their pleopods before swimming through the suspended material, or by ploughing through the surface layers with their antennae and transferring re-suspended material to their filtering basket with their periopods. These important reports appear to have been little commented upon until recent observations of *E. superba* actively feeding at abyssal depths by Clarke and Tyler (2008). Stomach content and stable isotope analyses also indicate that benthic feeding is a common trait in both Northern krill and Antarctic krill (Chapter 5).

3. 'TETHERED' SHRIMPS

One way of overcoming the problems presented by pelagic animals in aquaria is through the use of a 'tethered shrimp' approach. As with all tethering methods, this has the advantage of allowing the observer to keep the individual organism in view whilst still allowing it to react to various stimuli. It has the rather obvious disadvantage that it restricts free and natural movement preventing any interaction with other individuals. Attempts to assess stress levels of tethered krill singly or in close proximity suggested that association with conspecifics reduced heart rate to some extent (Ritz *et al.*, 2003). Kils (1982) investigated the flow field of the water produced by *M. norvegica* during swimming by gluing the animal to thin cord and measuring the flow of drifting bodies resulting from swimming activity. Hamner (1988) used tethered animals to allow close investigation of the feeding mechanism of *E. superba* and was able to dismiss previous suggestions that food entered the feeding basket from behind or laterally, rather than from the anterior.

Frank and Widder (1994a), who had noted the profound disorientation experienced by deep-sea shrimps when encountering any surface, developed a form of tethering that allowed observation and a degree of quantification of pelagic crustacean behaviour. Their solution was to attach a small stainless steel swivel to the dorsal carapace of the animal with the other end attached to a rigid plastic rod. Using this arrangement, the shrimp could rotate freely in both horizontal and vertical planes without reaching the walls of the aquarium. Monitoring the tilt or flexion of the body and changes in swimming activity (determined by counting beats of the pleopods) enabled the authors to find a threshold sensitivity to test flashes of light. The sensitivities of various species of decapods shrimps to near-UV and blue-green light were established using this approach followed by video analysis (Frank and Widder, 1994a,b). Abrahamsen *et al.* (2010) used tethered *M. norvegica* to examine their feeding behaviour. They found that the krill generally detected copepods that came within 7–23 mm in a hemisphere centred on the head of the animal such that the main detection volume is ventral and lateral. They also noted that the perceptual distance of the krill was higher in light than dark conditions. This, together with other work on feeding from aquarium-based studies, reinforces the concept of the importance of multi-sensory approaches to understand pelagic plankton responses to stimuli (Weissburg and Browman, 2005).

Others have gone a step further and attached crustaceans to force-transducers (e.g. Lenz and Hartline, 1999) or recorded the hydrodynamic signal generated by tethered krill using a microphone (Wiese and Ebina, 1995). One major advantage to this type of method is that swimming activity and behavioural responses can be recorded in such a way as to diminish the degree of subjectivity that may creep into analyses of video recordings. Another is that swimming activity and responses can easily be measured on several scales from instantaneous events such as escape responses (Lenz and Hartline, 1999) to changes in rate of response to longer term diurnal changes. One attempt to monitor long-term activity of krill using a combination of video and a carousel successfully recorded a single specimen of *M. norvegica* swimming 5.7 km in 24 h, despite the hindrance of a 10 cm wire attached to a simple swivel (Johnson and Thomasson, unpublished data). The krill was swimming at about 23 cm s^{-1} which is towards the upper end of swimming speed estimates from other sources (Kils, 1982).

Propulsion in krill is generated by highly efficient metachronal movements of the pleopods (Alben *et al.*, 2010; Kils, 1982). The swimming movements of *Euphausia superba* were described by Kils (1982) who paid particular attention to the extension and unfolding of the biramous pleopods at the onset of the power-stroke and the folding down and contraction at the beginning of the return stroke. These movements have the effect of a slight anterio-posterior jerk in the movement in the animal as it swims which has been recorded using pressure microphones for *E. superba* (Fig. 8.1) and a pendulum for *M. norvegica* (Thomasson *et al.*, 2003; Wiese and Ebina, 1995).

In order to record swimming behaviour and activity over long and short terms, Thomasson *et al.* (2003) used a virtually friction free rotational transducer as a pendulum to which *M. norvegica* was attached along the dorsal mid-line of the carapace (Fig. 8.2). This restricted the animal to movements along the main axis of their body. Swimming movements of

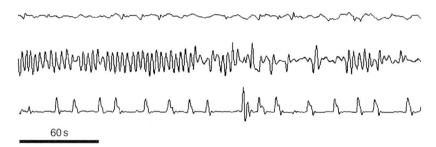

Figure 8.1 Example traces recorded from krill swimming while attached to a pendulum. As the animal swims the pendulum is displaced (increase on the Y axis). The top trace represents constant swimming, the middle cyclic swimming and the lower periodic swimming. Periodic swimming is found most often in satiated krill that, *in situ*, would be gliding downwards (Tarling and Johnson, 2006; reproduced with the permission of the publisher).

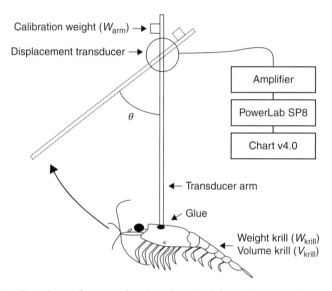

Figure 8.2 Experimental set-up showing the principles of the pendulum technique. Krill were glued to the transducer arm. Movement of the transducer arm caused by the animal swimming was measured in mV and recorded on a PowerLab SP8, software Chart 4.0. The mV readings were translated into angular displacement (θ), which was used to calculate swimming capacity in Newtons (Thomasson *et al.*, 2003; reproduced with the permission of the publisher).

the krill resulted in a signal from the transducer that was recorded to a computer via an analogue/digital convertor. Any thrust generated by the animal swimming pushed the pendulum forward until the weight of the animal plus a calibration weight exactly compensated for that thrust.

In the absence of any other weight, if the thrust generated by an animal was sufficient to allow it to hover then it would move a pendulum arm to 90° from vertical. Half that amount of thrust would move it to 30°. Thomasson et $al.$ (2003) connected transducers to an interface so that the exact angle that the animal was maintaining at any one time was known and the weight lifted by the animal (F, in N) could be calculated from:

$$F = [W_{krill} - (\rho H_2 O \times V_{krill}) + W_{arm}] \times g \times \sin\theta$$

where W_{krill} is the wet weight of krill in kg (total mass); $\rho H_2 O$, specific density of seawater in kg l^{-1} (generally taken as 1.03); V_{krill}, volume of the krill; W_{arm}, calibration weight of the transducer; g, acceleration due to gravity in m s^{-2} (9.82) and θ is the angle through which the transducer arm is displaced. In this manner, the relative swimming capacity of $M.$ $norvegica$ of different sizes, sexes and moult stages was determined. The swimming capacity was shown to increase with the size of the animal with no differences between male and female krill. There was a significant difference in swimming capacity between moult stages, with newly-moulted animals being weaker swimmers (Thomasson et $al.$, 2003). This apparatus was further modified for use at sea through developing control systems able to account for external motion. This allowed researchers to work on recently captured animals (Johnson and Tarling, 2008; Tarling and Johnson, 2006).

Ideally, it would be possible to calculate energy expended by an individual animal from the displacement of the arm. Because an animal swimming at a constant speed and attached to the pendulum is not moving, it is difficult to calculate 'work done' directly from Newton's 2nd law so we have to use the fact that it is the krill that is moving the water. If we assume a constant flow of water:

$$F = \frac{d(mv)}{d(t)} = r \times v$$

where m is the mass (kg); v, velocity in (m s^{-1}); t, time (s) and r is the mass of water moved per second.

We know F from the angle through which the pendulum is displaced but in order to estimate the power output and energy consumed by a swimming krill it is necessary to know how much water it is displacing to produce thrust. Yen et $al.$ (2003) estimated the volume of fluid displaced by $Euphausia$ $pacifica$ as being about 18 times the volume of the animal. The volume of a single $M.$ $norvegica$ can be estimated from Kils (1982) as:

$$\text{Volume} = 3.67 \times 10^{-6} L^{3.16}$$

where volume is in ml and length (L) is in mm. The mass of the water displaced per second (r) can be estimated using:

$$r = \text{PBR} \times \rho(\text{Vol} \times 18)$$

where PBR is the pleopod beat rate; ρ, the specific gravity of water and Vol $\times$ 18 is the estimated volume of water displaced during each pleopod beat cycle relative to the volume of the krill. Average pleopod beat frequency decreases with increasing length and can be estimated from Thomasson *et al.* (2003):

$$\text{PBR} = -0.107L + 10.18$$

where PBR is the pleopod beat rate in Hz and L, the total length of a krill. It should be noted that the pleopod beat rate varies significantly during the moult cycle of krill (Thomasson *et al.*, 2003). From the data obtained from the pendulum system (F and PBR), power in theory could be estimated as:

$$\frac{1}{2}F^2(18 \times \text{PBR} \times \rho \times \text{Vol})$$

Nevertheless, further parameters using this approach still need refinement. We do not know, for example if there are differences in the volume of water moved at different swimming speeds or by different sizes animals or how Reynolds numbers affect different sizes/species of krill in water bodies of different densities. Generally, the most efficient krill would move a large volume of water slowly. We also do not know how efficient pleopod-based swimming is and it is possible that a significant amount of energy is wasted in displacing water laterally. However the theory suggests that there is an exponential increase in the energy required for swimming with increasing size, something that undoubtedly limits the maximum size of negatively buoyant animals such as krill. These factors, based on observations of individual animals, may shed some light on subtle, larger scale ecological impacts of changes in local water density or reduced calcification rates induced by climate change.

The pendulum arrangement allows researchers to ask a broad range of ethological and physiological questions of individual krill of known state with regard to sex, size, moult stage and stomach fullness, and in response to olfactory, visual and physical stimuli or environmental variables such as light intensity, time, food availability, temperature or salinity. On occasion, observations of tethered animals can lead to interesting links between behaviour and physiology or anatomy. Tarling *et al.* (2009) linked swimming behaviour to egg release in *E. superba* and demonstrated that it was a slow and steady occurrence, where the swimming behaviour during spawning may explain some of the sorting evident in schools of this species. In mesopelagic shrimps,

Shelton *et al.* (2000) linked the eye-blink of mesopelagic decapods with the escape response and anatomy to demonstrate that previously unexplained tapetal distributions were linked to camouflage and bioluminescence. Gaten and Johnson (unpublished) found that brief flashes of dim blue light (ca. 10^{13} photons m^{-2} s^{-1}) were sufficient to cause dramatic changes in pleopod beat rate. Because swimming patterns may be affected by individual state and the artificial situation of the animal, as with simpler observational techniques, careful interpretation of short term responses may still be required (Johnson and Tarling, 2008; Thomasson *et al.*, 2003). Tarling and Johnson (2006) noted that there was a difference in swimming activity and swimming style depending on the stomach fullness of krill. Those that were satiated tended to be much less active than those that had empty stomachs. It was suggested that this behaviour could have significant implications for the carbon budget of the southern oceans as it was a mechanism that could carry biological material to depth more quickly than passive sinking of detritus. This work has implications for the standard tenet of biological oceanography that assumes that mesopelagic organisms migrate only once each day.

Using the pendulum system, comparison of swimming activities between species reveals some interesting differences between *M. norvegica* and *E. superba* (Thomasson *et al.*, 2003; Johnson and Tarling, 2008). Generally *M. norvegica* swam with a faster pleopod beat than equivalent sized *E. superba* and it was noted that in the latter species the males swam with a slower, stronger beat. Such a swimming style would be ideal as an honest signal (Zahavi and Zahavi, 1999) to female krill looking for a mate that is large (which is costly in the case of a negatively buoyant animal like krill), and would generate strong vortices that females could use to their advantage (see below). It is interesting that when we compared pleopod beat rates between moult stages of krill from the two species, a slightly different picture emerged. Soft, newly moulted *M. norvegica* had a low pleopod beat rate, suggesting that their strategy in this vulnerable condition was to sink quietly (Tarling *et al.*, 1999). *E. superba* had the highest pleopod beat rates when their carapace was soft which suggests that they may be using an increased beat rate to compensate for a lower output per beat (Johnson and Tarling, 2006). This would suggest that their strategy mandates staying with the school—a very different approach compared to *M. norvegica*, which may be indicative of important differences in lifestyles between the two species.

4. Swarming and Swimming

The understanding of swimming and swarming behaviour (and the underlying mechanisms) was limited in krill by the difficulty in maintaining these animals under experimental conditions that are conducive to these activities. However, the seminal experiments of Kils (1982) provided real insight into the behaviour and physiology of Antarctic krill, *E. superba*, particularly with respect to their swimming behaviour. The reasons suggested for swarming behaviour in krill and other organisms include reduction in energy expenditure, protection from predators, increased possibility of locating either food or a mate and improved decision making (List, 2004; Parrish and Hamner, 1997; Ritz, 1994; Ritz *et al.*, 2001). Strand and Hamner (1990) showed that swarming would occur in relatively large aquaria, provided they were free from any contrasting visual stimuli or external disturbance. Swarm formation can now be readily induced in *E. superba* at the Australian Antarctic Division research aquaria in Tasmania (Kawaguchi *et al.*, 2010). As Watson (2000) points out, the degree to which krill swarm can be of great ecological and commercial importance through knock-on effects on predator species.

There have been many *in situ* studies of swarm behaviour of *M. norvegica* that have revealed something of their spatial and temporal occurrence (Aitken, 1960; Nicol, 1986; Tarling *et al.*, 1998, 1999). Aspects that have been of interest to behavioural physiologists looking at the level of the individual have included the mechanisms that may promote swarm formation and maintenance and the possibility of advantages to individuals in terms of reduced costs of swimming (Swadling *et al.*, 2005; Wiese and Ebina, 1995).

The propulsion jet flow of *M. norvegica* and *E. superba* may qualify as a communication signal. To maintain a constant position in the krill formation and for fast re-assemblage even in darkness or after predator invasions, individuals have to be able to (A) perceive and analyse the flow field produced by the locomotion of their forerunners and (B) to synchronically beat their own pleopods. Three conditions are important for this task: the pleopods must generate a flow signal with specificity, the antennular sensor system must analyse the three-dimensional flow field, and a sensory-motor reflex from antennules to pleopods must exist (Abrahamsen *et al.*, 2010; Patria and Weise, 2004; Yen *et al.*, 2003).

In the flow jets produced by euphausiids the specificity rests in the spatial and temporal properties. The spatial properties determine the most favourable position energetically for an individual to swim within the formation, which is most certainly the contact point of the turbulent fringes of two adjacent propulsion jets (Wiese, 1996). The temporal component is characterised by the frequency spectrum of flow turbulence. Metachronal beats of the pleopods modulate the propulsion jet flow at 6 Hz in *M. norvegica* (Patria and Weise, 2004) and 3 Hz in *E. superba* (Kils, 1982), and thus provide a rhythmic mechanosensory input to the antennules of the following krill. The specificity of modulation of the flow signal renders it suitable for the purpose of communication (Markl, 1983).

The jetflow, which powers propulsion, may be as fast as 10 cm s^{-1} (Kils, 1983; Patria and Weise, 2004). The flow of water constitutes the carrier; it is directional and has a substantial range, considering the relatively small size of the krill (max. 50 mm). This range is determined first by the attenuation of flow velocity with distance (x) from the source, which is proportional to $1/x$ in jet flows of circular cross-section (Schlichting, 1982). Secondly, the range depends on the sensitivity of the receiving sensory organs, which is assumed to be 1 mm s^{-1} in the antennular flow receptors of tethered swimming euphausiids (Patria and Weise, 2004; Wiese and Marschall, 1990). Under these conditions, an assumed velocity of flow of 10 cm s^{-1} at a position 1 cm to the rear of the pleopods (Kils, 1982) would attenuate to 1 mm s^{-1} in a distance of 10 cm. Even in less densely organised schools of krill the nearest neighbour distance is estimated to be not larger than 50 cm and hence well within range of the flow signals from forerunners.

A modulated flow signal is produced by the propulsion jet pump of *M. norvegica* and *E. superba*. Kils (1982) has described in detail the action of the pleopods of *E. superba* (considered roughly equivalent to that of *M. norvegica*), especially the extension and unfolding of the biramous appendage at the onset of power-stroke and the folding down and contraction at the beginning of the return stroke. This rectifier effect in the periodic oscillation of pleopods leads to the occurrence of pressure jerks (pressure because a pressure sensitive microphone has been used in recording) during build up and decay of pressure peaks (Fig. 8.3; Wiese and Ebina, 1995).

As a consequence, the power spectrum of such a signal comprises a base frequency and three harmonics in fixed multiples of the base frequency (Wiese and Ebina, 1995). This means in the case of *M. norvegica* 6 Hz and about 12, 18 and 24 Hz; in the case of *E. superba* 3 Hz and about 6, 9 and 12 Hz. The three harmonics derive from the pressure jerks seen in the oscillating time course of pressure (Fig. 8.3), probably due to the rectifier effect in the pleopod action. The higher the order of the harmonic, the earlier this harmonic disappears from the signal with distance from the source. Provided that this rule applies to low frequency turbulent flow as well, and provided that the receiver system is able to encode them, harmonics may be used to indicate roughly the distance of the detector from the signal source (Wiese, 1996). Bleckmann *et al.* (1991) proposed that oscillations of water between 1 and 40 Hz are monitored very thoroughly by crustaceans as potential signals indicating the presence of prey or of conspecifics, whereas events comprising oscillation frequencies 40–100 Hz often result in escape responses.

Flow in euphausiids is sensed by one-sidedly hinged flagella of the antennules, with the proprioceptors of the hinge serving as transducers. The antennular flagella often break in captivity. Levers, in contrast to membranes, are suitable devices to work under various pressure conditions and therefore various water depths. The threshold curve of sensitivity of antennular proprioceptors has been measured both in *E. superba* (Wiese and Marschall, 1990) and in *M. norvegiva* (Fig.8.5; Patria and Weise, 2004) by

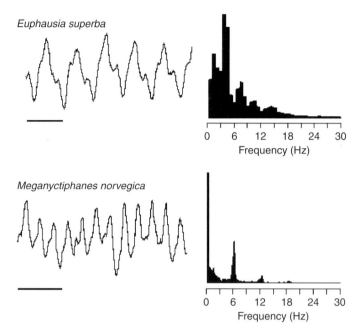

Figure 8.3 Recording of the water pressure generated by the action of swimmerets of *Euphausia superba* (upper trace) and *Meganyctiphanes norvegica* (lower trace) recorded using a pressure sensitive microphone. For *E. superba* on average the beat frequency is around 3 Hz (scale bar = 250 ms) and for *M. norvegica* around 6 Hz (scale bar = 500 ms). Pressure fluctuations show regular rhythmic peaks and troughs separated by interspersed pressure jerks which are interpreted as reflections of the unfolding and collapse (mechanical rectifier) of the pleopod. The FFT spectra show pressure jerks in the signal marking the presence of harmonics in the spectrum of the signal. In *E. superba* the harmonics are 6, 9, 12 Hz; in *M. norvegica* they are 12, 18 and 24 Hz. The low frequency character of at least the first of the harmonics suggests that some of the harmonics are encoded by antennular proprioceptors. The harmonics characterise the flow jet of *M. norvegica* and may help conspecifics to identify it (after Wiese and Ebina (1995)).

recording sensory cell activity in the basipodite of the antennule in response to quantitatively controlled vibrations of water applied to the vertically moving flagellum (Wiese *et al.*, 1980). A sensitivity threshold of around 0.5 mm s^{-1} was noticed (frequency range 1–40 Hz), with higher sensitivity in *E. superba* than *M. norvegica*. The reason for the encountered difference in sensitivity is not known although it may be explained in part by the differences in swarming and schooling activity between the two species. *M. norvegica* females appear unable to swim when newly moulted and hence tend to sink (Thomasson *et al.*, 2003), whereas female *E. superba* retain greater rigidity in their pleopods which enables them to maintain their position in the swarm (Johnson and Tarling, 2008).

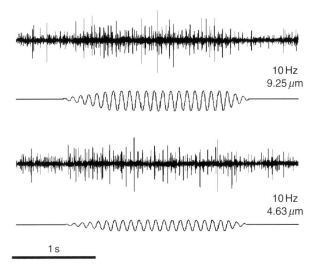

Figure 8.4 Recordings of evoked sensory activity from the antennular nerve of *M. norvegica* at a site 5 mm proximal to the flagellar hinge (upper traces) in response to water vibration stimuli (lower traces)(see Patria and Weise (2004) for details of the recording method). The extracellular recording shows several nerve axons responding (deduced by the different amplitudes of the action potentials). Two different axons often respond per sine cycle of the stimulus. This fact is interpreted as a sign of sensitivity to both flexion and extension of the flagellar hinge.

A sample of an actual recording of sensory nerve activity (Fig. 8.4) shows two different action potentials (different amplitude in the extracellular recording) within one cycle of the water oscillation. Both bending and stretching in the antennular flagellar hinge produce a signal in the proprioceptive sensory cells.

Systems of communication very much depend on the range through which the generated signals extend from the source to the prospective receiver. In this context, flow velocities generated by the pleopod jet pump have been measured for *E. superba* using kinematography (Kils, 1982) and for *M. norvegica* using computer assisted video analysis (Patria and Weise, 2004). According to these investigations, the range of generated flow velocities extends from 5–10 cm s^{-1} near the pleopods. Using the sensitivity threshold of 0.5 mm s^{-1} in velocity of flow (Fig. 8.5), the rule of Schlichting (1982) (see above) predicts a range of signals of modulated flow as produced by *M. norvegica* and *E. superba* of roughly 100 cm. That is, the dispersedly swimming krill more or less depend on accidental encounters with conspecifics to recruit new individuals to a potential formation.

Divers swimming close to formations of *E. superba* (Hamner, 1984) reported that, in formation, individuals swim at the maximum force available and are apparently sorted by size. Tarling (personal communication) has

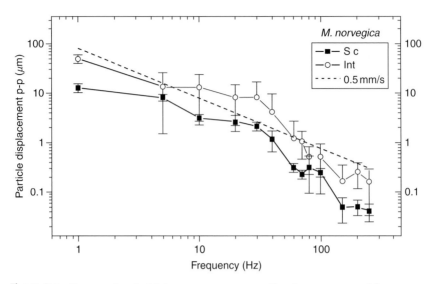

Figure 8.5 Sensory thresholds in response to water vibration as measured from axons of the antennular nerve (Sc) or in interneurons (Int) in the ventral nerve cord in *M. norvegica*. The dashed line represents a particle flow of 0.5 mm s^{-1}. Stimulus amplitude is given in micrometers peak to peak at threshold, the frequency range considered is 1–150 Hz (see Patria and Weise (2004) for details of the recording method).

measured inter-individual distances in krill formations and reported typically 20–25 cm between forerunner and follower. Measurements using tethered krill suggest that the position of the follower shrimp in the formation is linked to the ring-vortex established around the jetflow of circular cross-section produced by krill swimming ahead (Wiese, unpublished observations). In tethered *M. norvegica* this ring vortex is detected about 10–12 cm behind the pleopod-pump (Patria and Weise, 2004).

5. Light and Daily Rhythms

Light is a key factor in experiments with deep water crustaceans, although this is often overlooked with the result that unreliable data may unwittingly be collected. The problems of retinal breakdown in response to exposure to unnaturally high light levels have been demonstrated in a range of crustaceans, from the Norway lobster in Scottish sea lochs (Shelton *et al.*, 1985) to hydrothermal vent shrimps (Herring *et al.*, 1999). Working with *E. superba*, Newman *et al.* (2003) noted that krill reacted to UVA and photosynthetically active radiation (PAR) by moving away from it.

They suggested that this was a mechanism to protect them from over-exposure to damaging UVB light.

The daylight levels normally experienced by *M. norvegica* have been estimated to lie between 2.8×10^{12} and 1.4×10^{14} photons m^{-2} s^{-1} in the Oslofjord, Norway (Onsrud and Kaartvedt, 1998) and between 9.9×10^{12} and 4.5×10^{13} photons m^{-2} s^{-1} in the Gullmarsfjord, Sweden (Gaten and Johnson, unpublished) where the krill are normally found in the deepest part of the fjord during the day. Preliminary experiments have shown that following 3 h exposure to light at 1.8×10^{19} photons m^{-2} s^{-1} the eyes of *M. norvegica* showed minor damage to the rhabdoms when compared to animals fixed in the dark. However, krill fixed after being caught at midday and exposed briefly to sunlight during capture (8.1×10^{20} photons m^{-2} s^{-1}) showed extensive rhabdom degeneration (Fig. 8.6; Gaten and Johnson, unpublished). This shows that the eyes of this species (and probably all euphausiids) are vulnerable to light-induced damage that, although reversible following brief exposure, usually proved permanent in decapod crustaceans from various depths exposed to similar levels of daylight (Gaten *et al.*, 1990).

There is little doubt that their behaviour is also affected by brief flashes of light as they have, in common with almost all euphausiids, the ability to produce light (Herring and Locket, 1978). Bioluminescence is produced in *M. norvegica* by 10 photophores, one on each eyestalk, four on the ventral thorax and four on the ventral abdomen (Herring and Locket, 1978). They have been shown to respond to various photic stimuli with flashes of light from the photophores (Mauchline, 1960). Although photophores, such as the ocular peduncle, have been proposed to illuminate food in the dark

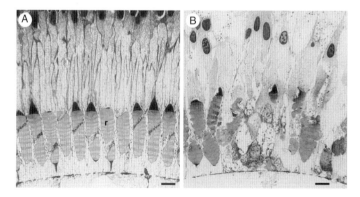

Figure 8.6 Cross sections through the retinula cell layer of *M. norvegica* show the result of a brief exposure to daylight. (A) In an eye fixed in the dark, rows of intact rhabdoms (*r*) are seen. (B) Exposure to daylight results in extensive breakdown of the rhabdoms and the surrounding cells. Scale bar = 10 μm.

(Macdonald, 1927), the photophores are mainly thought to provide adaptive benefits through camouflage involving counter shading that matches the downwelling radiation (Clarke, 1963). Recent laboratory-based work on tethered krill has suggested that they may form the basis of a system of communication using light flashes (Fregin and Wiese, 2002).

Euphausiids lack statocysts so it is probable that their only way of orientating themselves to the vertical is via the direction of light. Land (1980) showed that this is probably correct in experiments on a mesopelagic euphausiid, *Nematoscelis atlantica*. He observed specimens that were restrained, but that could freely rotate, and monitored their reaction to a small, moving light source. It was clear that the tethered animals rotated their eyes through 180° to track the light and that the ventral photophores rotated in synchrony with the movement of the eyes. This strongly suggests that the eyes tracked the downwelling light, allowing the animal to orientate itself vertically, whilst the photophores simultaneously rotated to remain pointing downwards (Land, 1980). The latter point lends support to the suggestion that the photophores are concerned primarily with counter-illumination, at least in this species. *M. norvegica* similarly has thoracic and abdominal photophores that can rotate through 180° (Hardy, 1962).

In addition to the reliance on light for vision and orientation, *M. norvegica* almost certainly uses the daily variation in light levels to control its circadian clock. Circadian rhythms have been found to be present in all organisms so far examined and they underlie the ability to predict daily variations in light and other factors. *M. norvegica* undertakes an extensive vertical migration, rising to the surface each night to feed and moving into deeper water during the daytime, probably to avoid visually guided pre-dators (but see Tarling and Johnson, 2006). The ability to predict the onset of day and night enables the animals to initiate their diel migrations at the appropriate time, irrespective of the ambient light conditions. Fine tuning of the migration depth then occurs in response to changes in the light climate, for instance at times of solar (Strömberg *et al.*, 2002) or lunar (Tarling *et al.*, 1999) eclipses.

The influence of a circadian rhythm on swimming activity of individual *M. norvegica* was clearly demonstrated in the laboratory by Velsch and Champalbert (1994) using both actographic analysis and infra-red video observations. After establishing that an activity rhythm was present under a light:dark cycle, with maximum activity during the night, they showed that an endogenous rhythm was present in total darkness with a period of less than 23 h. They concluded that light variations and the endogenous rhythm both had a role in vertical migration in *M. norvegica* (Velsch and Champalbert, 1994).

An adapted version of this actograph, using 12 experimental chambers, has been used with Antarctic krill in a ship-board experiment (Gaten *et al.*, 2008). The activity monitor comprised 12 vertical tubes, each containing

Figure 8.7 The krill activity monitor contains 12 vertical tubes with infra-red barriers close to the top and bottom of each tube. The illumination is provided by blue LEDs in the lid that are covered by neutral density material to provide light conditions close to ambient. The dimensions of the box are 84 × 58 cm with a depth of 61 cm.

5 l of seawater, retained upright in a light-tight box (Fig. 8.7). Each tube had infra-red barriers 5 cm from the top and bottom of the tube and the output from these barriers was recorded continuously. The lighting within the apparatus was adjusted to that experienced by the krill at normal daytime depths and the temperature in the cool-room kept at the level of the seawater supply. The krill were not fed during the experiment. Activity levels of the 12 individuals were recorded for five days under a light:dark cycle and then for a further five days in total darkness. The locomotor activity patterns showed a degree of complexity that suggested that the light: dark cycle was not the only *Zeitgeber* (external cue) in this species (Fig. 8.8). The presence of a secondary 12 h rhythm in 68% of the rhythmic animals is consistent with field observations of a 12 h component in vertical migration that becomes predominant when food is scarce (Godlewska, 1996). In addition, ultradian components in the activity rhythm are thought to reflect the stress caused by the isolation of these extremely social animals (Gaten *et al.*, 2008). Together these findings highlight the necessity of taking into account the effects of lack of food and of social cues when undertaking experiments on krill.

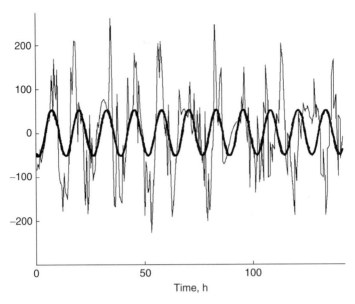

Figure 8.8 The presence of a 12 h activity rhythm in *Euphausia superba* is demonstrated by the filtered output from the activity monitor. The animal was in total
darkness throughout, so the trace is evidence of an endogenous bimodal circadian
rhythm. The data are plotted as relative activity as a function of time, with a super-
imposed sinusoid derived from the computed period of the oscillation (12.6 h). For
further details see Gaten *et al.* (2008).

The eyes of *M. norvegica* have been well described, both in terms of their
morphology (Hallberg and Nilsson, 1983) and their optics (Land *et al.*,
1979). The eyes are of the refracting superposition type and, in contrast to
the 'double eyes' found in many euphausiids, are spherically symmetrical.
Arguably, the most important property of an eye is the minimum resolvable
angle as this can be used to define the resolution of the eye under ideal
conditions. Land *et al.* (1979) used the inter-receptor angle ($\Delta\varphi$) to estimate
the resolution of the eye of *M. norvegica*:

$$\Delta\varphi = 57.3d/f$$

where d is the maximum diameter of the receptor and f is the focal length
(or posterior nodal distance). This gives a value of around 2.9° which is
much lower resolution than is usually found in the upper eyes of euphausiids
(1.2°—Land *et al.*, 1979) or mysids (1.5°—Gaten *et al.*, 2002) from deeper
water. This is assuming ideal optics and adequate lighting—in reality the
resolution would be much poorer due to imperfections in the optics. What
this means in practical terms is that *M. norvegica,* given the low light levels at

which these animals normally operate, is unlikely to be predominantly a visual predator. The higher predation rates on *Metridia longa* by *M. norvegica* noted by Torgersen (2001) is thus much more likely to be due to the fact that this prey species swims faster and further than the other copepod used in this experiment and is hence more likely to encounter the predator (Browman, 2005).

6. Conclusion

Observations of *Meganyctiphanes norvegica* and other species of krill in aquaria or on tethers will never be as satisfactory in principle as *in situ* observations of individual animals would be. There are some techniques developing that allow individual animals to be tracked in the medium term (e.g. McGehee and Jaffe, 1996) but they will probably never allow a full understanding of the state and circumstance of individual organisms. Aquarium-based studies have huge potential to answer fundamental questions about krill biology, ecology and behaviour, as long as the limitations are recognised and care is taken over the conclusions drawn (Kawaguchi *et al.*, 2010).

One aspect that must be considered is that krill are highly social animals that do not function in a normal fashion when isolated in an aquarium (Gaten *et al.*, 2008; Strand and Hamner, 1990). In addition, it is of paramount importance in designing experiments in aquaria to consider the light environment carefully. Light is a key determinant of behaviour of mesopelagic organisms and they tend to be specifically adapted to cope with a particular light intensity, in addition to defined spectral, spatial and temporal distributions of light (Gaten *et al.*, 1990, 1992, 2002; Johnson *et al.*, 2000a,b, 2002). Failure to consider these important factors is likely to lead to anomalous behaviour, physiological damage and erroneous conclusions.

ACKNOWLEDGEMENTS

We would like to thank Dr Andrew Mehta, University of Liverpool Physics Department, for his advice with regard to the pendulum equations.

REFERENCES

Abrahamsen, M., Browman, H. I., Fields, D. M., and Skiftesvik, A. B. (2010). The three-dimensional prey field of the northern krill, *Meganyctiphanes norvegica*, and the escape responses of their copepod prey. *Mar. Biol.* **157,** 1251–1258.

Aitken, J. J. (1960). Swarming in *Meganyctiphanes norvegica* (M. Sars) in Strangford lough, Co. Down. *Ir. Nat. J.* **13,** 140–142.

Alben, S., Spears, K., Garth, S., Murphy, D., and Yen, J. (2010). Coordination of multiple appendages in drag-based swimming. *J. R. Soc. Interface,* 10.1098/rsif.2010.0171.

Berkes, F. (1975). Some aspects of feeding mechanisms of Euphausiid crustaceans. *Crustaceana* **29,** 266–270.

Bleckmann, H., Breithaupt, T., Blickhan, R., and Tautz, J. (1991). The time course and frequency content of hydrodynamic events caused by moving fish, frogs and crustaceans. *J. Comp. Physiol. A* **168,** 749–757.

Browman, H. I. (2005). Applications of sensory biology in marine ecology and aquaculture. *Mar. Ecol. Prog. Ser.* **287,** 263–307.

Buchholz, F. (2003). Experiments on the physiology of southern and northern krill, *Euphausia superba* and *Meganyctiphanes norvegica,* with emphasis on moult and growth – A review. *Mar. Freshw. Behav. Physiol.* **36,** 229–247.

Clarke, W. D. (1963). Function of bioluminescence in mesopelagic organisms. *Nature* **198,** 1244–1246.

Clarke, A., and Tyler, P. (2008). Adult antarctic krill feeding at abyssal depths. *Curr. Biol.* **18,** 282–285.

Frank, T. M., and Widder, E. A. (1994a). Evidence for behavioural sensitivity to near-UV light in the deep-sea crustacean *Systellaspis debilis. Mar. Biol.* **118,** 279–284.

Frank, T. M., and Widder, E. A. (1994b). Comparative study of behavioural sensitivity thresholds to near-UV and blue-green light in deep-sea crustaceans. *Mar. Biol.* **121,** 229–235.

Fregin, T., and Wiese, K. (2002). The photophores of *Meganyctiphanes norvegica* (M. Sars) (Euphausiacea): Mode of operation. *Helgol. Mar. Res.* **56,** 112–124.

Gaten, E., Shelton, P. M. J., Chapman, C. J., and Shanks, A. M. (1990). Depth-related variation in structure and functioning of the compound eyes of the Norway lobster *Nephrops norvegicus. J. Mar. Biol. Assoc.* **70,** 343–355.

Gaten, E., Shelton, P. M. J., and Herring, P. J. (1992). Regional morphological variation in the compound eyes of mesopelagic decapods in relation to their habitat. *J. Mar. Biol. Assoc.* **72,** 61–75.

Gaten, E., Herring, P. J., and Shelton, P. M. J. (2002). Eye morphology and optics of a double-eyed mysid shrimp, *Euchaetomera typica. Acta Zool.* **83,** 221–230.

Gaten, E., Tarling, G., Dowse, H., Kyriacou, C. P., and Rosato, E. (2008). Is vertical migration in Antarctic krill (*Euphausia superba*) influenced by an underlying circadian rhythm? *J. Genet.* **87,** 473–483.

Godlewska, M. (1996). Vertical migrations of krill (*Euphausia superba* Dana). *P. Arch. Hydrobiol.* **43,** 9–63.

Hallberg, E., and Nilsson, D.-E. (1983). The euphausiid compound eye—a morphological re-investigation (Crustacea: Euphausiacea). *Zoomorphology* **103,** 59–66.

Hamner, W. M. (1984). Aspects of schooling in *Euphausia superba. J. Crustac. Biol.* **4,** 67–74.

Hamner, W. M. (1988). Biomechanics of filter feeding in the Antarctic krill *Euphausia superba*: review of past work and new observations. *J. Crustac. Biol.* **8,** 149–163.

Hamner, W. M., and Hamner, P. P. (2000). Behavior of Antarctic krill (*Euphausia superba*): Schooling foraging, and antipredatory behavior. *Can. J. Fish. Aquat. Sci.* **57,** 192–202.

Hardy, M. G. (1962). Photophore and eye movement in the Euphausiid *Meganyctiphanes norvegica* (G. O. Sars). *Nature* **196,** 790–791.

Harris, R. P., Wiebe, P. H., Lenz, J., Skjoldal, H. R., and Huntley, M. (2005). Zooplankton Methodology Manual. Elsevier, Amsterdam.

Herring, P. J., and Locket, N. A. (1978). The luminescence and photophores of euphausiid crustaceans. *J. Zool.* **186,** 431–462.

Herring, P. J., Gaten, E., and Shelton, P. M. J. (1999). Are vent shrimps blinded by science? *Nature* **398,** 116.

Johnson, M. L., and Tarling, G. A. (2008). Influence of individual state on swimming capacity and behaviour of Antarctic krill *Euphausia superba*. *Mar. Ecol. Prog. Ser.* **366,** 99–110.

Johnson, M. L., and Tarling, G. A. (2011). Krill. *In* "Crustacean Fieldwork Methods" (S. De Grave and J. W. Martin, eds.). Cambridge University press.

Johnson, M. L., Shelton, P. M. J., and Gaten, E. (2000a). Temporal resolution in the eyes of marine decapods from coastal and deep-sea habitats. *Mar. Biol.* **136,** 243–248.

Johnson, M. L., Shelton, P. M. J., Gaten, E., and Herring, P. (2000b). Relationship of dorsoventral eyeshine distributions to habitat depth and animal size in mesopelagic decapods. *Biol. Bull.* **199,** 6–13.

Johnson, M. L., Gaten, E., and Shelton, P. M. J. (2002). Spectral sensitivities of five marine decapod crustaceans and a review of spectral sensitivity variation in relation to habitat. *J. Mar. Biol. Assoc.* **82,** 835–842.

Kawaguchi, S., King, R., Meijers, R., Osborn, J. E., Swadling, K. M., Ritz, D. A., and Nicol, S. (2010). An experimental aquarium for observing the schooling behaviour of Antarctic krill (*Euphausia superba*). *Deep Sea Res. II* **57,** 683–692.

Kils, U. (1982). The swimming behaviour, swimming performance and energy balance of Antarctic krill, *Euphausia superba*. *Biomass Kiel Sci. Res. Ser.* **3,** 1–121.

Kils, U. (1983). Swimming and feeding of Antarctic krill, *Euphausia superba*—some outstanding energetics and dynamics, some unique morphological details. *Berichte zur Polarforschung, Sonderheft* **4,** 130–155.

Kunzig, R. (2000). Mapping the Deep The Extraordinary Story of Ocean Science. Sort of books, London.

Land, M. F. (1980). Eye movements and the mechanism of vertical steering in euphausiid Crustacea. *J. Comp. Physiol. A* **137,** 255–265.

Land, M. F., Burton, F. A., and Meyer-Rochow, V. B. (1979). The optical geometry of euphausiid eyes. *J. Comp. Physiol. A* **130,** 49–52.

Lenz, P. H., and Hartline, D. K. (1999). Reaction times and force production during escape behaviour of a calanoid copepod, *Undinula vulgaris*. *Mar. Biol.* **133,** 249–258.

List, C. (2004). Democracy in animal groups: a political science perspective. *Trends Ecol. Evol.* **19,** 168–169.

Macdonald, R. (1927). Food and habits of *Meganyctiphanes norvegica*. *J. Mar. Biol. Assoc.* **14,** 753–784.

Markl, H. (1983). Vibrational communication. *In* "Neuroethology and Behavioural Physiology" (F. Huber, H. Markl, F. Huber and H. Markl, eds.), pp. 332–353. Springer, Berlin Heidelberg, New York.

Mauchline, J. (1960). The biology of the euphausiid *Meganyctiphanes norvegica* (M. Sars). *Proc. R. Soc. Edinb.* **67B,** 141–179.

Mauchline, J. (1980). The biology of euphausiids. *Adv. Mar. Biol.* **18,** 373–623.

Mauchline, J. (1989). Functional morphology and feeding of euphausiids. *In* "Functional morphology of feeding and grooming in Crustacea" (B. E. Felgenhauer, L. Watling and A. B. Thistle, eds.), pp. 173–184. Balkema, Rotterdam, Netherlands.

Mauchline, J., and Fisher, L. R. (1969). The biology of Euphausiids. *Adv. Mar. Biol.* **7,** 1–421.

McClatchie, S. (1985). Feeding behaviour in *Meganyctiphanes norvegica* (M. Sars) (Crustacea: Euphausiacea). *J. Exp. Mar. Biol. Ecol.* **86,** 271–284.

McGehee, D., and Jaffe, J. S. (1996). Three-dimensional swimming behaviour of individual zooplankters: Observations using the acoustical imaging system FishTV. *ICES J. Mar. Sci.* **53,** 363–369.

Mills, E. (1989). Biological Oceanography, an early history, 1870-1960. Cornell University Press, Ithaca.

Newman, S. J., Ritz, D., and Nicol, S. (2003). Behavioural reactions of Antarctic krill (*Euphausia superba* Dana) to ultraviolet radiation and photosynthetically active radiation. *J. Exp. Biol.* **297,** 203–217.

Nicol, S. (1984). Population structure of daytime surface swarms of the euphausiid *Meganyctiphanes norvegica* in the Bay of Fundy. *Mar. Ecol. Prog. Ser.* **18**, 241–251.

Nicol, S. (1986). Shape, size and density of daytime surface swarms of the euphausiid *Meganyctiphanes norvegica* in the Bay of Fundy. *J. Plankton Res.* **8**, 29–39.

Onsrud, M. S. R., and Kaartvedt, S. (1998). Diel vertical migration of krill *Meganyctiphanes norvegica* in relation to physical environment, food and predators. *Mar. Ecol. Prog. Ser.* **171**, 209–219.

Parrish, J. K., and Hamner, W. M. (1997). Animal Groups in Three Dimensions Cambridge University Press, Cambridge.

Patria, M., and Weise, K. (2004). Swimming in formation in krill (Euphausiacea), a hypothesis: Dynamics of the flow field, properties of antennular sensor systems and a sensory motor link. *J. Plankton Res.* **26**, 1315–1325.

Price, H. J. (1989). Swimming behaviour of krill in response to algal patches: A mesocosm study. *Limnol. Oceanogr.* **34**, 649–659.

Ritz, D. A. (1994). Social aggregation in pelagic invertebrates. *Adv. Mar. Biol.* **30**, 155–216.

Ritz, D. A. (2000). Is social aggregation in aquatic crustaceans a strategy to conserve energy? *Can. J. Fish. Aquat. Sci.* **57**, 59–67.

Ritz, D. A., Foster, E. G., and Swadling, K. M. (2001). Benefits of swarming: Mysids in larger swarms save energy. *J. Mar. Biol. Assoc. UK* **81**, 543–544.

Ritz, D. A., Cromer, L., Swadling, K. M., Nicol, S., and Osborn, J. (2003). Heart rate as a measure of stress in Antarctic krill, *Euphausia superba*. *J. Mar. Biol. Assoc.* **83**, 329–330.

Ross, R. M., and Quetin, L. B. (2003). Working with living krill – the people and the places. *Marine and Freshwater Behaviour and Physiology* **36**, 207–228.

Schlichting, H. (1982). Grenzschicht-Theorie. Braun Verlag, Karlsruhe, pp. 754.

Shelton, P. M. J., Gaten, E., and Chapman, C. J. (1985). Light and retinal damage in *Nephrops norvegicus* (L.) (Crustacea). *Proc. R. Soc. Lond. B* **226**, 217–236.

Shelton, P. M. J., Gaten, E., Johnson, M. L., and Herring, P. J. (2000). The 'eye-blink' response of mesopelagic Natantia, eyeshine patterns and the escape reaction. *Crustac. Issues* **12**, 253–260.

Strand, S. W., and Hamner, W. M. (1990). Schooling behaviour of Antarctic krill (*Euphausia superba*) in laboratory aquaria: reactions to chemical and visual stimuli. *Mar. Biol.* **106**, 355–359.

Strömberg, J.-O., Spicer, J. I., Liljebladh, B., and Thomasson, M. A. (2002). Northern krill, *Meganyctiphanes norvegica*, come up to see the last eclipse of the millennium? *J. Mar. Biol. Assoc. UK* **82**, 919–920.

Swadling, K. M., Ritz, D. A., Nicol, S., Osborn, J. E., and Gurney, L. J. (2005). Respiration rate and cost of swimming for Antarctic krill, *Euphausia superba*, in large groups in the laboratory. *Mar. Biol.* **146**, 1169–1175.

Tarling, G. A., and Johnson, M. L. (2006). Satiation gives that sinking feeling. *Curr. Biol.* **16**, R83–R84.

Tarling, G. A., Matthews, J. B. L., Saborowski, R., and Buchholz, F. (1998). Vertical migratory behaviour of the euphausiid, *Meganyctiphanes norvegica*, and its dispersion in the Kattegat Channel. *Hydrobiologia* **375**, (376), 331–341.

Tarling, G. A., Buchholz, F., and Matthews, J. B. L. (1999). The effect of a lunar eclipse on the vertical migration behaviour of *Meganyctiphanes norvegica* (Crustacea: Euphausiacea) in the Ligurian Sea. *J. Plankton Res.* **21**, 1475–1488.

Tarling, G. A., Cuzin-Roudy, J., Wootton, K., and Johnson, M. L. (2009). Egg-release behaviour in Antarctic krill. *Polar Biol.* **32**, 1187–1194.

Thomasson, M. A., Johnson, M. L., Strömberg, J.-O., and Gaten, E. (2003). Variations in swimming power output of *Meganyctiphanes norvegica* in relation to sex, size and moult stage. *Mar. Ecol. Prog. Ser.* **250**, 205–213.

Torgersen, T. (2001). Visual predation by the euphausiid *Meganyctiphanes norvegica*. *Mar. Ecol. Prog. Ser.* **209,** 295–299.

Velsch, J.-P., and Champalbert, G. (1994). Swimming activity rhythms in Meganyctiphanes norvegica. *C.R. Acad. Sci. Ser. III* **317,** 857–862.

Watson, J. (2000). Aggregation and vertical migration. *In* "Krill Biology" (I. Everson, ed.), pp. 80–102. Ecology and Fisheries. Blackwell Science, Oxford.

Weissburg, M. J., and Browman, H. I. (2005). Sensory biology: Linking the external and internal ecologies of marine organisms. *Mar. Ecol. Prog. Ser.* **287,** 263–307.

Wiese, K. (1996). Sensory capacities of Euphausiids in the context of schooling. *J. Mar. Freshw. Behav. Physiol.* **28,** 183–194.

Wiese, K., and Ebina, Y. (1995). The propulsion jet of *Euphausia superba* (Antarctic krill) as a potential communication signal among conspecifics. *J. Mar. Biol. Assoc.* **75,** 43–54.

Wiese, K., and Marschall, H.-P. (1990). Sensitivity to vibration and turbulence of water in context with schooling in Antarctic krill. *Euphausia superba. In* "Front. Crustac. Neurobiol." (K. Weise, W.-D. Krenz, J. Tautz, H. Reichert and B. Mulloney, eds.), pp. 121–130. Basel, Birkhauser Verlag.

Wiese, K., Wollnik, F., and Jebram, D. (1980). The protective reflex of *Bowerbankia* (Bryozoa): Calibration and use to indicate movement below a capillary surface wave. *J. Comp. Physiol. A* **137,** 297–303.

Yen, J., Brown, J., and Webster, D. R. (2003). Analysis of the flow field of the krill *Euphausia pacifica. Mar. Freshw. Behav. Res.* **36,** 307–319.

Zahavi, A., and Zahavi, A. (1999). The Handicap Principle. Oxford University Press, NewYork.

CHAPTER NINE

DIEL VERTICAL MIGRATION BEHAVIOUR OF THE NORTHERN KRILL (*MEGANYCTIPHANES NORVEGICA* SARS)

Stein Kaartvedt

Contents

Abstract

The prototype of *Meganyctiphanes norvegica* diel vertical migration (DVM) behaviour comprises ascent around dusk, feeding near the surface at night, and descent at dawn, explained as a trade-off between feeding and predator avoidance in an environment where both food and risk of predation is highest near surface. Light is the proximate cue, and daytime distribution is deeper in clear waters and sunny weather and nocturnal distributions deeper in moonlight. However, both internal state and external factors further affect and modify the diel migration pattern. While *Meganyctiphanes* migrates in synchrony to the surface at sunset, part of the population may descend soon after the ascent with individuals re-entering upper layers throughout the night. This has been explained with hungry individuals being prone to take larger risks and hence stay shallower, while satiated individuals seek shelter at depth. Females migrate closer to the surface than males of equivalent size, possibly due to their greater demand for energy to fuel egg production. Freshly moulted *M. norvegica* remain at depth throughout the diel cycle. This has been related

King Abdullah University of Science and Technology, Thuwal, Saudi Arabia

Advances in Marine Biology, Volume 57
ISSN 0065-2881, DOI: 10.1016/S0065-2881(10)57009-3

to the fact that that krill do not feed during moulting, to reduced swimming capacity, and as a mechanism to avoid cannibalism whilst in a vulnerable condition. In some locations large parts of the population remain at depth at night. Such behaviour may incur access to demersal food sources, provide avoidance of predators, or can be a means to avoid horizontal transport to adjacent, unfavourable areas.

Environmental gradients can arrest migrations of *M. norvegica*, yet the effect of physics is not always distinguished from associated biological properties, like subsurface maxima of phytoplankton located at pycnocline boundaries. Deeper nocturnal distribution when predators were abundant has been reported, and krill may adjust their distribution upwards when exposed to deep-living predators. Instantaneous escape to approaching predators is a common component of the anti-predator repertoire of *Meganyctiphanes*. Occasionally reported schooling behaviour that overrides normal DVM behaviour may serve anti-predation purposes, as well as being related to reproduction.

M. norvegica can remain within confined areas, often defined by the bottom topography, even when exposed to strong currents. Behaviourally mediated retention may be accomplished by vertical migration in depth-stratified flows, but evidence for active use of DVM for the purpose of retention is so far circumstantial among *M. norvegica*. In several instances, large aggregations of krill that repeatedly occur in the same location appear to be accidental consequences of krill vertical migration behaviour interacting with the mean circulation and bottom topography, rather than representing active retention behaviour.

1. INTRODUCTION

At the time of the review by Mauchline in 1980, it was well known that *Meganyctiphanes norvegica* carries out diel vertical migrations (DVMs). The established DVM pattern comprised ascent from depth around dusk, feeding near the surface during the night and descent around dawn. The review focused on average depth day and night and amplitudes of migration. Ontogenetic differences in vertical distributions were established, with larvae and small individuals living shallower than adults or large individuals. Light was suggested to play an essential role. It had been observed that degree of cloud cover affected daytime distribution of krill with shallower distribution in heavy overcast and rainy weather (Mauchline and Fisher, 1969) and that moonlight caused deeper nocturnal distributions. The effect of light also included the notion that krill followed isolumes during their vertical migrations. The potential effects of physical gradients (pycnoclines) for hampering vertical migrations were discussed. The distribution of food was believed to play a role for migration patterns. The potential effect of predators, a predominant theme in current plankton research, was not addressed by Mauchline (1980), but evasion from visually searching predators as an explanation for descent at daytime had been referred to in the preceding review on krill biology by Mauchline and Fisher (1969).

Studies of DVM have been central in both freshwater and marine pelagic ecology since 1980. This has provided much new insight in both behavioural variations and driving forces behind the DVM behaviour. Increased knowledge and understanding of the behaviour of *M. norvegica* thus can be derived both from enhanced knowledge of this species in particular as well as from knowledge of DVM behaviour in general.

Much work on krill behaviour during the last decades has been carried out using echo sounders. This gives a bias towards studies of the large size classes, as the commonly used frequency when addressing krill (120 kHz) is most efficient in unveiling organisms larger than ∼20 mm and animals aggregating in acoustic scattering layers (SLs). To my knowledge, there has been no recent work on the vertical distribution and migration behaviour of larval stages of *M. norvegica*. This review, therefore, is restricted to post-larval individuals.

As in 1980, *M. norvegica* stays at subsurface waters during the day; it ascends at dusk and descends at dawn. However, there is much variation imposed on this generalised picture, and there has been considerable progress in unveiling the nature of this variation and its driving forces during the last three decades. The main aim of this paper is to summarise this new knowledge on *M. norvegica* DVM behaviour, yet some reference will also be made to other recent findings on the natural behaviour of this important species.

2. THE ROLE OF PREDATORS

It is now generally accepted that the main advantage to restricting the time spent in upper waters to dark periods is that risks of predation from visual predators such as fish are reduced. On the downside, these daily vertical movements are at the cost of leaving shallow food resources or advantageous temperature during the day; although *M. norvegica* may also feed in the daytime habitat (Kaartvedt et al., 2002; Onsrud and Kaartvedt, 1998; Sameoto, 1980). Already McLaren (1963) suggested that escape from predators was the reason for descent by day rather than by night among dielly migrating zooplankton, although the focus in his model is on metabolic consequences of DVM in waters with surplus food and vertical temperature gradients. Subsequently, Zaret and Suffern (1976) provided experimental evidence that DVM can be adaptive as a means of avoiding predation. This was followed by documentation of genotypically imprinted DVM among copepods which had been exposed to fish for many generations compared to non-migrating populations in fish-less environments (Gliwicz, 1986). Later, there has been much developments related to unveiling phenotypic flexibility in behaviour related to absence, presence and abundance of predators (fish). For a variety of taxa from different environments there is now evidence that proportions of populations and amplitudes of DVM increase with increasing predation pressure, plankton often assessing the danger of predation by some chemical cue (e.g. Cohen and Forward, 2009; Dawidowicz and Loose, 1992).

Much of the work on flexible DVM in relation to predators has been done with relatively small organisms (in the order of 1 to several millimeters). This may be because organisms of this size are easily handled in the aquaria, and also because plankton in this size range have much to gain at less risk by not performing a diurnal descent. There is limited information on flexible migration behaviour as a predator avoidance mechanism in *M. norevegica*. However, Onsrud and Kaartvedt (1998) reported that *M. norvegica* appeared to have a deeper nocturnal distribution when predators were abundant. Predation pressure is also included as a central driving force in models that evaluate the DVM pattern of *M. norvegica* (Tarling, 2003; Tarling *et al.*, 2000). Yet empirical data still are few. Studies of other euphausiid species provide circumstantial evidence that predation pressures affect krill vertical distribution. Klevjer *et al.* (2010) found that Antarctic krill were deeper distributed close to land than in the open ocean, which they attributed to populations of land-based predators. Bollens *et al.* (1992) observed variable DVM in larval *Euphausia pacifica* which they ascribed to variable presence of planktivorous fish. DVM patterns among juveniles and adults were, however, invariant.

A considerable proportion of *M. norvegica* may occasionally remain in relatively deep water also at night. Even within the same fjord Giske *et al.* (1990) observed that the population of *M. norvegica* behaves differently in different locations, with the majority of individuals migrating to the surface layer in one basin, while being concentrated below 50 m at night in adjacent waters. Kaartvedt and Svendsen (1990) made similar observations in another fjord system. And while Greene *et al.* (1988) report on dielly migrating *M. norvegica* off Georges Bank, USA, the most striking result in their case is the huge concentrations of krill that occur at several hundred metres depth, even at night. Deep concentrations were also observed by Youngbluth *et al.* (1989). The reasons for such distributions are not clear. They may relate to submergence at shallowing topography where the krill can feed on demersal food sources like detritus (e.g. Youngbluth *et al.*, 1989), these may be habitats where the krill is heavily susceptible to predation so that the nocturnal ascent is suppressed, or such behaviour may hamper nocturnal transport into shallower habitats where susceptibility to visual predators assumingly will be severe during the day (Kaartvedt, 1993). Baliño and Aksnes (1993) observed that krill in one such location was foraging on overwintering copepods *Calanus finmarchicus* compared to detritus else-where, while echograms suggest that *M. norvegica* in this location was attacked by fish in mid-waters. Unveiling the role of predators, food or other reasons behind this trait of *M. norvegica* behaviour would be a perti-nent question for future studies.

A usual assumption in analysis of vertical distribution and DVM patterns appears to be 'the deeper the better'. However, krill also are vulnerable to attack from deeper-living fish adapted to search for prey at low light intensities. Such planktivores will themselves have predators searching by sight, so that their vertical distribution is constrained to deeper layers during daytime. Kaartvedt *et al.* (1996) observed that the vertical distribution of krill (mainly *Thysanoessa inermis*) shallowed in accordance with increased predation pressure from below. Kaartvedt *et al.* (2005) argued that *M. norvegica* benefited from the presence of piscivorous fish since they hampered the predation threat of deep, krill-eating predators. According to these arguments, some weak light would be better than even weaker light, that is mid-waters better than deeper depths.

Instantaneous escape reactions represent another way of avoiding pre-dators, and krill can perform rapid backward 'tail-flips' as an escape response (Kils, 1981). Onsrud *et al.* (2004) reported on voids in acoustic SLs of *M. norvegica* around fish. While not often verified in the literature, use of submerged echo sounders recurrently documents such behaviour (e.g. Fig. 9.1), suggesting that an instantaneous escape from approaching pre-dators is a common component within the antipredator repertoire of *Meganyctiphanes*.

3. THE ROLE OF HUNGER AND SATIATION

The hunger/satiation hypothesis (Pearre, 1979, 2003) states that ver-tically migrating individuals return to deeper waters once fed. Satiated individuals will return to depth in the course of the night while hungry individuals will reenter the surface, resulting in asynchronous forage migra-tions. Field documentation of the hunger-satiation hypothesis in *M. norve-gica* is derived from observations of migration patterns combined with stomach investigations. *Meganyctiphanes* normally appears to migrate together to the surface at sunset, but part of the population starts descending soon after the ascent. Stomach investigations have revealed increased level of gut fluorescence in deep waters shortly after sunset as well as individuals with empty stomachs in upper waters late at night (Onsrud and Kaartvedt, 1998; Sourisseau *et al.*, 2008). The latter observation is interpreted as hungry krill reentering the surface layers, and a corresponding reorganisation of an acoustic krill layer in the morning is reported by Sourisseau *et al.* (2008).

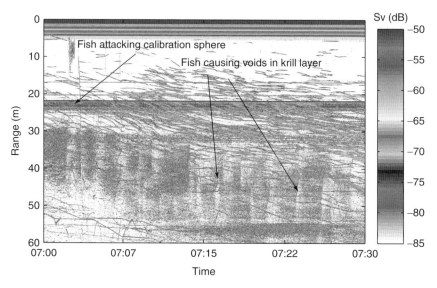

Figure 9.1 Echogram from a submerged (30 m) 120 kHz echosounder (Masfjorden, Norway; 5 November 2004). Vertically swimming fish caused marked voids in a descending scattering layer of *M. norvegica*. A calibration sphere—sometimes attacked by fish—is seen at 22 m, range (i.e. 52 m depth) the targets passing the sphere on their morning descent are the light fish *Maourlicus muelleri*.

Also studies of other krill species have found evidence supporting the hunger satiation hypothesis (Gibbons, 1993; Simard *et al.*, 1986a; Tarling and Johnson, 2006).

4. THE ROLE OF INTERNAL STATE AND MOTIVATION

Both theoretical arguments and field studies suggest that the benefit of migrating to the surface depends on the body condition of individuals, with organisms with more energy reserves being more inclined to remain in the safer waters at depth even at night (Fiksen and Carlotti, 1998; Hays *et al.*, 2001). Likewise, individuals having more to lose by exposing themselves to increased risks of predation (like egg bearing copepods) tend to remain in deep water throughout the diel cycle (Bollens and Frost, 1991). On the contrary, individuals with higher energy needs may be more inclined to seek prosperous, but dangerous feeding grounds. Tarling (2003) reported on sex-specific variations in DVM among *M. norvegica*, with females migrating closer to the surface at night than males of equivalent size. As predation risk is probably a function of the light available to visual predators, the females thus took greater risks than the males. The author's interpretation was that the greater demand for energy to fuel egg production was driving females to undertake a riskier DVM than males.

Tarling *et al.* (1999a) observed that freshly moulted *M. norvegica* remained at depth throughout the diel cycle. They suggested that moulting at night in the deepest layers could be a mechanism to avoid cannibalism whilst in a vulnerable condition. However, also the fact that krill do not feed during moulting (Thomassen *et al.*, 2003) would evidently imply reduced motivation for vertical feeding migrations. Thomassen *et al.* (2003) added swimming capacity to the explanatory variables, showing that newly moulted *Meganyctiphanes* had reduced swimming capacity.

5. THE ROLE OF PHYSICAL GRADIENTS

The role of temperature has a central place in the literature on DVM as the temperature affects metabolic rates, and therefore utilisation of energy and growth. Both cold and warm waters have been deemed beneficial (or detrimental). Cold waters make animals conserve energy, which is an advantage in times of food shortage. They also grow slower and may reach a larger size, which has been deemed advantageous due to higher fecundity (McLaren, 1963). However, in most instances, it is considered that the increased growth rate resulting from higher temperatures increases fitness and there is much documentation across a variety of taxa of the advantage of entering/disadvantage of leaving warm waters during DVM (Dawidowicz and Loose, 1992; Wurtsbaugh and Neverman, 1988).

In Mauchline (1980) and the subsequent decade, the literature on *M. norvegica* to a large extent focused on temperature gradients as physical barriers for vertical migrations, rather than their impact on fitness in terms of growth rates. Avoidance of surface layers has in several cases been ascribed to temperature gradients and avoidance of waters with temperatures above ~ 15 °C (Bergström and Strömberg, 1997; Buchholz *et al.*, 1995). Yet *M. norvegica* has a very broad geographic distribution and lives in habitats spanning large temperature ranges, both seasonally and geographically, and later studies have added information on krill ascending into temperatures near 20 °C at night (Kaartvedt *et al.*, 2002). 'Avoidance' of upper waters furthermore occurs even without physical gradients (Giske *et al.*, 1990; Onsrud and Kaartvedt, 1998). Therefore, while temperature plays a significant role in the physiology (Saborowski *et al.*, 2000, 2002) and probably in the habitat choice of *M. norvegica*, it likely does not often act as an impenetrable barrier for vertical migration. Nonetheless, it is not uncommon to observe that *M. norvegica* stops their ascent in the evening when encountering the thermocline/pycnolcine (own unpublished results).

This would likely have another explanation than the pycnocline functioning as a physical barrier for further ascent; for example, subsurface maxima of phytoplankton are often located at the pycnocline boundary and ascending krill may reach an adequate threshold level for feeding before reaching the surface (Andersen and Nival, 1991; Mauchline and Fisher, 1969).

Kaartvedt and Svendsen (1990) reported that *M. norvegica* avoided a brackish surface layer in a strongly freshwater influenced fjord system. However, in most of their normal habitat it does not seem that salinity constrain DVM in *M. norvegica*.

Low oxygen concentrations at depth may in some cases represent a 'false' bottom, constraining krill vertical distribution (Fig. 9.2). The example presented here is representative for several years of studies in the Oslofjord, Norway, which suggest that the lower tolerable oxygen concentrations at temperatures of ~ 7 °C is ~ 0.5 ml O_2 l^{-1} (saturation of about 7 %). Accordingly, Onsrud and Kaartvedt (1998) showed that oxygen concentrations of 1.3 ml l^{-1} did not constrain krill daytime depth. However, such low oxygen levels may still affect the biology of *Meganyctiphanes norvegica*.

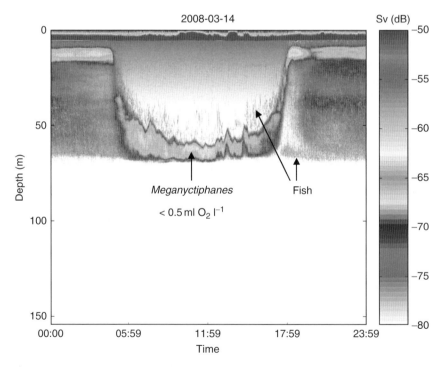

Figure 9.2   Echogram from an upward-looking 200 kHz echo sounder mounted on the bottom in the inner Oslofjord (14 March 2008). Low oxygen content (0.5 ml O_2 l^{-1}) constrained the lower daytime distribution of *Meganyctiphanes norvegica*. Time is GMT.

Spicer *et al.* (1999) concluded that *M. norvegica* that made excursions into regions of severe hypoxia during the day shifted to anaerobic metabolisms and needed to migrate to upper layers at night to pay their oxygen debt. There was high mortality among krill which was prevented from carrying out DVM (stocked in net cages at their daytime depth). Swimming is a major contributor in the bioenergetic budget of krill (Torress and Childress, 1983), and it is likely that *M. norvegica* will reduce their activity level as well when inhabiting hypoxic waters. Hypoxia also will affect feeding and digestion (e.g. Jordan and Steffensen, 2007), with possible implications for krill DVM patterns.

6. VERTICAL DISTRIBUTION AND HORIZONTAL TRANSPORT

Vertical distribution affects horizontal transport in regions with depth stratified flow. The interaction between zooplankton vertical distribution and the current profile and the role of vertical migrations for advection and retention is an old theme (e.g. Hardy and Gunther, 1935). There are basically two interpretations with respect to retention: Plankton may exploit vertical currents to remain within a preferred habitat, or retention and accumulation is an incidental result of the prevailing currents and the zooplankton behaviour. The first process has been particularly addressed in estuarine environments where plankton population maintenance in the phase of tidal flushing may be particularly striking (e.g. Cronin and Forward, 1979; Kimmerer and McKinnon, 1987). Also krill may remain within restricted areas (often defined by bottom topography), even when exposed to strong currents. The occurrence of behaviourally mediated retention within localised habitats has been suggested for *M. norvegica* (e.g. Buchholz *et al.*, 1995; Emsley *et al.*, 2005; Kaartvedt, 1993; Tarling *et al.*, 1998) and may involve both vertical migrations and horizontal swimming. Active retention-behaviour by vertical migrations (both ontogenetic and diel) has furthermore been suggested for other species of krill in upwelling regions (e.g. Barange and Pillar, 1992).

Advection and accumulation of krill (including *M. norvegica*) has been much studied in the St. Lawrence Estuary, Canada. The ultimate goal in this case has been to understand how the vertical distribution of krill interacts with the vertical current profile and the local topography to cause horizontal patchiness, with consequent ecosystem effects (transfer of food to predators). It has been concluded that large aggregations of krill result from negative phototaxis of the animals interacting with the upwelling and mean circulation in the area (Simard *et al.*, 1986b). The krill tend to stay below a barrier isolume, to which they react by actively swimming downwards at maximum speed (Cotté and Simard, 2005). There are no peculiarities in the DVM pattern of *M. norvegica* in this system compared to what has been established elsewhere (Simard *et al.*, 1986a,b; Sourisseau *et al.*, 2008), and the accumulation appears to be accidental/unintentional rather than representing active retention behaviour.

7. SWARMING AND SCHOOLING

Aggregations of *Meganyctiphanes norvegica* that form acoustical SLs are commonly reported to hold concentrations on the order of 1 (to 10) individuals per cubic metre (Cochran *et al.*, 1994; Greene *et al.*, 1992; Onsrud and Kaartvedt, 1998; Tarling *et al.*, 1998), while reports on behaviourally mediated swarms or schools are seldom. However, swarming or schooling involving much higher concentrations does occur, and such behaviour may override normal DVM patterns. Nicol (1984) reported on daytime surface aggregations of *M. norvegica* during consecutive summers, with concentrations up to 600,000 individuals per cubic metre. The krill were schooling when moving through the water and swarming when stationary (the term swarm used, unless uniform horizontal orientation was noted). Individual swarms had highly skewed sex ratios; both male dominated and female dominated swarms were encountered, so it was concluded that these swarms were in some way related to reproductive activity (Nicol, 1984).

Swarming of *M. norvegica* in another setting was observed by Kaartvedt *et al.* (2005) who reported on associations between oceanic krill swarms and large piscivorous fish in the Norwegian Sea. Swarms occurred at ~200–100 m depth both during day and the light summer night at this high latitude study site. 'Stacks' of large fish were found below each krill swarm, and the authors suggested that piscivores use krill swarms as feeding grounds in their hunt for planktivorous fish attracted by the swarms. For the individual krill, such patrols of large piscivores would add to the generally accepted anti-predator benefit provided by the swarming behaviour.

While the examples above involve social behaviour, high concentrations of *Meganyctiphanes norvegica* can also occur in cases where krill become physically aggregated. The case of horizontal aggregations in the Gulf of St. Lawrence was dealt with above; vertical aggregations occur as well. Stevick *et al.* (2008) observed that internal waves resulted in upward movement and concentration of *M. norvegica* at small, offshore banks through a coupling of physical processes and krill behaviour. This resulted in surface swarms with concentrations of >1500 m^{-3} in the top 1 m. Greene *et al.* (1988) reported on deep aggregations, the concentration mechanisms in this case included a presumed funnelling effect in deep canyons.

8. THE EFFECT OF LIGHT

The role of light for the distribution and behaviour of krill has long been acknowledged (cf. references in Mauchline, 1980). Basically, more light produces a deeper distribution. This relates to the differences between daytime and nighttime, but also to the actual light conditions both during day and night. Onsrud and Kaartvedt (1998) referred to deviating vertical distribution between fjords with different water clarity, *M. norvegica* remaining deeper in clear waters with deeper light penetration. The deepest daytime distribution of *M. norvegica* is accordingly reported from the relatively clear waters of the Mediterranean (Andersen and Nival, 1991). During daytime, this population may inhabit water between 400 m and 800 m (Sardou *et al.*, 1996; Panigada *et al.*, 1999; further references in Tarling *et al.*, 1999b), with the mean depth decreasing as chl-a concentration (shading) increases (Andersen *et al.*, 1998). Similarly, Frank and Widder (2002) reported that the daytime vertical distribution of several taxa, including *M. norvegica*, responded to an influx of turbid water which significantly decreased downwelling irradiance. The responding species ascended over 100 m in the water column during the influx and returned to their pre-influx depths once the influx had ceased and the turbidity declined. *In situ* light measurements demonstrated that each of these species was associated with the same irradiance levels during the influx as they were under pre-and post-influx conditions (Frank and Widder, 2002).

All cases referred to above seem to be responses to changes in absolute light levels, as are reports of responses of *M. norvegica* to infrequent and unpredictable (for animals) incidents like solar (Strömberg *et al.*, 2002) and lunar (Tarling *et al.*, 1999b) eclipses. Nevertheless, there is evidence that *M. norvegica* to some extent may adapt to the prevailing light conditions, as Myslinski *et al.* (2005) found that the photosensitivities of *M. norvegica* collected in clearer waters were lower than those of individuals collected from more turbid waters.

The importance of the light level is normally interpreted in terms of predator avoidance, but some light would be beneficial if *Meganyctiphanes norvegica* themselves act as visual predators. Based upon laboratory experiments (Torgersen, 2001) and field studies (Kaartvedt *et al.*, 2002) it has been suggested that visual predation may be the case for *M. norvegica*. This was disputed by Browman (2005) who concluded that mechanoreception is the main sensory modality involved in prey detection by *M. norvegica*, yet Abrahamsen *et al.* (2010) found that vision increased the sampling volume, making vision an important aspect of *M. norvegica*'s feeding ecology. Unveiling the searching mechanisms for prey has implications for interpretation of krill DVM patterns; for example Kaartvedt *et al.* (2002) suggested visual predation as an explanation for why *M. norvegica* in winter left abundant overwintering copepods (*Calanus*) in deep water to enter more scanty populated upper waters during the nighttime. The argument was that visual predation on this deep food source of dormant copepods might be possible during day, but not by night.

While the DVM behaviour is acknowledged to be related to evasion from visually searching predators, which aspect of the light field acts as a proximate cue for DVMs is still debated. There are several aspects of the light field which potentially can serve as cues. These have been discussed by Forward (1988) and summarised by Frank and Widder, 1997 as follows: (i) changes in underwater spectra; (ii) changes in the polarisation pattern; (iii) changes in absolute light intensity; (iv) the relative rate of change in intensity. According to Frank and Widder (1997), the most viable potential cues for mesopelagic species are a 'threshold' value of absolute light intensity, and/or a relative rate of change in intensity. In analysing data for vertically migrating *M. norvegica* in the Gulf of Maine, they concluded that "from the data presented here, it appears that the maximum rate of change in irradiance is not triggering the migration of these *M. norvegica* (Frank and Widder, 1997)." So apparently the results would favour the threshold hypothesis.

Some have proposed that organisms simply follow a preferred light-level (isolume) as it moves up and down the water column over the diel cycle (the preferendum hypothesis). For instance, Onsrud and Kaartvedt (1998) found that the upper fringe of migrating SLs of *M. norvegica* followed isolumes during ascent in the evening (Fig. 9.3). They measured surface light and estimated light at depth. Widder and Frank (2001) stressed the need for measuring underwater light as light in deeper waters changes faster due to narrowing of the spectral field with depth. Their results were also in accordance with the preferendum hypothesis. However, even while the light level associated with an individual may be referred to as an isolume, the light levels associated with a population would encompass a much broader range (Myslinski *et al.*, 2005; Widder and Frank, 2001). One assumption is that individuals within a population may follow different isolumes due to differences in size, sex or different trade-offs between foraging and predator avoidance. The most recent general review on the role of light in governing DVM (Cohen and Forward, 2009) provides a thorough evaluation for a variety of taxa, but does not shed additional light on the exogenous role of light in the DVM of *M. norvegica* specifically.

9. INDIVIDUAL SWIMMING BEHAVIOUR

Swimming behaviour was not a topic in the 1980 review. However, knowledge of individual behaviour is essential in analyses of marine ecosystems. The success of food capture depends on the search pattern (Løkkeborg and Fernø, 1999), metabolic rates increase exponentially with the swimming speed of an animal (Torress and Childress, 1983), and swimming is essential for predator–prey relationships both in relation to

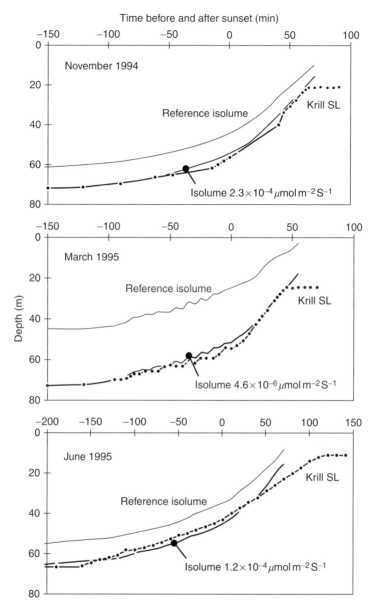

Figure 9.3 Positions of isolumes (●) and upper boundary of krill scattering layers (◆) during the transition from day to night at three different times of year (Oslofjord, Norway). Positions of the 10^{-3} μmol m^{-2} s^{-1} isolume are given as reference (from Onsrud and Kaartvedt, 1998).

search volume and encounter rates with predators (e.g. Buskey *et al.*, 1993; Gerritsen and Strickler, 1977). Also, *Meganyctiphanes norvegica* is negatively buoyant and has to swim constantly to avoid sinking (Kils, 1981).

Studies of krill swimming have been done both in aquaria and *in situ*, including analyses of day–night patterns. Velsch and Champalbert (1994) observed an endogenous swimming rhythm in *Meganyctiphanes norvegica* kept in total darkness, with increased swimming near the surface of an aquarium during the nighttime. The authors concluded that endogenous rhythms may play a role in krill DVM. Experimental studies by Thomassen *et al.* (2003) showed that the swimming capacity of *M. norvegica* increases with the size of the animal, and that newly moulted krill were weaker swimmers. The authors argued that such results may be useful in interpreting vertical migration studies, that is questions like range and speed of migrations being explained in terms of swimming capacity.

New methods have now been developed for *in situ* observations on krill swimming behaviour. De Robertis *et al.* (2000, 2003) exploited a submerged, specialised echosounder (Fish TV) to study *in situ* behaviour of *Euphausia pacifica*. In this way, both size-specific timings of vertical migrations (De Robertis *et al.*, 2000) and *in situ* three-dimensional swimming path and speed of the krill (De Robertis *et al.*, 2003) could be established. De Robertis *et al.* (2000) observed that smaller bodied *Euphausia pacifica* consistently ascended earlier in the afternoon and descended later in the morning than the adults. They concluded that the timing of vertical migration reflects how the size-dependent risk of attack by visual predators alters the tradeoff between feeding and predator avoidance.

Klevjer and Kaartvedt (2003, 2006) took advantage of split-beam echo sounders submerged into assemblages of *M. norvegica* to establish their swimming pattern and speed at their daytime depths using acoustic target tracking (Fig. 9.4). These first approaches have shown that target tracking of krill *in situ* indeed is possible by using conventional echo sounders. The first results revealed swimming speeds with a mode around 4 cm s^{-1} at daytime, being slightly above one body length per second (Klevjer and Kaartvedt, 2003). Swimming paths are both relatively straight and in tight loops (Klevjer and Kaartvedt, 2003, 2006), the relative frequency of different types of behaviour and their respective explanation remain to be established. This approach also would apply to studies of individual krill behaviour during DVMs, as done for small mesopelagic fish by Kaartvedt *et al.* (2008).

10. NEW METHODS/APPROACHES

Since the review by Mauchline (1980), new methods and approaches have been included in addressing the behaviour, vertical distribution, DVM patterns and their consequences for *M. norvegica* and other species of krill.

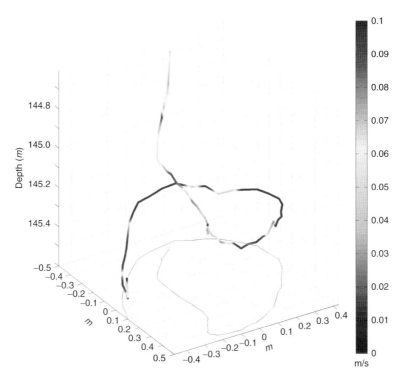

Figure 9.4 *In situ* swimming path and speed of an individual *Meganyctiphanes norvegica* at 145 m depth as revealed by acoustic target tracking over 56 s (06:30 h UTC, 20 April 2006). The results are from an upward-looking 120 kHz Simrad EK60 echsounder, mounted on the bottom in the Oslofjord, Norway, and cabled to shore for long-term studies. The average Target Strength (a proxy for size) for this krill was − 74.2 dB, and the average swimming speed 8 cm s^{-1}. Scale in metre. Colour scale depicts variations in swimming speed. Figure by Thor A. Klevjer.

Observations from submersibles were available also in 1980, but new documentation includes deep concentrations of *M. norvegica* (Greene *et al.*, 1988; Youngbluth *et al.*, 1989), krill bioluminescence and patchiness (Greene *et al.*, 1992; Widder *et al.*, 1992) and assessments of responses to downwelling light (Frank and Widder, 1997, 2002). Modelling is now among the methods applied in evaluating DVM behaviour (Andersen and Nival, 1991; Burrows and Tarling, 2004; Tarling *et al.*, 2000), as well as for assessing interactions between the DVM patterns of *M. norvegica* and the vertical current field (Emsley *et al.*, 2005; Sourisseau *et al.*, 2006).

There has been much improvement in video equipment during the last decades and video has proven a useful tool in experimental studies of krill behaviour (e.g. Abrahamsen *et al.*, 2010; Browman, 2005; Price, 1989; Tarling and Johnson, 2006; Torgersen, 2001). In field studies, cameras mounted on ROVs and other underwater instruments have been used in documenting the presence of krill (e.g. Clarke and Tyler, 2008) and verification of acoustic targets (e.g. Benfield *et al.*, 1998; Lawson *et al.*, 2008; Wiebe *et al.*, 2010). The application for *in situ* behavioural studies has been hampered by restricted range and the need of artificial light, which affects krill behaviour, yet some behavioural information has been obtained (Clarke and Tyler, 2008). Molecular methods are now used successfully in studies of krill biology (e.g. Papetti *et al.*, 2005; Vestheim and Jarman, 2008), also in the context of topics relevant to this chapter (Seear *et al.*, 2009).

Comprehensive advances have been made in ways of applying acoustic tools for assessing the behaviour of *M. norvegica*. Access to ADCPs has opened new avenues, including studies (sometimes long-term) of DVM patterns from moored systems (e.g. Buchholz *et al.*, 1995; Cochran *et al.*, 1994; Liljebladh and Thomasson, 2001; Sourisseau *et al.*, 2006), assessments of vertical swimming speeds and directions (Sourisseau *et al.*, 2006; Tarling *et al.*, 2002) and in evaluating interactions between krill vertical distribution and the vertical current profile (Liljebladh and Thomasson, 2001; Tarling *et al.*, 1998).

Submergence of traditional echo sounders for better resolution in deep water is yet another innovation. This procedure gives enhanced resolution on fine scale structures in deep water. Greene *et al.* (1992) quantified fine scale patchiness within an acoustic SL of *M. norvegica*, structures which could not be detected from a ship-borne echo sounder. As mentioned in the previous section, submergence of echo sounders into SLs also allows the detection of individual krill, which provide new opportunities for studies of individual, *in situ* behaviour (De Robertis *et al.*, 2000, 2003; Klevjer and Kaartvedt, 2003, 2006).

We now are at the verge of establishing large-scale cabled observatories on the bottom of the world's oceans that are equipped with an array of environmental sensors (e.g. Malakoff, 2004). Acoustic studies from preliminary observatories have provided detailed and new information on DVM patterns and the swimming behaviour of small mesopelagic fish (e.g. Kaartvedt *et al.*, 2009). This approach also promises for studies of *M. norvegica* and other species of krill. New platforms like AUVs and gliders are becoming increasingly available (e.g. Baumgartner and Fratantoni, 2008; Davis *et al.*, 2008), and have provided new information on Antarctic krill (Brierley *et al.*, 2002). Multibeam sonars have recently been exploited for characterising Antarctic krill school structure (Korneliussen *et al.,* 2009). Even the use of low frequency acoustics to characterise distribution of Antarctic krill at the mesoscale level has been suggested (Jagannathan *et al.*, 2009), but Antarctic krill represents a special case of a dominant, schooling species which does not strictly apply to *M. norvegica*.

It appears that the scientific community is now armed with an array of methods that can be used for further unveiling the behaviour of

Meganyctiphanes norvegica as well as other species of krill. However, as much as technology, the future direction and success of krill research depends on the human capital. The output from new and innovative methods and approaches depends on qualified researchers who know the possibilities and limitations of new techniques, who can pose the right questions and know how to address them.

REFERENCES

Abrahamsen, M. B., Browman, H. I., Fields, D. M., and Skiftesvik, A. B. (2010). The three-dimensional prey field of the northern krill, *Meganyctiphanes norvegica*, and the escape response of their copepod prey. *Mar. Biol.* **157**, 1251–1258.

Andersen, V., and Nival, P. (1991). A model of the diel vertical migration of zooplankton based on euphausiids. *J. Mar. Res.* **49**, 153–175.

Andersen, V., Francois, F., Sardou, J., Picheral, M., Scotto, M., and Nival, P. (1998). Vertical distributions of macroplankton and micronekton in the Ligurian and Tyrrhenian Seas (northwestern Mediterranean). *Oceanol. Acta* **21**, 655–676.

Baliño, B. M., and Aksnes, D. L. (1993). Winter distribution and migration of the sound scattering layers, zooplankton and micronekton in Masfjorden, western Norway. *Mar. Ecol. Prog. Ser.* **102**, 35–50.

Barange, M., and Pillar, S. C. (1992). Cross-shelf circulation, zonation and maintenance mechanisms of *Nyctiphanes capensis* and *Euphausia hanseni* (Euphausiacea) in the northern Benguela upwelling system. *Cont. Shelf Res.* **12**, 1027–1042.

Baumgartner, M. F., and Fratantoni, D. M. (2008). Diel periodicity in both sei whale vocalization rates and the vertical migration of their copepod prey observed from ocean gliders. *Limnol. Oceanogr.* **53**, (5, part 2), 2197–2209.

Benfield, M. C., Wiebe, P. H., Stanton, T. K., Davis, C. S., Gallager, S. M., and Greene, C. H. (1998). Estimating the spatial distribution of zooplankton biomass by combining Video Plankton Recorder and single-frequency acoustic data. *Deep Sea Res. II* **45**, 1175–1199.

Bergström, B., and Strömberg, J.-O. (1997). Behavioural differences in relation to pycno-clines during vertical migration of the euphausiids *Meganyctiphanes norvegica* (M. Sars) and *Thysanoessa raschii* (M. Sars). *J. Plankton Res.* **19**, 255–261.

Bollens, S. M., and Frost, B. W. (1991). Ovigerity, selective predation, and variable diel vertical migration in *Euchaeta elongata* (Copepoda: Calanoida). *Oecologia* **87**, 155–161.

Bollens, S. M., Frost, B. W., and Lin, T. S. (1992). Recruitment, growth and diel vertical migration of *Euphausia pacifica* in a temperate fjord. *Mar. Biol.* **114**, 219–228.

Brierley, A. S., Fernandes, P. G., Brandon, M. A., Armstrong, F., Millard, N. W., McPhail, S. D., Stevenson, P., Pebody, M., Perrett, J., Squires, M., Bone, D. G., and Griffiths, G. (2002). Antarctic krill under sea ice: Elevated abundance in a narrow band just south of ice edge. *Science* **295**, 1890–1892.

Browman, H. I. (2005). Applications of sensory biology in marine ecology and aquaculture. *Mar. Ecol. Prog. Ser.* **287**, 263–307.

Buchholz, F., Buchholz, C., Reppin, J., and Fischer, J. (1995). Diel vertical migration of *Meganyctiphanes norvegica* in the Kattegat: Comparison of net catches and measurements with Acoustic Doppler Current Profilers. *Helgol. Meeresunters* **49**, 849–866.

Burrows, M. T., and Tarling, G. (2004). Effects of density dependence on diel vertical migration of populations of northern krill: A genetic algorithm model. *Mar. Ecol. Prog. Ser.* **277**, 209–220.

Buskey, E. J., Coulter, C., and Strom, S. (1993). Locomotory patterns of microzooplankton: Potential effects on food selectivity of larval fish. *Bull. Mar. Sci.* **53**, 29–43.

Clarke, A., and Tyler, P. A. (2008). Adult Antarctic krill feeding at abyssal depths. *Curr. Biol.* **18**, 282–285.

Cochran, N. A., Sameoto, D. D., and Belliveau, D. J. (1994). Temporal variability of euphausiid concentrations in a Nova Scotia shelf basin using a bottommounted acoustic Doppler current profiler. *Mar. Ecol. Prog. Ser.* **107**, 55–66.

Cohen, J. H., and Forward, R. B. Jr. (2009). Zooplankton vertical migration – A review of proximate control. *Oceanogr. Mar. Biol. Annu. Rev.* **2009** (47), 77–110.

Cotté, C., and Simard, Y. (2005). Formation of dense krill patches under tidal forcing at whale feeding hot spots in the St Lawrence Estuary. *Mar. Ecol. Prog. Ser.* **288**, 199–210.

Cronin, T. W., and Forward, R. B. Jr. (1979). Tidal vertical migration: An endogenous rhythm in estuarine crab larvae. *Science* **205**, 1020–1022.

Davis, R. E., Ohman, M. D., Rudnick, D. L., and Sherman, J. T. (2008). Glider surveillance of physics and biology in the southern California Current System. *Limnol. Oceanogr.* **53**, (5, part 2), 2151–2168.

Dawidowicz, P., and Loose, C. J. (1992). Metabolic costs during predator-induced vertical migration of *Dafnia*. *Limnol. Oceanogr.* **37**, 1589–1595.

De Robertis, A., Jaffe, J. S., and Ohman, M. D. (2000). Size-dependent visual predation risk and the timing of vertical migration in zooplankton. *Limnol. Oceanogr.* **45**, 1838–1844.

De Robertis, A., Schell, C., and Jaffe, J. S. (2003). Acoustic observations of the swimming behaviour of the euphausiid *Euphausia pacifica* Hansen. *ICES J. Mar. Sci.* **60**, 885–898.

Emsley, S. M., Tarling, G. A., and Burrows, M. T. (2005). The effect of vertical migration strategy on retention and dispersion in the Irish Sea during spring–summer. *Fish. Oceanogr.* **14**, 161–174.

Fiksen, Ø., and Carlotti, F. (1998). A model of optimal life-history and diel vertical migration in *Calanus finmarchicus*. *Sarsia* **83**, 129–147.

Forward, R. B. Jr. (1988). Diel vertical migration: Zooplankton photobiology and behaviour. *Oceanogr. Mar. Biol. Annu. Rev.* **26**, 361–393.

Frank, T. M., and Widder, E. A. (1997). The correlation of downwelling irradiance and staggered vertical migration patterns of zooplankton in Wilkinson Basin Gulf of Maine. *J. Plankton Res.* **19**, 1975–1991.

Frank, T. M., and Widder, E. A. (2002). Effects of a decrease in downwelling irradiance on the daytime vertical distribution patterns of zooplankton and micronekton. *Mar. Biol.* **140**, 1181–1193.

Gerritsen, J., and Strickler, J. R. (1977). Encounter probabilities and community structure in zooplankton: A mathematical model. *J. Fish. Res. Board Can.* **34**, 73–82.

Gibbons, M. J. (1993). Vertical migration and feeding of *Euphausia lucens* at two 72 h stations in the southern Benguela upwelling region. *Mar. Biol.* **116**, 257–268.

Giske, J., Aksnes, D. L., Balino, B., Kaartvedt, S., Lie, U., Nordeide, J. T., Salvanes, A. G. V., Wakili, S. M., and Aadnesen, A. (1990). Vertical distribution and trophic interactions of zooplankton and fish in Masfjorden, Norway. *Sarsia* **75**, 65–81.

Gliwicz, M. Z. (1986). Predation and the evolution of vertical migration in zooplankton. *Nature* **320**, 746–748.

Greene, C. H., Wiebe, P. H., Burczynski, J., and Youngbluth, M. J. (1988). Acoustical detection of high-density krill demersal layers layers in the submarine canyons off Georges Bank. *Science* **241**, 359–361.

Greene, C. H., Widder, E. A., Tamse, M., and Johnson, G. E. (1992). The migration behaviour, fine structure, and bioluminescent activity of krill sound-scattering layers. *Limnol. Oceanogr.* **37**, (1992), 650–658.

Hardy, A. C., and Gunther, E. R. (1935). The plankton of the South Georgia Whaling grounds and adjacent waters. 1926-1927. *Discov. Rep.* Vol XI, 1–456.

Hays, G. C., Kennedy, H., and Frost, B. W. (2001). Individual variability in diel vertical migration of a marine copepod: Why some individuals remain at depth when others migrate. *Limnol. Oceanogr.* **46,** 2050–2054.

Jagannathan, S. *et al.* (2009). Ocean Acoustic Waveguide Remote (OAWRS) of marine ecosystems. *Mar. Ecol. Prog. Ser.* **395,** 137–160.

Jordan, A. D., and Steffensen, J. F. (2007). Effects of ration size and hypoxia on specific dynamic action in the cod. *Physiol. Biochem. Zool.* **80,** 178–185.

Kaartvedt, S. (1993). Drifting and resident plankton. *Bull. Mar. Sci.* **53,** 154–159.

Kaartvedt, S., and Svendsen, H. (1990). Advection of euphausiids in a Norwegian fjord system subject to altered freshwater input by hydro-electric power production. *J. Plankton Res.* **12,** 1263–1277.

Kaartvedt, S., Melle, W., Knutsen, T., and Skjoldal, H. R. (1996). Vertical distribution of fish and krill beneath water of varying optical properties. *Mar. Ecol. Prog. Ser.* **136,** 51–58.

Kaartvedt, S., Larsen, T., Hjelmseth, K., and Onsrud, M. S. R. (2002). Is the omnivorous krill *Meganyctiphanes norvegica* primarily a selectively feeding carnivore? *Mar. Ecol. Prog. Ser.* **228,** 193–204.

Kaartvedt, S., Røstad, A., Fiksen, Ø., Melle, W., Torgersen, T., Breien, M. T., and Klevjer, T. (2005). Piscivorous fish patrol krill swarms. *Mar. Ecol. Prog. Ser.* **299,** 1–5.

Kaartvedt, S., Torgersen, T., Klevjer, T. A., Røstad, A., and Devine, J. A. (2008). Behaviour of individual mesopelagic fish in acoustic scattering layers of Norwegian fjords. *Mar. Ecol. Prog. Ser.* **360,** 201–209.

Kaartvedt, S., Røstad, A., Klevjer, T. A., and Staby, A. (2009). Use of bottom-mounted echo sounders in exploring behaviour of mesopelagic fishes. *Mar. Ecol. Prog. Ser.* **395,** 109–118.

Kils, U. (1981). Swimming behaviour, swimming performance and energy balance of Antarctic krill *Euphausia superba*. *BIOMASS Sci. Ser.* **3,** 121.

Kimmerer, W. J., and McKinnon, A. D. (1987). Zooplankton in a marine bay II. Vertical migration to maintain horizontal distributions. *Mar. Ecol. Prog. Ser.* **41,** 53–60.

Klevjer, T. A., and Kaartvedt, S. (2003). Split-beam target tracking can be used to study the swimming behaviour of deep-living plankton *in situ*. *Aquat. Liv. Res.* **16,** 293–298.

Klevjer, T. A., and Kaartvedt, S. (2006). In situ target strength and behaviour of northern krill (*Meganyctiphanes norvegica*). *ICES J. Mar. Sci.* **63,** 1726–1735.

Klevjer, T. A., Tarling, G. A., and Fielding, S. (2010). Swarm characteristics of Antarctic krill *Euphausia superba* relative to the proximity of land during summer in the Scotia Sea. *Mar. Ecol. Prog. Ser.* **409,** 157–170.

Korneliussen, R. J., Heggelund, Y., Eliassen, I. K., Øye, O. K., Knutsen, T., and Dalen, J. (2009). Combining multibeam-sonar and multifrequencyechosounder data: Examples of the analysis and imaging of large euphausiid schools. *ICES J.Mar. Sci.* **66,** 991–997.

Lawson, G. L., Wiebe, P. H., Stanton, T. K., and Ashjian, C. J. (2008). Euphausiid distribution along the western Antarctic Peninsula. A. Development of robust multi-frequency acoustic techniques to identify euphausiid aggregations and quantify euphausiid size, abundance, and biomass. *Deep Sea Res. II* **55,** 412–431.

Liljebladh, B., and Thomasson, M. A. (2001). Krill behaviour as recorded by acoustic Doppler current profiler in the Gullmarsfjord. *J. Mar. Syst.* **27,** 301–313.

Løkkeborg, S., and Fernø, A. (1999). Diel activity pattern and food search behaviour in cod, *Gadus morhua*. *Environ. Biol. Fish.* **54,** 345–353.

Malakoff, D. (2004). John Delaney profile Marine geologist hopes to hear the heartbeat of the planet. *Science* **303,** 751–752.

Mauchline, J. (1980). The biology of mysids and euphausiids. *Adv. Mar. Biol.* **18,** 1–681.

Mauchline, J., and Fisher, L. R. (1969). The biology of euphausiids. *Adv. Mar. Biol.* **7,** 1–454.

McLaren, I. A. (1963). Effects of temperature on growth of zooplankton and the adaptive value of vertical migration. *J. Fish. Res. Board Can.* **20,** 685–727.

Myslinski, T. J., Frank, T. M., and Widder, E. A. (2005). Correlation between photosensitivity and downwelling irradiance in mesopelagic crustaceans. *Mar. Biol.* **147,** 619–629.

Nicol, S. (1984). Population structure of daytime surface swarms of the euphausiid *Meganyctiphanes norvegica* in the Bay of Fundy. *Mar. Ecol. Prog. Ser.* **18,** 241–251.

Onsrud, M. S. R., and Kaartvedt, S. (1998). Diel vertical migration of the krill *Meganyctiphanes norvegica* in relation to the physical environment, food and predators. *Mar. Ecol. Prog. Ser.* **171,** 209–219.

Onsrud, M. S. R., Kaartvedt, S., Røstad, A., and Klevjer, T. A. (2004). Vertical distribution and feeding patterns in fish foraging on the krill *Meganyctiphanes norvegica*. *ICES J. Mar. Sci.* **61,** 1278–1290.

Panigada, S., Zanardelli, M., Canese, S., and Jahoda, M. (1999). How deep can baleen whales dive? *Mar. Ecol. Prog. Ser.* **187,** 309–311.

Papetti, C., Zane, L., Bortolotto, E., Bucklin, A., and Patarnello, T. (2005). Genetic differentiation and local temporal stability of population structure in the euphausiid *Meganyctiphanes norvegica*. *Mar. Ecol. Prog. Ser.* **289,** 225–235.

Pearre, S. Jr. (1979). Problems of detection and interpretation of vertical migration. *J. Plankton Res.* **1,** 29–44.

Pearre, S. Jr. (2003). Eat and run? The hunger satiation hypothesis in vertical migration: History, evidence and consequences. *Biol. Rev.* **78,** 1–79.

Price, H. (1989). Swimming behaviour of krill in response to algal patches: A mesocosm study. *Limnol. Oceanogr.* **34,** 649–659.

Saborowski, R., Salomon, M., and Buchholz, F. (2000). The physiological response of krill *(Meganyctiphanes norvegica)* to temperature gradients in the Kattegat. *Hydrobiologia* **426,** 157–160.

Saborowski, R., Brohl, S., Tarling, G. A., and Buchholz, F. (2002). Metabolic properties of northern krill, *Meganyctiphanes norvegica*, from different climatic zones I. Respiration and excretion. *Mar. Biol.* **140,** 547–556.

Sameoto, D. D. (1980). Relationships between stomach contents and vertical migration in *Meganyctiphanes norvegica*, *Thysanoessa raschii* and *T. inermis* (Crustacea Euphausiacea). *J. Plankton Res.* **2,** 129–143.

Sardou, J., Etienne, M., and Andersen, V. (1996). Seasonal abundance and vertical distributions of macroplankton and micronekton in the Northwestern Mediterranean Sea. *Oceanol. Acta* **19,** 645–656.

Seear, P., Tarling, G. A., Teschke, M., Meyer, B., Thorne, M. A. S., Clark, M. S., Gaten, E., and Rosato, E. (2009). Effects of simulated light regimes on gene expression in Antarctic krill (*Euphausia superba* Dana). *J. Exp. Mar. Biol. Ecol.* **381,** 57–64.

Simard, Y., Lacroix, G., and Legendre, L. (1986a). Diel vertical migrantion and nocturnal feeding of a dense coastal krill scattering layer (*Thysanoessa raschii* and *Meganyctiphanes norvegica*) in stratified surface waters. *Mar. Biol.* **91,** 93–105.

Simard, Y., de Ladurantaye, R., and Therriault, J.-C. (1986b). Aggregation of euphausiids along a coastal shelf in an upwelling environment. *Mar. Ecol. Prog. Ser.* **32,** 203–215.

Sourisseau, M., Simard, Y., and Saucier, F. J. (2006). Krill aggregation in the St. Lawrence system, and supply of krill to the whale feeding grounds in the estuary from the gulf. *Mar. Ecol. Prog. Ser.* **314,** 257–270.

Sourisseau, M., Simard, Y., and Saucier, F. J. (2008). Krill diel vertical migration fine dynamics, nocturnal overturns, and their roles for aggregation in stratified flows. *Can. J. Fish. Aquat. Sci.* **65,** 574–587.

Spicer, J. I., Thomasson, M. A., and Strömberg, J.-O. (1999). Possessing a poor anaerobic capacity does not prevent the diel vertical migration of Nordic krill *Meganyctiphanes norvegica* into hypoxic waters. *Mar. Ecol. Prog. Ser.* **185,** 181–187.

Stevick, P. T., Incze, L. S., Kraus, S. D., Rosen, S., Wolff, N., and Baukus, N. (2008). Trophic relationships and oceanography on and around a small offshore bank. *Mar. Ecol. Prog. Ser.* **363,** 15–28.

Strömberg, J.-O., Spicer, J. I., Liljebladh, B., and Thomasson, M. A. (2002). Northern krill, *Meganyctiphanes norvegica*, come up to see the last eclipse of the millennium? *J. Mar. Biol. Assoc. UK* **82,** 919–920.

Tarling, G. A. (2003). Sex-dependent diel vertical migration in northern krill *Meganyctiphanes norvegica* and its consequences for population dynamics. *Mar. Ecol. Prog. Ser.* **260,** 173–188.

Tarling, G. A., and Johnson, M. L. (2006). Satiation gives krill that sinking feeling. *Curr. Biol.* **16,** 83–84.

Tarling, G. A., Matthews, J. B. L., Saborowski, R., and Buchholz, F. (1998). Vertical migratory behaviour of the euphausiid *Meganyctiphanes norvegica*, and its dispersion in the Kattegat Channel. *Hydrobiologia* **375,** (376), 331–341.

Tarling, G. A., Cuzin-Roudy, J., and Buchholz, F. (1999a). Vertical migration behaviour in the northern krill *Meganyctiphanes norvegica* is influenced by moult and reproductive processes. *Mar. Ecol. Prog. Ser.* **190,** 253–262.

Tarling, G. A., Buchholz, F., and Matthews, J. B. L. (1999b). The effect of a lunar eclipse on the vertical migration of *Meganyctiphanes norvegica* (Crustacea: Euphausiacea) in the Ligurian Sea. *J. Plankton Res.* **21,** 1475–1488.

Tarling, G. A., Matthews, J. B. L., Burrows, M., Saborowski, R., Buchholz, F., Bedo, A., and Mayzaud, P. (2000). An optimisation model of the diel vertical migration of 'Northern krill'(*Meganyctiphanes norvegica*) in the Clyde Sea and Kattegat. *Can. J. Fish. Aquat. Sci.* **57,** (Suppl 3), 38–50.

Tarling, G. A., Jarvis, T., Emsley, S. M., and Matthews, J. B. L. (2002). Midnight sinking behaviour of *Calanus finmarchicus*: A response to satiation or krill predation. *Mar. Ecol. Prog. Ser.* **240,** 183–194.

Thomassen, M. A., Johnson, M. L., Strømberg, J.-O., and Gaten, E. (2003). Swimming capacity and pleopod beat rate as a function of sex, size and moult stage in Northern krill *Meganyctiphanes norvegica*. *Mar. Ecol. Prog. Ser.* **250,** 205–213.

Torgersen, T. (2001). Visual predation by the euphausiid *Meganyctiphanes norvegica*. *Mar. Ecol. Prog. Ser.* **209,** 295–299.

Torress, J. J., and Childress, J. J. (1983). Relationship of oxygen consumption to swimming speed in *Euphausia pacifica*. 1. Effects of temperature and pressure. *Mar. Biol.* **74,** 79–86.

Velsch, J.-P., and Champalbert, G. (1994). Rythmes d'activite natatoire chez *Meganyctiphanes norvegica* (Crustacea, Euphausiacea). *C. R. Acad. Sci. Paris Sci. Life Sci.* **317,** 857–862.

Vestheim, H., and Jarman, S. N. (2008). Blocking primers to enhance PCR amplification of rare sequences in mixed samples – a case study on prey DNA in Antarctic krill stomachs. *Front. Zool.* **5,** 12. doi:10.1186/1742-9994-5-12.

Widder, E. A., and Frank, T. M. (2001). The speed of an isolume: A shrimp's eye view. *Mar. Biol.* **138,** 669–677.

Widder, E. A., Greene, C. H., and Youngbluth, M. J. (1992). Bioluminescence of sound-scattering layers in the Gulf of Maine. *J. Plankton Res.* **14,** 1607–1624.

Wiebe, P. H., Chu, D., Kaartvedt, S., Hundt, A., Melle, W., Ona, E., and Batta-Lona, P. (2010). The Acoustic properties of *Salpa thompsoni*. *ICES J. Mar. Sci.* **67,** 583–593.

Wurtsbaugh, W. A., and Neverman, D. (1988). Post-feeding thermotaxis and daily vertical migration in larval fish. *Nature* **333,** 846–848.

Youngbluth, M. J., Bailey, T. G., Davoll, P. J., Jacoby, C. A., Blades-Eckelbarger, P. I., and Griswold, C. A. (1989). Fecal pellet production and diel migratory behaviour by the euphausiid *Meganyctiphanes norvegica* effect benthic-pelagic coupling. *Deep Sea Res.* **36,** 1491–1501.

Zaret, T. M., and Suffern, J. S. (1976). Vertical migration in zooplankton as a predator avoidance mechanism. *Limnol. Oceanogr.* **21,** 804–813.

CHAPTER TEN

Predation on Northern Krill (*Meganyctiphanes norvegica* Sars)

Yvan Simard*,† *and* Michel Harvey*

Contents

Abstract

We consider predation as a function of prey concentration with a focus on how this interaction is influenced by biological–physical interactions, and wider oceanographic processes. In particular, we examine how the anti-predation behaviour of Northern krill interacts with ocean-circulation process to influence its vulnerability to predation. We describe how three-dimensional (3D) circulation interacts with *in situ* light levels to modulate predator–prey interactions from small to large scales, and illustrate how the stability of the predator–prey system is sometimes perturbed as a consequence. Northern krill predators include a wide range of species from the pelagic and benthic strata, as well as birds. Many exhibit adaptations in their feeding strategy to take advantage of the dynamic physical–biological processes that determine the distribution, concentration and vulnerability of Northern krill. Among them, baleen whales appear to have developed particularly efficient predation strategies. A literature search indicates that Northern krill are a major contributor to ecosystem function throughout its distributional range, and a key species with respect to the flow of energy to

* Maurice Lamontagne Institute, Fisheries and Oceans Canada, Mont-Joli, Québec, Canada
† Marine Science Institute, University of Québec at Rimouski, Rimouski, Québec, Canada

Advances in Marine Biology, Volume 57
ISSN 0065-2881, DOI: 10.1016/S0065-2881(10)57010-X

upper trophic levels. A list of future research needed to fill gaps in our under-standing of Northern krill predator–prey interaction is provided.

1. INTRODUCTION

Predation is the basic mechanism that controls the flow of energy through the ecosystem, from the initial foraging of herbivores on primary producers through to the apex predators. Predation plays a key role in the functioning and multidimensional-structuring of ecosystems over a broad continuum of time and space scales. The pressures of predation have determined the contours of species niches, species morphology, camouflage and chemical characteristics, behavioural adaptations, community structure and ultimately, species biodiversity through evolutionary processes. Preda-tion on pelagic prey, such as Northern krill *Meganyctiphanes norvegica*, is thought to be the main reason behind widespread behaviours such as diel vertical migration (DVM), where animals seek refuge from visual predators in darker depths during daytime (see Chapter 9), schooling/swarming behaviour, which may reduce predation success and capture risk for indi-vidual prey (Pitcher and Parrish, 1993), and bioluminescence, used to escape or disturb attacking predators (Widder *et al.*, 1992). All these partic-ular adaptations of *M. norvegica* will affect its vulnerability to predation and its contribution to ecosystem function through trophic exchange.

Basic predation dynamics are described by a functional response (f), which relates the predation or consumption rate (y) to prey concentration (x):

$$y = f(x)$$

The functional response, $f(x)$, can be expressed by a number of different formulations, often containing several terms to take into account relevant components like the prey encounter rate, prey evaluation, pursuit, attack and digestion times, predator success rate and satiation (see review by Jeschke *et al.*, 2002). In all cases, the most important variable is the prey concentration, x, which is strongly structured in the horizontal and vertical planes in pelagic marine ecosystems. If the prey is dispersed in large 3D ocean volumes, there can be little energetic gain to the predator whatever the functional response may be, and the energy transfer to upper trophic levels is inefficient. Baleen whales, which rely on large quantities of small prey such as krill for their survival, are clear living proof of this paradigm since they must focus their foraging towards areas where their prey are densely concentrated (Acevedo-Gutiérrez *et al.*, 2002; Brodie *et al.*, 1978; Croll *et al.*, 2001). *Predation is therefore more likely to be significant in areas where the prey concentrates* and where the predator–prey interactions are intensified.

M. norvegica is a good species to use to try and typify the more general features of predation on euphausiid species given our detailed knowledge of its behavioural characteristics and its distribution in space and time.

2. PREDATION EFFECTS ON *M. NORVEGICA*

2.1. Anti-predation behaviour

M. norvegica undergo predation mortality similar to other *r*-breeding species with a planktonic developmental phase, where cohort success is determined very early in the life cycle (Houde, 2002). In the early free drifting stages, *M. norvegica* larvae are a minor contributor to a large and diverse low-mobility epipelagic planktonic community. Their relative concentration, size, shape and other passive characteristics (e.g. detectability, buoyancy and chemical properties) are the main factors controlling the predation on them through encounter rate, edibility and digestibility. With growth, the larvae gradually acquire mobility that allows some degree of control over its horizontal and vertical position and the development of anti-predator behaviours. Among them, the control of its detectability by visual predators through vertical swimming towards darker depths is acquired very early, as soon as the swimming appendages develop (Lacroix, 1961; Mauchline, 1959). Initially, the refuge depth below the "barrier isolume" (the avoided level of light (Simard *et al.*, 1986a)) is shallow and then deepens with age as the visual detectability increases with increasing body size (Cohen and Forward, 2009). This negative phototactism and DVM behaviour plays an important role in determining the 3D distribution, aggregation and concentration of *M. norvegica* (and other krill species) through their interaction with prevailing circulation regimes (Lavoie *et al.*, 2000; Simard and Lavoie, 1999; Sourisseau *et al.*, 2006). The upper limit to the depth distribution of adult krill is controlled by the *in situ* light level in the blue wavelengths, towards which krill eyes have peak sensitivity (Frank and Widder, 1999). The degree of penetration of these wavelengths through the water column depends on surface light levels and the local absorption profile. Trends and variability in blue light penetration are highly predictable; so are the most probable depth horizons of krill and their predators.

One consequence of this strong DVM behaviour is the formation of deep layers of krill, known as deep scattering layers (SLs) and detectable by high-frequency echosounders. These SLs appear to be another anti-predator strategy, as krill concentrated in them avoid capture by plankton nets to a significant degree (Simard and Sourisseau, 2009). How this avoidance is actually triggered is still unclear but visual cues appear to be involved, and possibly, bioluminescent communication signals propagating through the SL (Greene *et al.*, 1992; Simard and Sourisseau, 2009; Wiebe *et al.*, 2004).

Temporary saturation of krill visual receptors with strobe light flashes was shown to be effective in reducing such avoidance (Sameoto et al., 1993; Simard and Sourisseau, 2009; Wiebe et al., 2004). Krill anti-predation behaviour is, to some extent, quite predictable, and bulk-consumption predators, such as baleen whales, have learnt how to exploit these predictable behaviours as described later.

2.2. Anti-predation behaviours coupled with structured flows:

Northern krill are confined to a deep SL between sunrise and sunset and are mostly distributed into epipelagic SLs of variable thickness and density during the night (Harvey et al., 2009; Onsrud et al., 2004; Sourisseau et al., 2008; Tarling et al., 2001). This cyclical vertical distribution pattern exposes them to different horizontal transport regimes depending on the particular vertical current profile they experience, particularly in regions with vertically stratified flows (Lavoie et al., 2000; Zhou et al., 2005). This was shown to be responsible for large-scale retention/dispersion patterns as well as for the generation of dense aggregations where tidal currents strongly interact with topographic structures (Cotte and Simard, 2005; Mackas et al., 1997; Simard, 2009; Simard et al., 1986a; Sourisseau et al., 2006; Sourisseau et al., 2008). This biophysical coupling not only influences krill aggregation patterns but also the intensity of predation on krill as the predators focus on these aggregations to maximise their foraging and capture efficiencies.

A particularly well-described mechanism which aggregates krill is shelf/slope-current combined with phototactism (Figs. 10.1(case 1) and 10.2; Simard et al., 1986a). In this mechanism, up-slope components of the tidal and mean current transport krill up through the water column towards the surface-lit layers. Meanwhile, the krill actively try to oppose this upward movement to avoid entering these layers and become overly exposed to visual predation. This mechanism has been both documented by empirical observations (Cotte and Simard, 2005) and its strong aggregation effect successfully replicated in ground-truthed simulation models incorporating DVM and 3D circulation fields within large-scale basins (Sourisseau et al., 2006). Wind-driven and strong tidal upwelling circulation cells often reinforce this up-slope concentrating current at depth (U, Fig. 10.1). Along-slope currents can then transport these krill aggregations towards topographic traps such as canyons carved in the shelf (Croll et al., 2005; Simard and Mackas, 1989), large basin meanders, diverticula or channel heads (Kulka et al., 1982; Simard and Lavoie, 1999; Fig. 10.1(case 2)). Other Northern krill traps can occur in small basins carved on continental shelves (Cochrane et al., 1991) or semi-closed embayments like fjords (e.g. Zhou et al., 2005; Fig. 10.1(case 3)). Seamounts and shallow banks (Fig. 10.1(case 4)) are also favourable aggregation areas because of the aforementioned mechanism and also other coupled circulation-DVM effects (Genin,

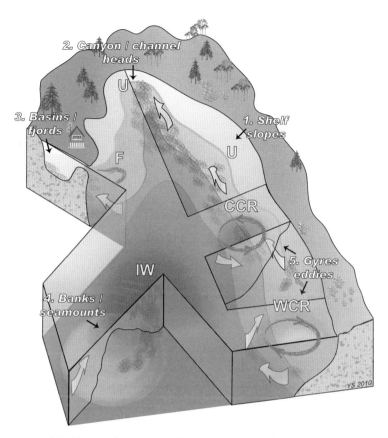

Figure 10.1 3D scheme of oceanographic processes involved in the governing of the interactions between predators and krill species including *M. norvegica*, by generating prey aggregations (red dots) at predictable sites and times, which predators have learnt to take advantage of and where intensive trophic exchanges occur. 1: Krill aggregation along slopes by upslope upwelling (U) currents and bathymetry guided currents; 2: Canyon and channel head krill entrapments by upstream flow at the krill scattering layer depth and intense upwelling at the topographic dead-end; 3: Krill entrapment in basins carved on continental shelves and in fjords; 4: Krill aggregation around banks and seamounts by 3D currents interacting with topography, often generating internal waves (IW); 5: Krill aggregation by cyclonic (CCR: cold core ring) and anti-cyclonic (WCR: warm core ring) eddies and gyres. Fronts (F) can sometimes concentrate krill by convergence or through the response of krill to environmental discontinuities. (See Colour Insert.)

2004). Internal waves (IW, Fig 10.1) can interact with topography to aggregate zooplankton (Jillett and Zeldis, 1985; Lennert-Cody and Franks, 1999; Warren *et al.*, 2003) and also cause rapid vertical displacements through the water column, altering availability to epipelagic predators. Gyres and mesoscale eddies (Fig. 10.1(case 5)) that persist for sufficient

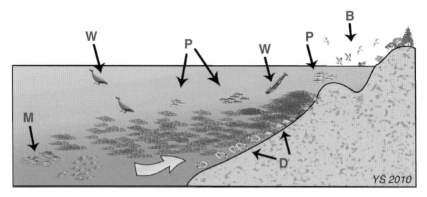

Figure 10.2   Schematic representation of typical predator–prey interactions along sloping topography where krill (red dots) is aggregating during daytime at latitudes inhabited by *M. norvegica*. D: Demersal predators where the krill SL horizon comes into contact with the slope; P: Pelagic predators aggregated over the neighbouring shallows or in the epipelagic surface-lit layer allowing visual predation; M: Mesopelagic and bathypelagic fish occupying deep layers in the water column. W: Baleen and other marine mammals diving to feed on the krill SL; B: Surface feeding and diving marine birds taking advantage of the available krill SL, notably at night or during upwelling events. (See Colour Insert.)

periods can generate 3D circulation cells that may also aggregate krill when combined with their depth-keeping behaviour (Wiebe, 1982; Wiebe and Flierl, 1983). For instance, cyclonic circulation cells create a doming of deep waters in the centre of the circulation ring (cold core ring, CCR, Fig. 10.1), thereby bringing krill closer to the surface-lit waters and favouring aggrega-tion in a similar fashion to that generated by the up-slope currents. Anti-cyclonic circulation cells (warm core ring, WCR, Fig. 10.1) could generate different concentration patterns depending on their interactions with krill DVM. Fronts generated by several mechanisms (topographic interactions, upwelling, plumes, jets, convergences, etc.; F, Fig. 10.1) can sometimes concentrate krill by passive convergence or by active responses to environ-mental discontinuities (Franks, 1992). Nevertheless, the extent to which this occurs, and becomes a major site for predation, is strongly dependent on whether krill are present at the right depths and times, and also the relative reactions of predators and prey to the frontal discontinuity. The relative contribution of fronts to the aggregation of krill is therefore less predictable than the shelf/slope-current phototactism mechanism. When fronts co-occur along topographic features, the aggregation process may become confounded with slope processes. Large-scale persistent frontal features such as water mass boundaries, coastal jets and shelf break currents also play a role in krill distribution patterns, notably trough concentration process at the tilted interface between the two water masses or by convergence in the upper water column where krill is mainly distributed during the night.

3. NORTHERN KRILL PREDATORS AND PREDATION STRATEGY

3.1. Predator types

Northern krill predators (Table 10.1) can be sorted into five functional groups depending on their distribution and feeding behaviour (Fig. 10.2): (a) epipelagic fish such as herring, capelin and mackerel (P, Fig. 10.2), which are members of the pelagic community, living and feeding in the upper ∼100 m of the water column and maintaining predation pressure on Northern krill SLs from above; (b) benthic and demersal species such as flat fishes, cod, redfish, pollock and crabs (D, Fig. 10.2), which live on the seabed or within ∼100 m of the bottom, and take advantage of Northern krill SLs shoaling on shelf-slopes; (c) mesopelagic and benthopelagic species such as squid, hake, whiting and Norway pout (M, Fig. 10.2), which can live in the mid water column as well as in the vicinity of the bottom; (d) air breathing predators such as birds (B, Fig. 10.2) and marine mammals (W, Fig. 10.2), which can dive into the krill SL at depth during daytime or when it is closer to the surface during its upward DVM during night-time or in response to some physical forcing under certain circumstances.

Among these predators, baleen whales are unique, since they do not feed on individuals but on bulk amounts of krill, targeting aggregations that they engulf in mouthfuls. The whales take advantage of krill swarming behaviour which is otherwise a beneficial anti-predation strategy (Pitcher and Parrish, 1993). Nevertheless, aggregation may not be completely ineffective against whale predation, since it at least minimises the spatial extent of the krill distribution and makes it harder for roaming predators to detect them. However, once a predator has detected the edge of an aggregation, the probability of it being of considerable biomass is high, since aggregations often extend over kilometre scales (Simard and Lavoie, 1999). It is likely that predator migration routes and fidelities to certain sites are traits that have adapted over time to take advantage of predictable Northern krill aggregation sites, such as those sketched in Fig. 10.1.

3.2. Predator–prey diel rules and tactics

Prey-detectability, encounter rate and attack success are the key variables that influence predator–prey interactions at smaller scales. Underwater light is the principal controlling factor for visual predation, and DVM behaviour is the main anti-predation response of Northern krill to this factor. The

Table 10.1 Diversity and distribution of Northern krill, *M. norvegica*, predators from a literature search

Predator group	Predator species	Sampled area (years)	Location – Reference
Actinopterygii (Ray-finned fishes)			
Gadiformes	Pollock	Bay of Fundy, Western Atlantic (1960–1961)	1 – (Steele, 1963)
	Pollachius virens	Georges Bank, Gulf of Maine (1977–1980)	2 – (Bowman *et al.*, 2000)
		Scotian Shelf and Bay of Fundy (1958–1967; 1996–2002)	3 – (Carruyhers *et al.*, 2005)
		Scottish waters (1975–1982)	4 – (Du Buit, 1991)
		Faroe Island (1996)	5 – (Hojgaard, 1999)
	Atlantic Cod	Balsfjorden, Northern Norway (1978)	6 – (Pearcy *et al.*, 1979)
	Gadus morhua	Icelandic subarctic waters (north and south-east of Iceland) (1983–1984)	7 – (Astthorsson and Palsson, 1987)
		Scotian Shelf (1998)	8 – (Sameoto *et al.*, 1994)
		Gulf of St. Lawrence (1998–2009)	9 – (Chabot, 2010)
		Gulf of Maine and Scotian Shelf (1977–1980)	2 – (Bowman *et al.*, 2000)
	Juvenile cod (age 0–2)	Barents Sea (1984–2002)	10 – (Dalpadado and Bogstad, 2004)
	Gadus morhua		
	Silvery pout	Raunefjorden Fjord, Western Norway (1975)	11 – (Mattson, 1981)
	Gadiculus argenteus	Algarve, Portugal (1996–1997)	12 – (Santos and Borges, 2001)
	Norway pout	Raunefjorden Fjord, Western Norway (1975)	11 – (Mattson, 1981)
	Trisopterus esmarkii	Oslofjord, Norway (1997–2001)	13 – (Onsrud *et al.*, 2004)
	Fourbeard rockling		
	Rhinonemus cimbrius	Raunefjorden Fjord, Western Norway (1975)	11 – (Mattson, 1981)
	Enchelyopus cimbrius	Lower St. Lawrence Estuary (2008)	14 – (Chabot, 2010)
	Silver hake	Scotian Shelf (1998)	8 – (Sameoto *et al.*, 1994)
	Merluccius bilinearis	Browns Bank, Scotian Shelf (1985)	15 – (Sherstyukov and Nazarova, 1991)
		Georges Bank, Gulf of Maine (1977–1980)	2 – (Bowman *et al.*, 2000)

European hake *Merluccius merluccius*	Balearic Islands, Mediterranean sea (2003–2004)	16– (Cartes *et al.*, 2009)
	Algarve, Portugal (1996–1997)	12 – (Santos and Borges, 2001)
	Tyrrhenian Sea, Mediterranean sea (1998–1999)	17 – (Sartor *et al.*, 2003)
Red hake *Urophycis chuss*	Georges Bank, Gulf of Maine (1977–1980)	2 – (Bowman *et al.*, 2000)
White hake *Urophycis tenuis*	Gulf of Maine (1977–1980)	2 – (Bowman *et al.*, 2000)
Blue whiting *Micromesistius poutassou*	Algarve, Portugal (1996–1997)	12 – (Santos and Borges, 2001)
English whiting *Merlangius merlangus*	Portuguese coast (1990–1992)	18 – (Cabral and Murta, 2002)
	Oslofjord, Norway (1997–2001)	13 – (Onsrud *et al.*, 2004)
Rock grenadier *Coryphaenoides rupestris*	Le Danois Bank, Cantabrian Sea, N Spain (2003–2004)	19 – (Preciado *et al.*, 2006)
Common grenadier *Nezumia bairdii*	Laurentian Chanel (2001)	20 – (Chabot, 2010)
Haddock *Melanogrammus aeglefinus*	Gulf of Maine and Scotian Shelf (1977–1980)	2 – (Bowman *et al.*, 2000)
Clupeiformes		
Atlantic herring *Clupea harengus*	Balsfjorden, northern Norway (1978)	6 – (Pearcy *et al.*, 1979)
	Coast of Norway (1991–1993)	21 – (Dalpadado, 1993)
	Coastal areas of Norway to the Atlantic and Arctic water masses (1994–1996)	22– (Dalpadado *et al.*, 1998)
	Western Norwegian Sea (1995–1996)	23 – (Gislason and Astthorsson, 2000)
	Oslofjord, Norway (1997–2001)	13 – (Onsrud *et al.*, 2004)
	Lower St. Lawrence Estuary (2008)	14 – (Chabot, 2010)
	Gulf of Maine, Scotian Shelf (1977–1980)	2 – (Bowman *et al.*, 2000)

(*continued*)

Table 10.1 (continued)

Predator group	Predator species	Sampled area (years)	Location – Reference
	Alewife *Alosa pseudoharengus*	Scotian Shelf (winter) and Bay of Fundy (summer) (1990–1991)	24 – (Stone and Jessop, 1994)
Perciformes	Horse mackerel *Trachurus trachurus*	Gulf of Maine and Scotian Shelf (1977–1980) Cantabrian Sea (north of Spain) (1998)	2 – (Bowman *et al.*, 2000) 25 – (Olaso *et al.*, 1999)
	Chub mackerel *Scomber japonicus*	Algarve, Portugal (1996–1997)	12 – (Santos and Borges, 2001)
		Algarve, Portugal (1996–1997)	12 – (Santos and Borges, 2001)
	Vahl's eelpout *Lycodes vahlii*	Laurentian Chanel (2001)	20 – (Chabot, 2010)
	Black sea bass *Centropristis striata*	Middle Atlantic (1977–1980)	26 – (Bowman *et al.*, 2000)
	Atlantic mackerel *Scomber scombrus*	Georges Bank (1977–1980)	2 – (Bowman *et al.*, 2000)
	Butterfish *Peprilus triacanthus*	Middle Atlantic (1977–1980)	26 – (Bowman *et al.*, 2000)
Pleuronectiformes	Gray sole *Glyptocephalus cynoglossus*	Raunefjorden Fjord, Western Norway (1975) Whole St. Lawrence marine system (2004)	11 – (Mattson, 1981) 27 – (Chabot, 2010)
	American plaice *Hippoglossoides platessoides*	Lower St. Lawrence Estuary (2004)	14 – (Chabot, 2010)
	Atlantic halibut *Hippoglossus hippoglossus*	Whole St. Lawrence marine system (1998–2009)	27 – (Chabot, 2010)
	Greenland halibut *Reinhardtius hippoglossoides*	Whole St. Lawrence marine system (1993–2009)	27 – (Chabot, 2010)
	Summer flounder *Paralichthys dentatus*	Gulf of Maine (1977–1980)	2 – (Bowman *et al.*, 2000)

Order	Common name / Species	Location (year)	Reference
Lampridiformes	Dealfish *Trachipterus arcticus*	Sjona, Norway (1981)	28 – (Straumfors, 1981)
Zeiformes	Boarfish *Capros aper*	Algarve, Portugal (1996–1997)	12 – (Santos and Borges, 2001)
Scorpaeniformes	Atlantic thornyhead *Trachyscorpia cristulata*	Le Danois Bank, Cantabrian Sea, N Spain (2003–2004)	19 – (Preciado et al., 2006)
	Redfishes *Sebastes* sp.	Whole St. Lawrence marine system (1993–1997)	27 – (Chabot, 2010)
	Deepwater redfish *Sebastes mentella*	Reykjanes Ridge, Iceland bassin (2003–2004)	30 – (Petursdottir et al., 2008)
	Acadian redfish *Sebastes fasciatus*	Georges Bank, Gulf of Maine and Scotian Shelf (1977–1980)	2 – (Bowman et al., 2000)
	Unidentified large piscivorous fish	Norwegian Sea (2004)	29 – (Kaartvedt et al., 2005)
Osmeriformes	Slickheads *Alepocephalus rostratus*	Le Danois Bank, Cantabrian Sea, N Spain (2003–2004)	19 – (Preciado et al., 2006)
	Capelin *Mallotus villosus*	Whole St. Lawrence marine system (2003, 2008)	27 – (Chabot, 2010)
	Atlantic argentine *Argentina silus*	Gulf of Maine and Scotian Shelf (1977–1980)	2 – (Bowman et al., 2000)
Aulopiformes	Shortnose *Chlorophthalmus agassizii*	Le Danois Bank, Cantabrian Sea, N Spain (2003–2004)	19 – (Preciado et al., 2006)
	White barracudina *Arctozenus risso*	Laurentian Chanel (2001)	20 – (Chabot, 2010)
Beryciformes	Mediterranean redfish *Hoplostethus mediterraneus*	Algarve, Portugal (1996–1999)	31 – (Pais, 2002)

(continued)

Table 10.1 (continued)

Predator group	Predator species	Sampled area (years)	Location – Reference
Chondrichthyes (Cartilaginous fishes)			
Carcharhiniformes	Blackmouth catshark	Le Danois Bank, Cantabrian Sea, N Spain (2003–2004)	1 – (Preciado *et al.*, 2006)
	Galeus melastomus	Raunefjorden Fjord, Western Norway (1975)	2 – (Mattson, 1981)
		Gulf of Genoa, Mediterranean sea (1975)	3 – (Relini Orsi and Wurtz, 1975)
	Bluntsnout grenadier	Cantabrian Sea, N. Spain (2003–2004)	4 – (Preciado *et al.*, 2009)
	Nezumia sclerorhynchus	Le Danois Bank, Cantabrian Sea, N. Spain (2003–2004)	1 – (Preciado *et al.*, 2006)
	Chain dogfish	Gulf of Maine (1977–1980)	10 – (Bowman *et al.*, 2000)
	Scyliorhinus rotifer		
Rajiformes	Thorny skate	Whole St. Lawrence marine system (2005, 2008)	9 – (Chabot, 2010)
	Amblyraja radiata		
	Smooth skate	Whole St. Lawrence marine system (2005, 2008)	9 – (Chabot, 2010)
	Malacoraja senta		
Squaliformes	Smooth lanternshark	Algarve, Portugal (1996–1997)	5 – (Santos and Borges, 2001)
	Etmopterus pusillus		
	Birdbeak dogfish	Le Danois Bank, Cantabrian Sea, N. Spain (2003–2004)	1 – (Preciado *et al.*, 2006)
	Deania calcea		
	Velvet belly	Algarve, Portugal (1996–1997)	5 – (Santos and Borges, 2001)
	Etmopterus spinax	Gulf of Genoa, Mediterranean sea (1977)	6 – (Relini Orsi and Wurtz, 1977)
		Southern Portugal (2003–2004)	7 – (Neiva *et al.*, 2006)
		Cantabrian Sea, N. Spain (2003–2004)	4 – (Preciado *et al.*, 2009)
		Norway (57°30′–57°31′ N 06°52′–07°03′ E) (2001)	8 – (Klimpel *et al.*, 2003)

Cephalopoda (Squids)			
Teuthida	Northern shortfin squid	Bay of Fundy (1979–1982)	1 – (Nicol and O'Dor, 1985)
	Illex illecebrosus	Continental shelf west and south-west of Ireland (1993–1997)	2 – (Lordan et al., 1998)
		Northwest Atlantic, Cape Hatteras – Scotian Shelf (1968–1981)	3 – (Froerman, 1984)
	Lesser flying squid	Continental shelf west and south-west of Ireland (1993–1997)	2 – (Lordan et al., 1998)
	Todaropsis eblanae		
	Veined squid	Northwest Atlantic, Cape Hatteras – Scotian Shelf (1991–1993)	4 – (Collins et al., 1994)
	Loligo forbesi		
Sepiolida	Bobtail squid	Central part of Gullmar Fjord, Sweden (1979–1982)	5 – (Bergstroem, 1985)
	Sepietta oweniana		
Malacostraca (Crustacea)			
Decapoda	Oplophorid shrimp	Deep Western Mediterranean sea (1988–1989)	1 – (Cartes, 1993)
	Acanthephyra pelagica		
	Pandalid shrimps	Alboran Sea, S. Spain, Mediterranean sea (2002)	2 – (Fanelli and Cartes, 2004)
	Plesionika gigliolii		
	P. heterocarpus		
	P. martia		
	Deep-sea shrimp	Balearic Islands, Mediterranean sea (2003–2004)	3 – (Cartes et al., 2008)
	Aristeus antennatus		
	Squat lobsters	UK continental shelf West of Shetland (2002)	4 – (Hudson and Wigham, 2003)
	Munida sarsi		
	Northern shrimp	Lower St. Lawrence Estuary (2008)	5 – (Chabot, 2010)
	Pandalus borealis		

(continued)

Table 10.1 (continued)

Predator group	Predator species	Sampled area (years)	Location – Reference
Ophiuroidea (Brittle star)			
Phrynophiurida	Euryhaline brittle stars	Oslofjord Norway (2003)	1 – (Rosenberg et al., 2005)
	Gorgonocephalus caputmedusae		
	G. arcticus		
	G. arcticus	Bay of Fundy (1982–1984, 1988)	2 – (Emson et al., 1991)
Mammalia (Mammals)			
Cetacea	Fin whale	Scotian shelf (1965)	1 – (Brodie et al., 1978)
	Balaenoptera physalus	Iceland Basin (1979–1989)	3 – (Sigurjónsson and Vikingsson, 1997)
		Iceland Basin (1967–1989)	4 – (Stefansson et al., 1997)
		West and Southwest of Iceland (1976–1989)	5 – (Vikingsson, 1997)
		Western Ligurian Sea, Mediterranean sea (1998)	6 – (Panigada et al., 1999)
	Minke whale	Northern Norway (1988–1990)	2 – (Nordoey and Blix, 1992)
	Balaenoptera acutorostrata	Iceland Basin (1979–1989)	3 – (Sigurjónsson and Vikingsson, 1997)
		Iceland Basin (1983–1993)	4 – (Stefansson et al., 1997)
		Southern Barents Sea (1998)	7 – (Harbitz and Lindstroem, 2001)
		Southern Barents Sea (May–June 1998)	8 – (Lindstroem and Haug, 2002)
		Gulf of Maine (USA) (1989–1994)	9 – (Gannon et al., 1998)
	Harbor porpoise	Bay of Biscay and western Channel, Northeast Atlantic French coast (1998–2003)	10 – (Spitz et al., 2006)
	Phocoena phocoena		

Carnivora	Harp seal *Pagophilus groenlandicus*	Lower St. Lawrence Estuary (1996)	11 – (Hammill, 2010)
Aves (Birds) *Ciconiiformes*	Thick-billed murres *Uria lomvia*	Western Greenland (1988–1989)	1 – (Falk and Durinck, 1993)
	Sooty shearwaters *Puffinus gravis* *P. griseus*	Brier Island, Bay of Fundy (1974)	2 – (Brown *et al.*, 1981)
	Leach's Storm Petrel *Oceanodroma leucorhoa*	Green Is., Gull Is., Eastern Newfoundland (1987)	3 – (Montevecchi *et al.*, 1992)

Locations numbered for each Class taxa are mapped in Fig. 10.4.

main SL during daytime is located within a safe *in situ* light level, of which the top margin is often bounded by patrolling pelagic fish (Kaartvedt *et al.*, 2005; Onsrud *et al.*, 2004). These fish can take advantage of fast changes in light levels during twilight when vertically migrating krill entering their patrolled layer may become detectable (Daytime 1, Fig. 10.3). Krill is also vulnerable to predation from below, especially where the SL approaches the bottom as it comes into contact with shelf-slope regions or shallow bathymetry (e.g. basins/fjords, Fig. 10.1). The bottom margin of SLs is also vulnerable to off-shelf predation from meso- and bathypelagic predators. The spatial separation between predator and prey, through their different preferred levels of *in situ* light, temperature, pressure, oxygen, etc., means that their interactions are generally limited to fringe regions where these preferred habitats may overlap. Northern krill in SLs have also developed efficient detection and avoidance mechanisms (Simard and Sourisseau, 2009; Wiebe *et al.*, 2004) to minimise further the rates of encounter and attack-success. They effectively detect nets and avoid being caught, most likely using visual cues including bioluminescent flashes, which become inefficient when strobe lighting is present.

Baleen whales efficiently exploit krill SLs during daytime. Their feeding-dive behaviour is now better known, owing to electronic tags that track their depth, orientation and speed underwater (Acevedo-Gutiérrez *et al.*, 2002; Calambokidis *et al.*, 2007; Croll *et al.*, 2001). Their predation strategy appears to exploit krill anti-predator behaviours that are otherwise effective against most other predators, such as the tendency to remain below a certain *in situ* light level and to escape downwards when under attack. Blue whales, which are thought to feed exclusively on krill, exhibit typical feeding dives that include several up-and-down excursions through krill SLs located in waters with relatively deep bathymetries (Daytime 1, Fig. 10.3). During the initial dive, the whale first descends through the SL after which it immediately turns around for a fast ascent, mouth open (Acevedo-Gutiérrez *et al.*, 2002; Calambokidis *et al.*, 2007) (arrowed yellow up path, Daytime 1, Fig. 10.3). The whale then rotates for another slow dive through the SL before resuming the fast ascent feeding behaviour. The process is repeated several times before surfacing again. The whale does not feed during the slow descent phase probably because of the effective downward escape response of krill to predators approaching from above. However, this downward escape response means that they are more likely to swim into the open mouth of the whale during its fast ascent phase, where the speed of the whale further minimises krill escape possibilities. This behaviour appears to be an efficient feeding strategy for whales feeding in deep water, but is not likely to be as efficient in exploiting aggregations in high-density shallow areas where the krill SL has shoaled and concentrated (Daytimes, Fig. 10.3).

Other daytime predators, such as harbour porpoises, can acoustically detect and exploit krill SLs with biosonar. The high-frequency clicks of

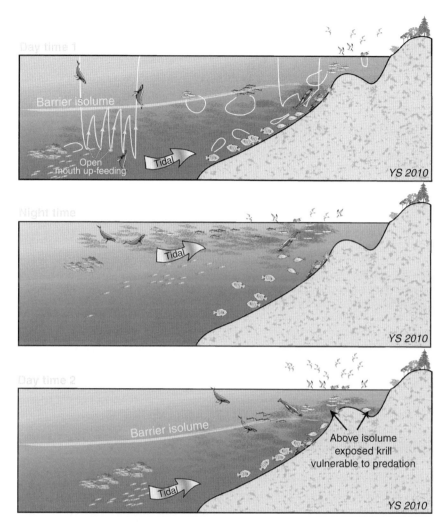

Figure 10.3 Schematic representation of predator–prey behaviours at a Northern krill aggregation site over the diel cycle and with a tidal current regime. Daytime 1: Typical predator–prey configuration along the sloping bottom, where the krill SL shoals and concentrates at the junction point between the barrier isolume and the bottom under forcing from up-slope currents at tidal frequencies. Predators from all types illustrated in Fig. 10.2 can have excursions (white lines) into the SLs or graze its top or bottom margins. Night time: During night, the krill SL is in the upper water column and the entire predator community changes its behaviour while tracking the prey community into the upper water column. The currents then transport the krill in directions determined by the local currents in the upper water column. Daytime 2: All the krill still in upper water column at dawn return to the depths when the sun rises and they track the barrier isolume downwards. Because of their transport during night by upper water column currents, the krill may descend onto bathymetries that are shallower than the depth of the barrier isolume. They then become visible to predators, their eyes are

these organisms (Au *et al.*, 1999; Mohl and Andersen, 1973) are in the optimal frequency band for detecting krill (Lavery *et al.*, 2007). Stomach contents indicate that they do sometimes feed on Northern krill (Tables 10.1, 10.2).

During the night-time, the light-forced spatial organisation of predators and prey becomes modified through DVM (Night-time, Fig. 10.3). The upper water column now becomes the focus of predator–prey interactions, where Northern krill is intensively feeding on phytoplankton and small meso-zoo-plankton (Simard *et al.*, 1986b; Sourisseau *et al.*, 2008; see Chapter 5) and where the predators are gathered. Northern krill does not appear to be as reactive to predators during night-time since net avoidance appears to be reduced (Simard and Sourisseau, 2009). However, their stay in this risky environment, where the presence of predators is mainly detectable through physical encounters and/or chemical tracers (Cohen and Forward, 2009), appears to be short, since the proportion that occupies the upper layers gradually decreases as the night goes on (Sourisseau *et al.*, 2008). This gradual descent out of the upper water column during night-time, termed "midnight sinking", is believed to be a physiological response to satiation, but it also serves as an anti-predation mechanism (Pearre, 2003). The time window where the highest krill concentration is found in the upper surface layer is the 1–2 h period after the arriving of the SL ascent at sunset (Sourisseau *et al.*, 2008). It is likely that some predators have learnt to exploit that highly predictable time window of high krill concentration. Likewise, the rapid reformation of the deep SL during night (Sourisseau *et al.*, 2008) results from krill descending to minimise their level of risk from occupation of the surface layers (Pearre, 2003).

At dawn, the remaining krill in the surface layers descend rapidly to join the rest of the population that are already within the deep SL. Because of the currents acting on the upper layers, where they spent the night, there is a risk that the descent is onto shallow bathymetry, shallower than the daytime depth of the barrier isolume (Daytime 2, Fig. 10.3). Visual predators are able to detect krill within such light levels and high levels of krill predation result, as the balance between predator and prey moves in favour of the predators. Such scenarios were originally revealed through the introduction of echosounders capable of tracking the DVM of SLs and predator distributions (Isaacs and Schwartzlose, 1965). Such trapping of krill is more likely to happen in shallow areas adjoining deep basins, where krill tend to concentrate against topo-graphic obstacles to the flow, such as channel heads or banks. There, tidal currents interacting with topography are sometimes strong enough to displace

---

exposed to high light levels and they no longer can escape by moving down. On the contrary, this reaction concentrates them and makes them still more vulnerable to all pelagic, demersal and benthic predators. With the usual anti-predation strategies fail-ing, krill may be found over the whole water column, even at the surface, often under strong tidal current interactions with the topography, to the benefit of avian predators. (See Colour Insert.)

Table 10.2 Stomach content (weight and occurrence) in Northern krill, *M. norvegica*, for different predators from a literature search

Class		Stomach content		
Order species	Region	Weight % or mean weight in g	Occurrence % or mean number per predator / N	Reference
Actinopterygii (Ray-finned fishes) *Gadiformes*				
Pollock	1	30.8–83.4%	–	Bowman *et al.* (2000)
	3	8–85%	–	Du Buit (1991)
	2	4%	3.2%	Hojgaard (1999)
Atlantic Cod	2	–	0.4–2N	Astthorsson and Palsson (1987)
	1	–	1–39%	Sameoto *et al.* (1994)
	1	0.5%	–	Chabot (2010)
	1	0–6.4%	–	Bowman *et al.* (2000)
Silvery pout	4	0.2 g	1.7 N	Mattson (1981)
	3	65%	57%	Santos and Borges (2001)
Norway pout	4	0.3 g	0.1 N	Mattson (1981)
	4	44–75%	1.5–2.2 N	Onsrud *et al.* (2004)
Fourbeard	4	0.1 g	0.4 N	Mattson (1981)
rockling	1	8.8%	–	Chabot (2010)
Silver hake	1	36–60%	–	Sameoto *et al.* (1994)
	1	60–100%	1–314 N	Sherstyukov and Nazarova (1991)
		3.7–62.0%	–	Bowman *et al.* (2000)
European hake	3	0.0–59.0 g	–	Cartes *et al.* (2009)
	3	26%	54%	Santos and Borges (2001)
	3	4.4%	2.8%	Sartor *et al.* (2003)
Red hake	1	0.4–37%	–	Bowman *et al.* (2000)
White hake	1	0–37%	–	Bowman *et al.* (2000)
Blue whiting	3	46%	64%	Santos and Borges (2001)
	3	24%	8%	Cabral and Murta (2002)
English whiting	4	34–80%	1.4–6 N	Onsrud *et al.* (2004)
Commun grenadier	1	14.5%	–	Chabot (2010)

(*continued*)

Table 10.2 *(continued)*

Class		Stomach content		
			Occurrence % or mean	
		Weight % or mean weight	number per predator /	
Order species	Region	in g	N	Reference
Haddock	1	0–61%		Bowman *et al.* (2000)
Clupeiformes				
Atlantic herring	4	0–10%	0–25%	Gislason and Astthorsson (2000)
	4	17–50%	1.5–5.1 N	Onsrud *et al.* (2004)
	1	11.0%	–	Chabot (2010)
	1	0–82%	–	Bowman *et al.* (2000)
Perciformes				
Horse mackerel	3	67%	41%	Santos and Borges (2001)
Chub mackerel	3	41%	69%	Santos and Borges (2001)
Vahl's eelpout	1	3.1%	–	Chabot (2010)
Black sea bass	1	14.8%	–	Chabot (2010)
Atlantic mackerel	1	10%	–	Chabot (2010)
Butterfish	1	2.6%	–	Chabot (2010)
Pleuronectiformes				
Gray sole	4	0.02 g	0.003 N	Mattson (1981)
	1	3.6%	–	Chabot (2010)
American plaice	1	4.6%	–	Chabot (2010)
Atlantic halibut	1	0.1%	–	Chabot (2010)
Greenland halibut	1	0.2%	–	Chabot (2010)
Summer flounder	1	0-6%	–	Bowman *et al.* (2000)
Zeiformes				
Boarfish	3	16%	12%	Santos and Borges (2001)
Scorpaeniformes				
Redfishes	1	1.4%	–	Chabot (2010)
Acadian redfish	1	0–90%	–	Bowman *et al.* (2000)
Osmeriformes				
Capelin	1	2.6%	–	Chabot (2010)
Atlantic argentine	1	4.8–100%	–	Bowman *et al.* (2000)

Table 10.2 *(continued)*

Class			Stomach content		
Order species	Region	Weight % or mean weight in g	Occurrence % or mean number per predator / N		Reference
Aulopiformes					
White barracudina	1	23.8%	–		Chabot (2010)
Beryciformes					
Mediterranean redfish	3	–	41.4%		Pais (2002)
Chondrichthyes (Cartilaginous fishes)					
Carcharhiniforme					
Blackmouth catshark	4	0.03 g	1.1 N		Mattson (1981)
	3		13–40%		Preciado *et al.* (2009)
Chain dogfish	1	0–12%	–		Bowman *et al.* (2000)
Rajiformes					
Thorny skate	1	0.5%	–		Chabot (2010)
Smooth skate	1	0.6%	–		Chabot (2010)
Squaliformes					
Smooth lanternshark	3	1%	4%		Santos and Borges (2001)
	3	50%	91%		Santos and Borges (2001)
Velvet belly	3	22.0%	3.3%		Neiva *et al.* (2006)
	3	–	24–60%		Preciado *et al.* (2009)
	4	–	2–12 N		Klimpel *et al.* (2003)
Cephalopoda (Squids)					
Teuthida					
Northern shortfin squid	1		6 N		Nicol and O'Dor (1985)
	3	–	9.5%		Lordan *et al.* (1998)
	1	–	10–86.5%		Froerman (1984)
Lesser Flying Squid	3	–	7.6%		Lordan *et al.* (1998)
Sepiolida					
Bobtail squid	4	–	56%		Bergstroem (1985)
Malacostraca (Crustacea)					
Decapoda					
Pandalid shrimp	3	22%	8%		Fanelli and Cartes
		20%	10%		(2004)
		10%	5%		

(continued)

Table 10.2   *(continued)*

Order species	Region	Weight % or mean weight in g	Occurrence % or mean number per predator / N	Reference
Class		**Stomach content**		
Deep-sea shrimp	3	–	16–40%	Cartes *et al.* (2008)
Ophiuroidea (Brittle star)				
Phrynophiurida				
Euryhaline brittle stars	1	–	1–5 N	Emson *et al.* (1991)
Mammalia (Mammals)				
Cetacea				
Fin whale	1	–	100%	Brodie *et al.* (1978)
	2	–	97%	Sigurjónsson and Víkingsson (1997)
	2	–	60–100%	Vikingsson (1997)
Minke whale	2	–	41%	Sigurjónsson and Víkingsson (1997)
Harbor Porpoise	3	0.2%	12.7%	Spitz *et al.* (2006)
Aves (Birds)				
Ciconiiformes				
Thick-billed murres	1	0–3.1%	–	Falk and Durinck (1993)
Leach's Storm Petrel	1	–	10%	Montevecchi *et al.* (1992)

1-North-West Atlantic (Canada and USA, Western Greenland); 2 – Iceland and Faroe Islands; 3 – Eastern Europe (United Kingdom, Ireland, France, Spain, Portugal, Italy; Mediterranean sea and Western Atlantic); 4 – North-East Atlantic (Norway, Sweden, Denmark, Finland).

SLs over the shallows (e.g. Simard *et al.* (2002)). Currents > 25 cm s^{-1} (~ 0.5 knots) are common within slope flows, internal waves and upwelling pulses (Cotte and Simard, 2005; Lavoie *et al.*, 2000; Simard *et al.*, 2002), and *M. norvegica* is unlikely to be able to oppose such currents over extended periods. High-energy shallow areas are therefore favourable places for predators which benefit from regular inputs of displaced deep-water prey, such as krill (e.g. Cotte and Simard, 2005; Simard *et al.*, 2002; Stevick *et al.*, 2008). In these locations, it is not uncommon to observe krill close to or at the surface during daytime, and corresponding surface feeding frenzies of fish, whales and birds, (e.g. Nicol and O'Dor, 1985; Stevick *et al.*, 2008).

Surface swarming during the daytime does not appear to be an adaptive strategy for krill. It may nevertheless be explained by several biological and physical processes. Among them, we find coordinated groups of visual predators such as schooling fish actively concentrating krill towards the surface by circling around krill swarms, as underwater video footage has documented for several predator–prey systems, including those of fish and krill. Rapid and large vertical displacements of krill trapped in internal waves and responses to 3D strong environmental gradients at fronts are other means by which krill may be transported to or constrained within the surface layers. In all cases, it is possible that the sensitive visual capabilities of Northern krill become temporarily or permanently dysfunctional because of exposure to high light levels, as is suggested by the enhancement of net catches by instruments equipped with artificial strobe-lighting (Wiebe *et al.*, 2004). Such particular physical–biological coupling may be involved in daytime surface swarming of *M. norvegica* that are often encountered in high energy areas such as the Bay of Fundy (Brown *et al.*, 1979; Nicol, 1986).

4. Diversity of Northern Krill Predators and their Diet

A literature search and unpublished institutional data identified 71 species whose stomach contents specifically included *M. norvegica* (Table 10.1). Fish accounted for 49 species, invertebrates for 14 species, mammals for 5 species and birds for 3 species. Although this represents a large number of species, the Northern krill predator community is undoubtedly much larger than this list since only a part of the potential predators were analysed and, among them, only those where the stomach contents were analysed to the species level were considered. Besides, this list takes into account only recent records and does not necessarily include all other predator species identified in the euphausiid monographs of Mauchline (1980), Mauchline and Fisher (1969). The list includes representatives from all functional groups previously introduced (Fig. 10.2). Their spatial distribution (Fig. 10.4) covers all areas where *M. norvegica* is distributed in the North Atlantic (see Chapter 1).

The stomach contents of several predator species have been found to be mainly composed of *M. norvegica* (Table 10.2). Among these species are pollock, pout, hake, whiting, grenadier, herring, mackerel, redfish, lantern shark, squids, shrimps and baleen whales. This is the minimum number of species that rely on *M. norvegica*, since reported gut contents of other predator species have not been documented to the species level. Nevertheless, Table 10.2 identifies major players in the energy transfer from *M. norvegica* to the higher levels of the food web, which includes several top predators such as large commercially exploited fish, and marine mammals, including other pinnipeds and odonto-cetes, which are not specifically mentioned in Tables 10.1 and 10.2.

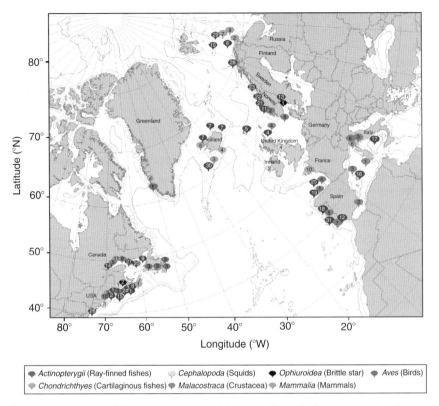

Figure 10.4 Map of the predators containing Northern krill, *M. norvegica*, from a literature search. Each taxonomic group is indicated by different colors. Numbers correspond to references in Table 10.1. (See Colour Insert.)

5. Perspectives and Further Research

M. norvegica is a successful member of the euphausiid community. The present rapid global warming of the planet is likely to affect its distributional range, particularly with regard to a poleward shift in its habitat. Competition with other species and predation from new predators at the fringes of its distribution will contribute to determine the success of its response to this potential change. Predators that depend on Northern krill for a large part of their diet will have to find alternative prey if they do not respond similarly to this habitat shift. Likewise, the predators relying on other forage species in newly colonised areas may find Northern krill to be an advantageous substitute for their usual resources which have otherwise migrated elsewhere and/or decreased.

In this general scenario of global change, the forces responsible for Northern krill concentration in particular areas where predators efficiently forage on them will still be the same. However, the changes in ocean circulation, ecosystem production and in the DVM response to a shift in latitudinal distribution, modifying the photoperiod the animals are exposed to, will result in new distributions of Northern krill aggregations. Present knowledge of Northern krill aggregation behaviour can be combined with ocean and ecosystem change scenarios to try and predict where these new aggregation sites will be. To improve our capacity to carry out such predictions and understand the predator–prey dynamics with regard to Northern krill, future research should include carefully designed *in situ* experiments to characterise the individual and collective responses of Northern krill to predators. These should include: (a) the visual, chemical and mechanical detection of predators coming from different directions, (b) the communication of the presence of predators to nearby conspecifics, through bioluminescent flashes or other signals, and their effective range, (c) the description of the corresponding avoidance reactions and (d) the interference of other behaviours such as feeding and reproduction on this predator–prey interaction. The adaptation of Northern krill to altered light levels and the potential overexposure of their eyes whilst in the surface layers are other areas where more information is needed for a comprehensive understanding of the role of light on predation on Northern krill. Finally, the specific contribution of Northern krill to the global diet of its diverse predators, notably that of baleen whales, still requires substantial efforts to improve our still very fragmentary present knowledge.

ACKNOWLEDGEMENTS

We thank Geraint Tarling, British Antarctic Survey, for his invitation to contribute this chapter, the initiation and coordination of this review on an important player in North Atlantic ecosystems, as well as for his help in providing references, edition and his patience in accommodating our busy agenda. We also specially thank Denis Chabot and Mike Hammill, Maurice Lamontagne Institute, Fisheries and Oceans Canada, for the access to unpublished data on the diet of several predators of *M. norvegica* in the Estuary and Gulf of St. Lawrence.

REFERENCES

Acevedo-Gutiérrez, A., Croll, D. A., and Tershy, B. R. (2002). High feeding costs limit dive time in the largest whales. *J. Exp. Biol.* **205**, 1747–1753.

Astthorsson, O. S., and Palsson, O. K. (1987). Predation on euphausiids by cod, *Gadus morhua*, in winter in Icelandic subarctic waters. *Mar. Biol.* **96**, 327–334.

Au, W. W. L., Kastelein, R. A., Rippe, T., and Schooneman, N. M. (1999). Transmission beam pattern and echolocation signals of a harbor porpoise (*Phocoena phocoena*). *J. Acoust. Soc. Am.* **106**, 3699–3705.

Bergstroem, B. I. (1985). Aspects of natural foraging by *Sepietta oweniana* (Mollusca, Cephalopoda). *Ophelia* **24**, 65–74.

NMFS-NEBowman, R. E., Stillwell, C. E., Michaels, W. L., and Grosslein, M. D. (2000). Food of northwest Atlantic fishes and two common species of squid NOAA. *Tech. Mem.* **155**, 1–137.

Brodie, P. F., Sameoto, D. D., and Sheldon, R. W. (1978). Population densities of euphausiids off Nova Scotia as indicated by net samples, whale stomach contents, and sonar. *Limnol. Oceanogr.* **23**, 1264–1267.

Brown, R. G. B., Barker, S. P., and Gaskin, D. E. (1979). Daytime surface swarming by *Meganyctiphanes norvegica* (M Sars) (Crustacea, Euphausiacea) off Brier Island, Bay of Fundy. *Can. J. Zool.* **57**, 2285–2291.

Brown, R. G. B., Barker, S. P., Gaskin, D. E., and Sandeman, M. R. (1981). The foods of great and sooty shearwaters *Puffinus gravis* and *P. griseus* in Eastern Canadian waters. *Ibis* **123**, 19–30.

Cabral, H. N., and Murta, A. G. (2002). The diet of blue whiting, hake, horse mackerel and mackerel off Portugal. *J. Appl. Ichthyol.* **18**, 14–23.

Calambokidis, J., Schorr, G. S., Steiger, G. H., Francis, J., Bakhtiari, M., Marshall, G., Oleson, E. M., Gendron, D., and Robertson, K. (2007). Insights into the underwater diving, feeding, and calling behavior of blue whales from a suction-cup-attached video-imaging tag (CRITTERCAM). *Mar. Technol. Soc. J.* **41**, 19–29.

Carruyhers, E. H., Neilson, J. D., Waters, C., and Perley, P. (2005). Long-term changes in the feeding of *Pollachius virens* on the Scotian Shelf: Responses to a dynamic ecosystem. *J. Fish Biol.* **66**, 327–347.

Cartes, J. E. (1993). Feeding habits of oplophorid shrimps in the deep Western Mediterranean. *J. Mar. Biolog. Assoc. U.K.* **73**, 193–206.

Cartes, J. E., Papiol, V., and Guijarro, B. (2008). The feeding and diet of the deep-sea shrimp *Aristeus antennatus* off the Balearic Islands (Western Mediterranean): Influence of environmental factors and relationship with the biological cycle. *Prog. Oceanogr.* **79**, 37–54.

Cartes, J. E., Hidalgo, M., Papiol, V., Massut, E., and Moranta, J. (2009). Changes in the diet and feeding of the hake *Merluccius merluccius* at the shelf-break of the Balearic Islands: Influence of the mesopelagic-boundary community. *Deep Sea Res.* **56**, 344–365.

Chabot, D. (2010). Unpublished data. Maurice Lamontagne Institute, Fisheries and Oceans Canada, 850 route de la Mer, Mont-Joli, Québec GOA-1N0, Canada.

Cochrane, N. A., Sameoto, D., Herman, A. W., and Neilson, J. (1991). Multiple-frequency acoustic backscattering and zooplankton aggregations in the inner Scotian Shelf basins. *Can. J. Fish. Aquat. Sci.* **48**, 340–355.

Cohen, J. H., and Forward, R. B. (2009). Zooplankton diel vertical migration: A review of proximate control. *Oceanogr. Mar. Biol. Annu. Rev.* **47**, 77–109.

Collins, M. A., De Grave, S., Lordan, C., Burnell, G. M., and Rodhouse, P. G. (1994). Diet of the squid *Loligo forbesi* Steenstrup (Cephalopoda: Loliginidae) in Irish waters. *ICES J. Mar. Sci.* **51**, 337–344.

Cotte, C., and Simard, Y. (2005). Formation of dense krill patches under tidal forcing at whale feeding hot spots in the St. Lawrence Estuary. *Mar. Ecol. Prog. Ser.* **288**, 199–210.

Croll, D. A., Acevedo-Gutiérrez, A., Tershy, B. R., and Urbén-Ramirez, J. (2001). The diving behavior of blue and fin whales: Is dive duration shorter than expected based on oxygen stores? *Comp. Biochem. Physiol., Part A Mol. Integr. Physiol.* **129**, 797–809.

Croll, D. A., Marinovic, B., Benson, S., Chavez, F. P., Black, N., Ternullo, R., and Tershy, B. R. (2005). From wind to whales: Trophic links in a coastal upwelling system. *Mar. Ecol. Prog. Ser.* **289**, 117–130.

Dalpadado, P. (1993). Some observations on the feeding ecology of Norwegian spring spawning herring *Clupea harengus* along the coast of Norway. *Counc. Meet. of the Int. Counc. for the, C.M./L* 47.

Dalpadado, P., and Bogstad, B. (2004). Diet of juvenile cod (age 0-2) in the Barents Sea in relation to food availability and cod growth. *Polar Biol.* **27**, 140–154.

Dalpadado, P., Ellertsen, B., Melle, W., and Dommasnes, A. (1998). Food and feeding conditions and prey selectivity of herring (*Clupea harengus*) through its feeding migrations from coastal areas of Norway to the Atlantic and Arctic watermasses of the Nordic Seas. "*Counc. Meet. of the Int. Counc. for the Exploration of the Sea, CM/R12.*

Du Buit, M. H. (1991). Food and feeding of saithe (*Pollachius virens* L.) off Scotland. *Fish. Res.* **12**, 307–323.

Emson, R. H., Mladenov, P. V., and Barrow, K. (1991). The feeding mechanism of the basket star *Gorgonocephalus arcticus*. *Can. J. Zool.* **69**, 449–455.

Falk, K., and Durinck, J. (1993). The winter diet of thick-billed murres, *Uria lomvia*, in western Greenland, 1988-1989. *Can. J. Zool.* **71**, 264–272.

Fanelli, E., and Cartes, J. E. (2004). Feeding habits of pandalid shrimps in the Alboran Sea (SW Mediterranean): Influence of biological and environmental factors. *Mar. Ecol. Prog. Ser.* **280**, 227–238.

Frank, T. M., and Widder, E. A. (1999). Comparative study of the spectral sensitivities of mesopelagic crustaceans. *J. Comp. Physiol. A* **185**, 255–265.

Franks, P. J. S. (1992). Sink or swim: Accumulation of biomass at fronts. *Mar. Ecol. Prog. Ser.* **82**, 1–12.

Froerman, Y. M. (1984). Feeding spectrum and trophic relationships of Short-finned Squid (*Illex illecebrosus*) in the Northern Atlantic. *NAFO Sci. Coun. Studies* **7**, 67–75.

Gannon, D. P., Craddock, J. E., and Read, A. J. (1998). Autumn food habits of harbor porpoises, *Phocoena phocoena*, in the Gulf of Maine. *Fish. Bull.* **96**, 428–437.

Genin, A. (2004). Bio-physical coupling in the formation of zooplankton and fish aggregations over abrupt topographies. *J. Mar. Syst.* **50**, 3–20.

Gislason, A., and Astthorsson, O. S. (eds.) (2000). The food of Norwegian Spring spawning Herring in the Western Norwegian Sea in relation to the annual cycle of zooplankton. *Counc. Meet. of the Int. Counc. for the Exploration of the Sea, CM/M 9.*

Greene, C. H., Widder, E. A., Youngbluth, M. J., Tamse, A., and Johnson, G. E. (1992). The migration behavior, fine-structure, and bioluminescent activity of krill Sound-Scattering layers. *Limnol. Oceanogr.* **37**, 650–658.

Harbitz, A., and Lindstroem, U. (2001). Stochastic spatial analysis of marine resources with application to minke whales (*Balaenoptera acutorostrata*) foraging: A synoptic case study from the southern Barents Sea. *Sarsia* **86**, 485–501.

Harvey, M., Galbraith, P. S., and Descroix, A. (2009). Vertical distribution and diel migration of macrozooplankton in the St. Lawrence marine system (Canada) in relation with the cold intermediate layer thermal properties. *Prog. Oceanogr.* **80**, 1–21.

Hojgaard, D. P. (1999). Food and parasitic nematodes of saithe, *Pollachius virens* (L.), from the Faroe Islands. *Sarsia* **84**, 473–478.

Houde, E. D. (2002). Mortality. In "The unique contributions of early life stages" (L. A. F.a. R. G. Werner, ed.), pp. 64–87. Blackwell Publishing, Oxford.

Hudson, I. R., and Wigham, B. D. (2003). In situ observations of predatory feeding behaviour of the galatheid squat lobster *Munida sarsi* (Huus, 1935) using a remotely operated vehicle. *J. Mar. Biol. Ass. U.K.* **83**, 1–2.

Isaacs, J. D., and Schwartzlose, R. A. (1965). Migrant sound scatterers: Interaction with the sea floor. *Science* **150**, 1810–1813.

Jeschke, J. M., Kopp, M., and Tollrian, R. (2002). Predator functional responses: Discriminating between handling and digesting prey. *Ecol. Monogr.* **72**, 95–112.

Jillett, J. B., and Zeldis, J. R. (1985). Aerial observations of surface patchiness of a planktonic crustacean. *Bull. Mar. Sci.* **37**, 609–619.

Kaartvedt, S., Røstad, A., Fiksen, Ø., Melle, W., Thomas Torgersen, T., Breien, M. T., and Klevjer, T. A. (2005). Piscivorous fish patrol krill swarms. *Mar. Ecol. Prog. Ser.* **299**, 1–5.

Klimpel, S., Palm, H. W., and Seehagen, A. (2003). Metazoan parasites and food composition of juvenile *Etmopterus spinax* (L., 1758) (Dalatiidae, Squaliformes) from the Norwegian Deep. *Parasitol. Res.* **89,** 245–251.

Kulka, D. W., Corey, S., and Iles, T. D. (1982). Community structure and biomass of euphausiids in the Bay of Fundy. *Can. J. Fish. Aquat. Sci.* **39,** 326–334.

Lacroix, G. (1961). Les migrations verticales journalières des euphausides à l'entrée de la baie des Chaleurs. *Nat. Can.* **88,** 257–317.

Lavery, A. C., Wiebe, P. H., Stanton, T. K., Lawson, G. L., Benfield, M. C., and Copley, N. (2007). Determining dominant scatterers of sound in mixed zooplankton populations. *J. Acoust. Soc. Am.* **122,** 3304–3326.

Lavoie, D., Simard, Y., and Saucier, F. J. (2000). Aggregation and dispersion of krill at channel heads and shelf edges: The dynamics in the Saguenay-St. Lawrence Marine Park. *Can. J. Fish. Aquat. Sci.* **57,** 1853–1869.

Lennert-Cody, C. E., and Franks, P. J. S. (1999). Plankton patchiness in high-frequency internal waves. *Mar. Ecol. Prog. Ser.* **186,** 59–66.

Lindstroem, U., and Haug, T. (2002). Feeding strategy and prey selectivity in common minke whales (*Balaenoptera acutorostrata*) foraging in the southern Barents Sea during early summer. *J. Cetacean Res. Manag.* **3,** 239–249.

Lordan, C., Burnell, G. M., and Cross, T. F. (1998). The diet and ecological importance of *Illex coindetii* and *Todaropsis eblanae* (Cephalopoda: Ommastrephidae) in Irish waters. *S. Afr. J. Mar. Sci.* **20,** 153–163.

Mackas, D. L., Kieser, R., Saunders, M., Yelland, D. R., Brown, R. M., and Moore, D. F. (1997). Aggregation of euphausiids and Pacific hake (*Merluccius productus*) along the outer continental shelf off Vancouver Island. *Can. J. Fish. Aquat. Sci.* **54,** 2080–2096.

Mattson, S. (1981). The food of *Galeus melastomus*, *Gadiculus argenteus thori*, *Trisopterus esmarkii*, *Rhinonemus cimbrius*, and *Glyptocephalus cynoglossus* (Pisces) caught during the day with shrimp trawl in a West-Norwegian Fjord. *Sarsia* **66,** 109–127.

Mauchline, J. (1959). The development of the euphausiacea (Crustacea) especially that of *Meganyctiphanes norvegica* (M. Sars). *Proc. Zool. Soc. Lond.* **132,** 627–639.

Mauchline, J. (1980). The biology of mysids and euphausiids. Academic Press, London.

Mauchline, J., and Fisher, L. R. (1969). The biology of euphausiids. Academis Press, London.

Mohl, B., and Andersen, S. (1973). Echolocation: High frequency component in the click of the harbour porpoise (*Phocoena* phocoena L.). *J. Acoust. Soc. Am.* **54,** 1368–1372.

Montevecchi, W. A., Birtfriesen, V. L., and Cairns, D. K. (1992). Reproductive energetics and prey Harvest of Leachs Storm-Petrels in the Northwest Atlantic. *Ecology.* **73,** 823–832.

Neiva, J., Coelho, R., and Erzini, K. (2006). Feeding habits of the velvet belly lanternshark *Etmopterus spinax* (Chondrichthyes: Etmopteridae) off the Algarve, southern Portugal. *J. Mar. Biol. Ass. U.K.* **86,** 835–841.

Nicol, S. (1986). Shape, size and density of daytime surface swarms of the euphausiid *Meganyctiphanes norvegica* in the bay of Fundy. *J. Plankton Res.* **8,** 29–39.

Nicol, S., and O'Dor, R. K. (1985). Predatory behaviour of squid (*Illex illecebrosus*) feeding on surface swarms of euphausiids. *Can. J. Zool.* **63,** 15–17.

Nordoey, E. S., and Blix, A. S. (1992). Diet of minke whales in the northeastern Atlantic. *Rep. Int. Whaling Comm.* **42,** 393–398.

Olaso, I., Cendrero, O., and Abaunza, P. (1999). The diet of the horse mackerel, *Trachurus trachurus* (Linnaeus 1758), in the Cantabrian Sea (north of Spain). *J. Appl. Ichthyol. / Z. Angew. Ichthyol.* **15,** 193–198.

Onsrud, M. S. R., Kaartvedt, S., Røstad, A., and Klevjer, T. A. (2004). Vertical distribution and feeding patterns in fish foraging on the krill *Meganyctiphanes norvegica*. *ICES J. Mar. Sci.* **61,** 1278–1290.

Pais, C. (2002). Diet of a deep-see fish, *Hoplostethus mediterraneus* from the south coast of Portugal. *J. Mar. Biol. Ass. U.K.* **82,** 351–352.

Panigada, S., Zanardelli, M., Canese, S., and Jahoda, M. (1999). How deep can baleen whales dive. *Mar. Ecol. Prog. Ser.* **187,** 309–311.

Pearcy, W. G., Hopkins, C. C. E., Gronvik, S., and Evans, R. A. (1979). Feeding habits of cod, capelin, and herring in Balsfjorden, northern Norway, July–August 1978: The importance of euphausiids. *Sarsia* **64,** 269–277.

Pearre, S. (2003). Eat and run? The hunger/satiation hypothesis in vertical migration: History, evidence and consequences. *Biol. Rev.* **78,** 1–79.

Petursdottir, H., Gislasona, A., Falk-Petersenb, S., Hopb, H., and Svavarssonc, J. (2008). Trophic interactions of the pelagic ecosystem over the Reykjanes Ridge as evaluated by fatty acid and stable isotope analyses. *Deep Sea Res. (II Top. Stud. Oceanogr.)* **55,** 83–93.

Pitcher, T. J., and Parrish, J. K. (1993). Functions of shoaling behaviour in teleosts. *In* "The behaviour of teleost fishes" (T. J. Pitcher. ed.), pp. 363–439. Chapman & Hall, London.

Preciado, I., Cartes, J., Velasco, F., Olaso, I., Serrano, A., Frutos, I., and Sanchez, F. (2006). The role of suprabenthic and epibenthic communities in the diet of a deep-sea fish assemblage (Le Danois Bank, Cantabrian Sea, N Spain). *In* "11th International Deep-Sea Biology Symposium, Southampton, UK," pp. 147–148.

Preciado, I., Cartes, J. E., Serrano, A., Velasco, F., Olaso, I., Sánchez, F., and Frutos, I. (2009). Resource utilization by deep-sea sharks at the Le Danois Bank, Cantabrian Sea, north-east Atlantic Ocean. *J. Fish Biol.* **75,** 1331–1355.

Relini Orsi, L., and Wurtz, M. (1975). Remarks on feeding of *Galeus melastomus* from ligurian bathyal grounds. *Ancona* **2,** 17–36.

Relini-Orsi, L., and Wurtz, M. (1977). Food habits of *Etmopterus spinax* (Chondrichthyes, Squalidae). *In* "8th National Congress of Italian Society of Marine Biology (Taormina, Italy)," pp. 74–75.

Rosenberg, R., Dupont, S., Lundälv, T., Nilsson Sköld, H., Norkko, A., Roth, J., Stach, T., and Thorndyke, M. (2005). Biology of the basket star Gorgonocephalus caputmedusae (L.). *Mar. Biol.* **148,** 43–50.

Sameoto, D., Cochrane, N., and Herman, A. (1993). Convergence of acoustic, optical, and net-catch estimates of euphausiid abundance: Use of artificial-light to reduce net avoidance. *Can. J. Fish. Aquat. Sci.* **50,** 334–346.

Sameoto, D., Neilson, J., and Waldron, D. (1994). Zooplankton prey selection by juvenile fish in Nova Scotlan Shelf basins. *J. Plankton Res.* **16,** 1003–1019.

Santos, J., and Borges, T. (2001). Trophic relationships in deep-water fish communities off Algarve, Portugal. *Fish Res.* **51,** 337–341.

Sartor, P., Carlini, F., and De Ranieri, S. (2003). Diet of young European hake (*Merluccius merluccius*) in the Northern Tyrrhenian Sea. *Biol. Mar. Mediterr.* **10,** 904–908.

Sherstyukov, A. I., and Nazarova, G. I. (1991). Vertical migration and feeding of 0-Group Silver Hanke (*Merluccius bilinearis*) on the Scotian Shelf, November 1985. *NAFO Sci. Coun. Studies* **15,** 53–58.

Sigurjónsson, J., and Víkingsson, G. A. (1997). Seasonal abundance of and estimated food consumption by cetaceans in Icelandic and Adjacent Waters. *J. Northw. Atl. Fish. Sci.* **22,** 271–287.

Simard, Y. (2009). The Saguenay-St. Lawrence Marine Park: Oceanographic processes at the basis of this unique forage site of northwest Atlantic whales. *Rev. Sci. Eau/J. Water Sci.* **22,** 177–197.

Simard, Y., and Lavoie, D. (1999). The rich krill aggregation of the Saguenay - St. Lawrence Marine Park: hydroacoustic and geostatistical biomass estimates, structure, variability, and significance for whales. *Can. J. Fish. Aquat. Sci.* **56,** 1182–1197.

Simard, Y., and Mackas, D. L. (1989). Mesoscale aggregations of euphausiid sound scattering layers on the continental shelf of Vancouver Island. *Can. J. Fish. Aquat. Sci.* **46,** 1238–1249.

Simard, Y., and Sourisseau, M. (2009). Diel changes in acoustic and catch estimates of krill biomass. *ICES J. Mar. Sci.* **66,** 1318–1325.

Simard, Y., De Ladurantaye, R., and Therriault, J.-C. (1986a). Aggregation of euphausiids along a coastal shelf in an upwelling environment. *Mar. Ecol. Prog. Ser.* **32,** 203–215.

Simard, Y., Lacroix, G., and Legendre, L. (1986b). Diel vertical migrations and nocturnal feeding of a dense coastal krill scattering layer (*Thysanoessa raschi* and *Meganyctiphanes norvegica*) in stratified surface waters. *Mar. Biol.* **91,** 93–105.

Simard, Y., Lavoie, D., and Saucier, F. J. (2002). Channel head dynamics: Capelin (*Mallotus villosus*) aggregation in the tidally driven upwelling system of the Saguenay - St. Lawrence marine park's whale feeding ground. *Can. J. Fish. Aquat. Sci.* **59,** 197–210.

Sourisseau, M., Simard, Y., and Saucier, F. J. (2006). Krill aggregation in the St. Lawrence system, and supply of krill to the whale feeding grounds in the estuary from the gulf. *Mar. Ecol. Prog. Ser.* **314,** 257–270.

Sourisseau, M., Simard, Y., and Saucier, F. J. (2008). Krill diel vertical migration fine dynamics, nocturnal overturns, and their roles for aggregation in stratified flows. *Can. J. Fish. Aquat. Sci.* **65,** 574–587.

Spitz, J., Rousseau, Y., and Ridoux, V. (2006). Diet overlap between harbour porpoise and bottlenose dolphin: An argument in favour of interference competition for food? *Estuar. Coast. Shelf Sci.* **70,** 259–270.

Steele, D. H. (1963). Pollock (*Pollachius virens* (L.)) in the Bay of Fundy. *J. Fish. Res. Bd. Can.* **20,** 1267–1314.

Stefansson, G., Sigurjonsson, J., and Vikingsson, G. A. (1997). On dynamic interactions between some fish resources and cetaceans off Iceland based on a simulation model. *J. Northw. Atl. Fish. Sci.* **22,** 357–370.

Stevick, P. T., Incze, L. S., Kraus, S. D., Rosen, S., Wolff, N., and Baukus, A. (2008). Trophic relationships and oceanography on and around a small offshore bank. *Mar. Ecol. Prog. Ser.* **363,** 15–28.

Stone, H. H., and Jessop, B. M. (1994). Feeding habits of anadromous alewives, *Alosa pseudoharengus*, off the Atlantic Coast of Nova Scotia. *Fish. Bull.* **92,** 157–170.

Straumfors, P. (1981). New record of *Trachipterus arcticus* (Brunnich) in Sjona, Nordland. *Fauna (Blindern)* **34,** 45.

Tarling, G. A., Matthews, J. B. L., David, P., Guerin, O., and Buchholz, F. (2001). The swarm dynamics of northern krill (*Meganyctiphanes norvegica*) and pteropods (*Cavolinia inflexa*) during vertical migration in the Ligurian Sea observed by an acoustic Doppler current profiler. *Deep Sea Res. (I Oceanogr. Res. Pap.)* **48,** 1671–1686.

Vikingsson, G. A. (1997). Feeding of Fin Whales (*Balaenoptera physalus*) off Iceland - Diurnal and Seasonal Variation and Possible Rates. *J. Northw. Atl. Fish. Sci.* **22,** 77–89.

Warren, J. D., Stanton, T. K., Wiebe, P. H., and Seim, H. E. (2003). Inference of biological and physical parameters in an internal wave using multiple-frequency, acoustic-scattering data. *ICES J. Mar. Sci.* **60,** 1033–1046.

Widder, E. A., Greene, C. H., and Youngbluth, M. J. (1992). Bioluminescence of sound-scattering layers in the Gulf of Maine. *J. Plankton Res.* **14,** 1607–1624.

Wiebe, P. H. (1982). Rings of the Gulf Stream. *Sci. Am.* **246,** 60–70.

Wiebe, P. H., and Flierl, G. R. (1983). Euphausiid invasion/dispersal in Gulf Stream cold-core rings. *Aust. J. Mar. Freshw. Res.* **34,** 625–652.

Wiebe, P. H., Ashjian, C. J., Gallager, S. M., Davis, C. S., Lawson, G. L., and Copley, N. J. (2004). Using a high-powered strobe light to increase the catch of Antarctic krill. *Mar. Biol.* **144,** 493–502.

Zhou, M., Zhu, Y. W., and Tande, K. S. (2005). Circulation and behavior of euphausiids in two Norwegian sub-Arctic fjords. *Mar. Ecol. Prog. Ser.* **300,** 159–178.

Taxonomic Index

SUBJECT INDEX

Geraint A. Tarling *et al.*, Figure 1.5 The pattern of photophore illumination in Northern krill (*Meganyctiphanes norvegica*) as viewed ventrally. The image was photographed in darkness, exposure time 40 s/400 ASA. The natural colour, a narrow band blue of 480 nm wavelength, is visible to the human eye with background irradiation dimmer than 10^{-2} μW/cm^2. Note that parts of the appendages are illuminated.

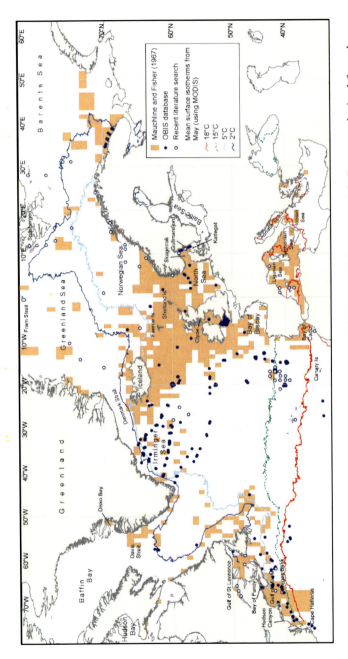

Geraint A. Tarling *et al*., Figure 1.6 The distribution of Northern krill (*Meganyctiphanes norvegica*). Data was obtained from three sources: shaded areas were transposed from Mauchline and Fisher (1967), including any records of larvae or point samples. Blue dots represent records of *M. norvegica* present on the Ocean Biogeographic Information System (OBIS—www.iobis.org) as of January 2010. Red dots represent further novel records obtained in a recent literature search by the present authors. Surface isotherms were extracted from data provided by the NASA Moderate Resolution Imaging Spectroradiometer (MODIS) representing a mean value for May (2003–2010): the 2 and 18 °C isotherms approximate the distributional limits, with a few exceptions; the 5 and 15 °C isotherms represent the notional breeding limits, following Einarsson (1945).

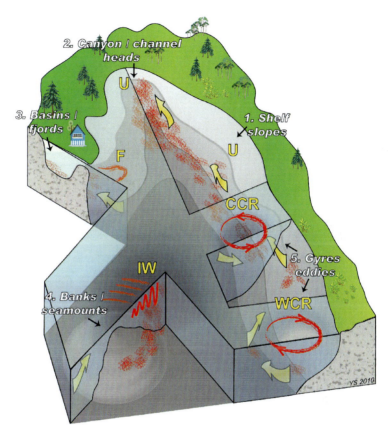

Yvan Simard and Michel Harvey, Figure 10.1 3D scheme of oceanographic processes involved in the governing of the interactions between predators and krill species including *M. norvegica*, by generating prey aggregations (red dots) at predictable sites and times, which predators have learnt to take advantage of and where intensive trophic exchanges occur. 1: Krill aggregation along slopes by upslope upwelling (U) currents and bathymetry guided currents; 2: Canyon and channel head krill entrapments by upstream flow at the krill scattering layer depth and intense upwelling at the topographic dead-end; 3: Krill entrapment in basins carved on continental shelves and in fjords; 4: Krill aggregation around banks and seamounts by 3D currents interacting with topography, often generating internal waves (IW); 5: Krill aggregation by cyclonic (CCR: cold core ring) and anti-cyclonic (WCR: warm core ring) eddies and gyres. Fronts (F) can sometimes concentrate krill by convergence or through the response of krill to environmental discontinuities.

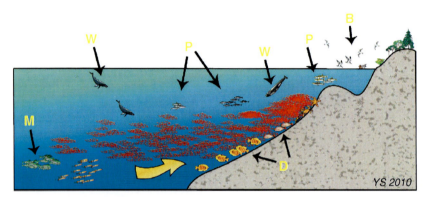

Yvan Simard and Michel Harvey, Figure 10.2 Schematic representation of typical predator–prey interactions along sloping topography where krill (red dots) is aggregating during daytime at latitudes inhabited by *M. norvegica*. D: Demersal predators where the krill SL horizon comes into contact with the slope; P: Pelagic predators aggregated over the neighbouring shallows or in the epipelagic surface-lit layer allowing visual predation; M: Mesopelagic and bathypelagic fish occupying deep layers in the water column. W: Baleen and other marine mammals diving to feed on the krill SL; B: Surface feeding and diving marine birds taking advantage of the available krill SL, notably at night or during upwelling events.

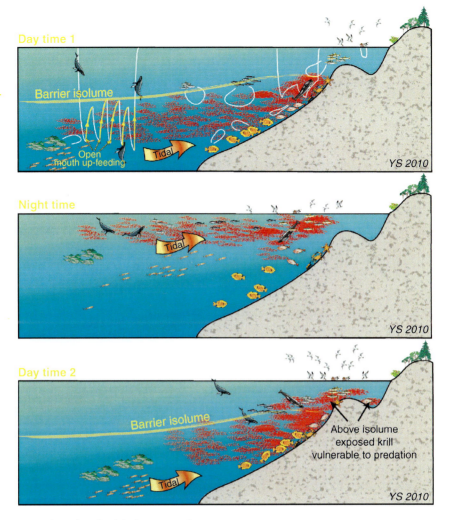

Yvan Simard and Michel Harvey, Figure 10.3 Schematic representation of predator–prey behaviours at a Northern krill aggregation site over the diel cycle and with a tidal current regime. Daytime 1: Typical predator–prey configuration along the sloping bottom, where the krill SL shoals and concentrates at the junction point between the barrier isolume and the bottom under forcing from up-slope currents at tidal frequencies. Predators from all types illustrated in Fig. 10.2 can have excursions (white lines) into the SLs or graze its top or bottom margins. Night time: During night, the krill SL is in the upper water column and the entire predator community changes its behaviour while tracking the prey community into the upper water column. The currents then transport the krill in directions determined by the local currents in the upper water column. Daytime 2: All the krill still in upper water column at dawn return to the depths when the sun rises and they track the barrier isolume downwards. Because of their transport during night by upper water column currents, the krill may descend onto bathymetries that are shallower than the depth of the barrier isolume. They then become visible to predators, their eyes are exposed to high light levels and they no longer can escape by moving down. On the contrary, this reaction concentrates them and makes them still more vulnerable to all pelagic, demersal and benthic predators. With the usual anti-predation strategies failing, krill may be found over the whole water column, even at the surface, often under strong tidal current interactions with the topography, to the benefit of avian predators.

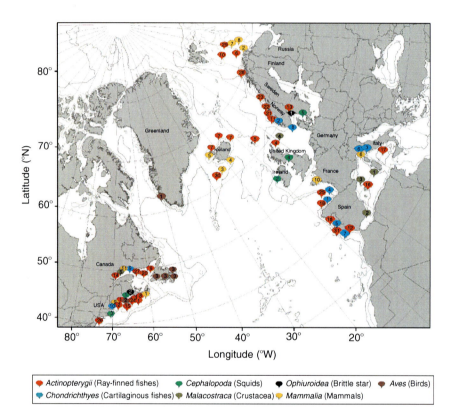

Yvan Simard and Michel Harvey, Figure 10.4 Map of the predators containing Northern krill, *M. norvegica*, from a literature search. Each taxonomic group is indicated by different colors. Numbers correspond to references in Table 10.1.